ERIC WIEBE

**Pathfinder Associative Networks:
Studies in Knowledge Organization**

ABLEX SERIES IN COMPUTATIONAL SCIENCES

Derek Partridge, University of Exeter

Series Editor

Pathfinder Associative Networks: Studies in Knowledge Organization
  Roger W. Schvaneveldt (Editor)

In Preparation

A New Guide to Artificial Intelligence
  Derek Partridge

Artificial Intelligence and Software Engineering
  Derek Partridge (Editor)

Binding Time — Six Studies in Programming Technology and Milieu
  Mark Halpern

Artificial Intelligence and Business Management
  Derek Partridge and K.M. Hussain

Intelligent Systems
  Eric Dietrich and Chris Fields

Database in Practice Rather Than in Theory
  K.G. Jeffrey

New Generation Architectures and Languages
  Stephen J. Turner

Computer Analysis of English: Lexical Semantics and Preference Semantics Analysis
  Brian Slator

Pathfinder Associative Networks: Studies in Knowledge Organization

Edited by

Roger W. Schvaneveldt
*Department of Psychology and
Computing Research Laboratory
New Mexico State University*

ABLEX PUBLISHING CORPORATION
Norwood, New Jersey 07648

Copyright © 1990 by Ablex Publishing Corporation.

All rights reserved. No part of this publication may be reproduced in any form, by photostat, microfilm, retrieval system, or any other means, without the prior permission of the publisher.

Printed in the United States of America.

Library of Congress Cataloging-in-Publication Data

Pathfinder associative networks : studies in knowledge organization / edited by Roger W. Schvaneveldt.
    p. cm. – (Ablex series in computational sciences)
    Includes bibliographical references.
    ISBN 0-89391-624-2
    1. Expert systems (Computer science). 2. Artificial intelligence. 3. Graph theory. 4. Knowledge acquisition (Expert systems).
    I. Schvaneveldt, Roger W.   II. Series.
QA76.76.E95P28   1989
006.3'3–dc20   89-18219
                                                                                                          CIP

ABLEX Publishing Corporation
355 Chestnut Street
Norwood, New Jersey 07648

Contents

List of Contributors		vii
Preface		ix
	Roger W. Schvaneveldt	
1	**Properties of Pathfinder Networks**	1
	Donald W. Dearholt & Roger W. Schvaneveldt	
2	**Graphs in the Social and Psychological Sciences: Empirical Contributions of Pathfinder**	31
	Francis T. Durso & Kathy A. Coggins	
3	**Fuzzy PFNETs: Coping with Variability in Proximity Data**	53
	Chris Esposito	
4	**Discriminating Between Degrees of Low or High Similarity: Implications for Scaling Techniques Using Semantic Judgments**	61
	Renate J. Roske-Hofstrand & Kenneth R. Paap	
5	**Assessing Structural Similarity of Graphs**	75
	Timothy E. Goldsmith & Daniel M. Davenport	
6	**A Graph-Theoretic Approach to Concept Clustering**	89
	Chris Esposito	
7	**Empirically Defined Semantic Relatedness and Category Judgment Time**	101
	Nancy Jaworski Cooke	
8	**Pathfinder Networks and Multidimensional Spaces: Relative Strengths in Representing Strong Associates**	111
	Russell J. Branaghan	
9	**Directed Graphs as Memory Representations: The Case of Rhyme**	121
	David C. Rubin	
10	**Proximities, Networks, and Schemata**	135
	Roger W. Schvaneveldt	

11 Using Pathfinder to Extract Semantic Information from Text 149
    James E. McDonald, Tony A. Plate, & Roger W. Schvaneveldt

12 Information Retrieval Using Pathfinder Networks 165
    Richard H. Fowler & Donald W. Dearholt

13 Using Pathfinder to Evaluate User and System Models 179
    Wendy A. Kellogg & Timothy J. Breen

14 Hypertext Perspectives:
    Using Pathfinder to Build Hypertext Systems 197
    James E. McDonald, Kenneth R. Paap, & Deborah R. McDonald

15 Expert Conceptual Structure:
    The Stability of Pathfinder Representations 213
    John G. Gammack

16 Using Pathfinder as a Knowledge Elicitation Tool:
    Link Interpretation 227
    Nancy Jaworski Cooke

17 A Structural Assessment of Classroom Learning 241
    Timothy E. Goldsmith & Peder J. Johnson

18 Representation of Problem Schemata 255
    Lisa A. Onorato

19 A Measure of the Knowledge Reorganization Underlying Insight 267
    Tom Dayton, Francis T. Durso, & Jack D. Shepard

References 279

Graph Theory and Pathfinder Primer 297

Glossary 301

Author Index 305

Subject Index 313

Contributors

Russell J. Branaghan
Department of Psychology and
Computing Research Laboratory
New Mexico State University
Las Cruces, NM 88003

Timothy J. Breen
Knowledge Systems Laboratory
Boeing Computer Services
P.O. Box 24346, MS 7L-64
Seattle WA 98124-0346

Kathy A. Coggins
Department of Psychology
University of Oklahoma
Norman, OK 73019

Nancy Jaworski Cooke
Department of Psychology
Rice University
Box 1892
Houston, TX 77251

Daniel M. Davenport
P. O. Box 5800
Sandia National Laboratories
Albuquerque, NM 87185

Tom Dayton
Department of Psychology
University of Oklahoma
Norman, OK 73019

Donald W. Dearholt
Department of Computer Science and
Computing Research Laboratory
New Mexico State University
Las Cruces, NM 88003

Francis T. Durso
Department of Psychology
University of Oklahoma
Norman, OK 73019

Chris Esposito
Boeing Advanced Technology Center
P.O. Box 24346 MS 7L-64
Seattle, WA 98124

Richard H. Fowler
Department of Mathematics and
Computer Science
Pan American University
Edinburg, TX 78539

John G. Gammack
MRC Applied Psychology Unit
15 Chaucer Road
Cambridge, CB2 2EF
England

Timothy E. Goldsmith
Department of Psychology
University of New Mexico
Albuquerque, NM 87131

Peder J. Johnson
Department of Psychology
University of New Mexico
Albuquerque, NM 87131

Wendy A. Kellogg
User Interface Institute
IBM Thomas J. Watson Research
Center
P. O. Box 704
Yorktown Heights, NY 10598

Deborah R. McDonald
Department of Psychology
New Mexico State University
Las Cruces, NM 88003

James E. McDonald
Department of Psychology and
Computing Research Laboratory
New Mexico State University
Las Cruces, NM 88003

Lisa A. Onorato
Department of Psychology
Hartwick College
Oneonta, NY 13820

Kenneth R. Paap
Department of Psychology and
Computing Research Laboratory
New Mexico State University
Las Cruces, NM 88003

Tony A. Plate
Department of Computer Science
University of Toronto
Toronto M5S 1A1
Canada

Renate J. Roske-Hofstrand
Aerospace Human Factors Research
Division
Nasa-Ames Research Center
Mailstop 239-21
Moffett Field, CA 94035

David C. Rubin
Department of Psychology
Duke University
Durham, NC 27706

Roger W. Schvaneveldt
Department of Psychology and
Computing Research Laboratory
New Mexico State University
Las Cruces, NM 88003

Jack D. Shepard
Department of Psychology
University of Oklahoma
Norman, OK 73019

Preface

The chapters in this book represent a sampling of theoretical, empirical, and applied work with *Pathfinder networks*. These networks began in 1981 (Schvaneveldt & Durso, 1981) as an attempt to develop a *network* model for *proximity* data. The intervening years have seen several developments of that original work. A theoretical paper relating Pathfinder networks to fundamental concepts in graph theory (Schvaneveldt, Dearholt, & Durso, 1988) grew out of a conference organized by Frank Harary and Keith Phillips. The chapters in this book represent a wide range of applications for network models.

The original motivation for developing Pathfinder grew out of our realization that although network representations abound in theoretical work in cognitive psychology and artificial intelligence, there were few methods for arriving at a network representation from empirical data. Proximity data offer a convenient starting point for networks. Indeed, proximity data serve as the building block for several interesting structural models such as multidimensional scaling (MDS) and cluster analysis. Essentially, Pathfinder networks are determined by identifying the proximities that provide the most efficient connections between the entities by considering the indirect connections provided by *paths* through other entities. The resulting networks have several interesting properties (see Chapter 1), and they have also proven to be useful in a variety of applications. There are now various algorithms for deriving PFNETs in several computer languages running on several different computers.[1]

There are a few features of this book that should be helpful to readers without much background knowledge of graph theory. A brief primer on graph theory and Pathfinder and a glossary can be found at the back of the book. References from all of the chapters are compiled in a single reference section at the back of the book. Chapter 1 reviews some definitions and properties of Pathfinder networks as well as some algorithms for deriving these networks from proximity data. Chapter 2 is a general review of empirical work with Pathfinder in cognitive modeling and an exploration of potential applications in social networks. The other chapters relate to several major themes.

Chapters 3, 4, and 5 address some methodological issues. Esposito (Chapter 3) develops and evaluates a version of Pathfinder that takes variability of proximity data into account. Roske-Hofstrand and Paap (Chapter 4) analyze some properties of proximity data obtained by ratings and the implications for Pathfinder networks. Goldsmith and Davenport (Chapter 5) present some measures of the similarity of two networks.

Chapters 6 through 10 report investigations of some basic phenomena in human memory. Esposito (Chapter 6) analyzes the relation between human judgments of the goodness of categories and various formal characteristics of graphs. Cooke (Chapter 7) examines the time required to judge that two concepts belong to the same category. Branaghan (Chapter 8) analyzes the ease with which lists of associations are learned. Rubin (Chapter 9) investigates the strategies people use to search memory. Schvaneveldt (Chapter 10) examines the representation of schemata in Pathfinder and connectionist style networks.

---

[1] Programs have been written in Pascal, C, LISP, and APL. Various versions of the programs run on IBM PC, Apple Macintosh, and SUN Microsystems computers. Information on obtaining programs is available from: Interlink, Inc., P.O. Box 4086 UPB, Las Cruces, NM 88003-4086.

Chapters 11 through 16 address applications of Pathfinder to problems in knowledge elicitation, information retrieval, and interface design. McDonald, Plate, and Schvaneveldt (Chapter 11) extract associative information from text and use this information to resolve word ambiguity. Fowler and Dearholt (Chapter 12) address the classic problem of retrieving information from large collections as in libraries. Kellogg and Breen (Chapter 13) compare the models of systems to mental models of users. McDonald, Paap, and McDonald (Chapter 14) attack the problem of establishing connections in Hypertext. Gammack (Chapter 15) analyzes the use of different techniques for eliciting proximity information from an expert. Cooke (Chapter 16) develops a method for identifying the nature of the relations between linked concepts in a network.

Chapters 17, 18, and 19 are concerned with still other aspects of knowledge representation. Goldsmith and Johnson (Chapter 17) investigate the use of networks and MDS spaces to assess classroom learning. Onorato (Chapter 18) analyzes the ways in which people organize information depending on the purpose of the information. Dayton, Durso, and Shepard (Chapter 19) examine the differences in the way solvers and nonsolvers organize problem-relevant information.

Obviously, there are many interrelations among the various chapters. As an aid to seeing these relations and as an initial illustration of the use of Pathfinder, I constructed Figure 1. This figure shows a Pathfinder network depicting the close associations among the chapters.

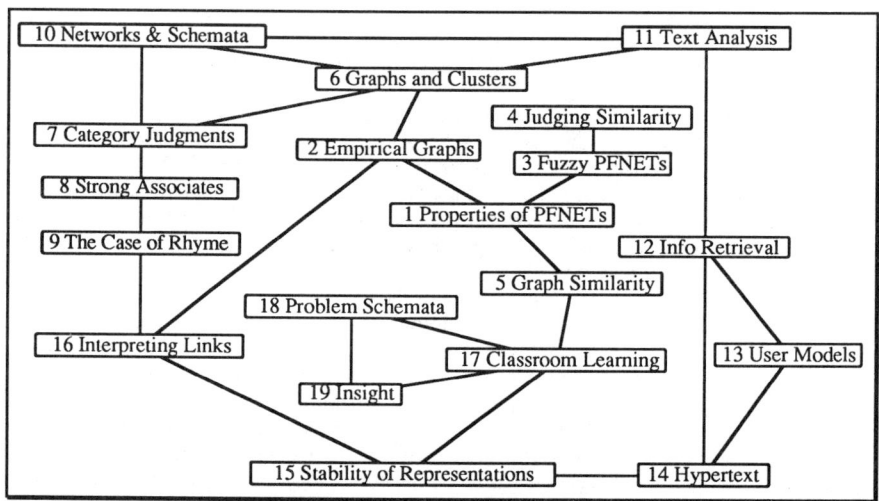

Figure 1. The modified PFNET($r = \infty$, $q = n–1$) for the chapters in this book.

To construct Figure 1, I first made a list of the three chapters most closely related to each chapter. This ordered list of associations was used to construct a matrix of proximity data where an entry was 1, 2, or 3 if the chapter on the column was the first, second, or third most associated with the chapter on the row. Other entries were treated as infinite. This matrix was non-symmetrical and the Pathfinder network that resulted from analyzing the matrix had directed links. However, I was not able to interpret the directions of the links so I made all of the links undirected as shown in the figure.

Preface

This figure can be used to find chapters that are closely related to other chapters. Several groups of interrelated chapters can be identified in addition to the one I used to order the chapters. It is obviously impossible to capture all of these relations in the linear ordering enforced on a book.

The development of Pathfinder and much of the research reported in this book have been supported by the National Science Foundation (IST-8506706), the Air Force Human Resources Laboratory (F33615-84-C-0072 and F33615-80-C-0004), Texas Instruments, Inc., and the National Aeronautic and Space Administration (NAG 2-453). Such support has been invaluable in the development of the methods and research.

I gratefully acknowledge the assistance of several people in assembling this book. Derek Partridge encouraged me to undertake the project in the first place. Most of the authors reviewed one or more chapters in addition to writing their own. Douglas Nelson and David Farwell also provided very helpful reviews. My associates here at New Mexico State were invaluable in their assistance with all of the details and in the defense of conceptual coherence. I am particularly grateful to Bob Fiegel, Tarra Fiegel, Rebecca Gomez, and Paula Moreland for their help. Thanks also to my wife, Ann, and daughter, Susan, for their love and support.

R. Schvaneveldt
April 11, 1989
Las Cruces, New Mexico

Chapter 1

Properties of Pathfinder Networks

Donald W. Dearholt and Roger W. Schvaneveldt

Network models have played important roles in various areas of cognitive science and computer science. In cognitive psychology and artificial intelligence, network representations of concepts stored in semantic memory have been used in models of memory retrieval and human performance (e.g., Anderson, 1983; Collins & Loftus, 1975; Collins & Quillian, 1969; Friendly, 1977, 1979; Meyer & Schvaneveldt, 1976; Rumelhart & McClelland, 1986), scene description and analysis (e.g., Brooks & Binford, 1980; Waltz, 1972; Winston, 1970), natural-language processing (e.g., Bobrow & Webber, 1980; Fillenbaum & Rapoport, 1971; Kintsch, 1974; Quillian, 1967, 1969; Schank, 1972; Woods, Kaplan, & Nash-Webber, 1972), and knowledge representation (e.g., Brachman, 1977, 1979; Fahlman, 1979; Fikes & Hendrix, 1977; Griffith, 1982; Novak, 1977; Schmolze & Lipkis, 1983; Sowa, 1984; Woods, 1975).

In database systems, a network data model often results in efficient representations of sets of concepts (Date, 1981; Ullman, 1982). Thus far, the network model incorporated in database systems has been constructed with two primary objectives: providing efficient data access for the anticipated user environment and making the most of the rather severe limitations imposed by present computer operations and architecture. Although the network data model used most frequently in database models (CODASYL, 1971) can support abstractions of essentially any type, there are constraints (for the purpose of modularity, simplicity of definition, and hardware support) that must be circumvented by artificial programming devices. Networks identifying relationships between data items have been proposed for designing the logical schema of a database system (e.g., see Martin, 1977, Chapter 6) by means of bubble charts. The bubble charts are used to indicate relationships between data items (e.g., functional dependencies, primary keys, and secondary keys). The bubble charts are usually viewed as an intermediate step in the development of a logical schema. Clustering strategies for data items have been investigated and proposed for improving expected retrieval time, based on the estimated likelihood of retrieval of data items contingent upon the retrieval of other data items (e.g., Navathe & Fry, 1976; Schkolnick, 1977).

Recently developed techniques from our laboratory and elsewhere allow researchers to derive networks from the same proximity data employed by multidimensional scaling (Dearholt, Schvaneveldt, & Durso, 1985; Hutchinson, 1989; Schvaneveldt, Dearholt, & Durso, 1988; Schvaneveldt & Durso, 1981; Schvaneveldt, Durso, & Dearholt, 1989).

Networks and Proximity Data

Hutchinson's NETSCAL procedure (Hutchinson, 1981, 1989), which makes *ordinal* data assumptions, is based on a theorem of Hakimi and Yau (1964) regarding the *distance* matrix of a *graph* and its realizability. The distance metric used by Hakimi and Yau is the sum of *edge weights* along a path, so that the distance between *nodes* is the (minimum) sum of the weights (distances) of the edges along a path between the nodes. This measure of *path length* is appropriate for *ratio*-scale data. Hutchinson, however, also used a distance metric in which the distance between nodes is the smallest maximum weight along any of the paths between the two nodes. This path-length measure is appropriate for ordinal as well as ratio-scale data. A serious shortcoming of Hutchinson's work is that his Corollary I considers *triangle inequalities* of only two-link paths. That is, the triangle inequality can be violated in paths having three or more links in Hutchinson's networks. This seems to be an unfortunate limitation, inappropriate for the scaling of data, and perhaps also for cognitive modeling, although psychological proximity may not always obey the triangle inequality (Tversky, 1977; Tversky & Gati, 1982).

The triangle inequality can be viewed in three different domains. The first, and the sparsest, is in euclidean space, as addressed by Hakimi and Yau (1964), in which the triangle inequality must always be satisfied. The second is the class of problems in which measures of similarity or "distance" are measured objectively by set intersections; for most such problems, there is no expectation of transitivity holding, so that there is likewise no expectation that the triangle inequality will be satisfied, either. That is, if we know the intersections of sets A and B, and of B and C, we do not generally know anything about the intersection of sets A and C. The information retrieval application to be discussed in detail in the chapter by Fowler and Dearholt (Chapter 12, this volume) is an example of such a problem in which the triangle inequality may be violated. The third domain is that of subjective estimates of similarity, in which data frequently show violations of the triangle inequality. Philosophically, it is attractive to use geodetic distance measures, in which the distance between each pair of nodes is considered to be the length of the shortest path available between those nodes; indeed, in graph theory this has been the usual definition of distance. Then, a violation of some triangle inequality is never a part of a path between any pair of nodes, because a shorter alternative path is always available. Thus the omission of the edges which violate some triangle inequality in a network assures the preservation of the (geodetic) distances between all pairs of nodes and provides a simpler structure which possesses precisely those edges which are responsible for the most economical paths (Schvaneveldt, Dearholt, & Durso, 1988).

The links that are omitted include those due to differences on two or more separable dimensions, in which the triangle inequality is expected to be violated, as discussed by Tversky and Gati (1982). That is, if A and B are judged to be similar because of feature x, and B and C are judged to be similar because of feature y, then A and C will normally be judged to be less similar than the triangle inequality would indicate. Thus if salient associations are linked in a graph paradigm, the absence of a *link* can denote a difference in the basis for judged similarity.

We have developed a procedure called Pathfinder (several equivalent procedures, actually) to generate a class of networks called PFNETs, which are based on estimates or measures of distances between pairs of entities. This procedure allows a spectrum of assumptions to be made about the data, including ordinal and ratio properties. The data required are either similarities or distances. Similarities can be obtained either from a subject's estimates of the similarity of each pair of entities in the set or from a measure of set intersections. Distances can be obtained in some domains by estimating or computing appropriate differences between all pairs of entities. The result of the Pathfinder procedure is a network which is either a directed graph (if the similarity or distance matrix is not symmetrical) or an undirected graph otherwise. Each entity in the set is represented as a node in the network, and each link that is entered in the network has a weight value determined

1 Properties of Pathfinder Networks

by the distance between the two entities so linked. Our network generation procedure incorporates two parameters. The first, the Minkowski r-metric, determines how distance between two nodes not directly linked is computed. The weight of a path with weights w_1, $w_2,..., w_k$ is:

$$W(P) = \left(\sum_{i=1}^{k} w_i^r \right)^{1/r}$$

For $r = 1$, the path weight is the sum of the link weights along the path; for $r = 2$, the path weight is computed as euclidean distance is computed; and for $r = \infty$, the path weight is the same as the maximum weight associated with any link along the path. We will use "distance" in this chapter to mean the Minkowski distance (geodetic), which depends upon the value of the r-metric. The second parameter is the q parameter, which is a limit on the number of links in the paths examined in constructing a network. Its value determines the maximum number of links in paths in which the triangle inequalities are guaranteed to be satisfied in the resulting network. Our procedure generates families of PFNETs, and we can generate Hutchinson's (1989) networks as a special case with $r = \infty$ and $q = 2$.

The links omitted from a PFNET are omitted because they violate a triangle inequality involving q or fewer links. These omissions preserve all (geodetic) distances from the original data, however, and because not all links are present in most PFNETs, structural features are easier to ascertain. If a distance between two nodes not directly linked must be computed from the PFNET, it is computed using the Minkowski metric, resulting in a computed distance less than that given explicitly in the original data.

Advantages of PFNETs include (1) the capability of directly modeling asymmetrical relationships (Hutchinson, 1981; Tversky, 1977), which is more difficult with multidimensional scaling (Constantine & Gower, 1978; Harshman, Green, Wind, & Lundy, 1982; Krumhansl, 1978); (2) the provision for a complementary alternative to multidimensional scaling which often provides a more accurate representation of local data relationships than does multidimensional scaling; since multidimensional scaling must move data points to minimize a global error criterion, the resulting relationships between neighboring points is often significantly different than the original data would indicate; (3) the fact that hierarchical constraints in most cluster analysis techniques do not apply to PFNETs; (4) the representation of the most "salient" relationships present in the data; (5) the provision for a new paradigm in studying models of classification; and (6) the provision for a more quantitative paradigm for some of the issues in which networks have been invoked qualitatively or designed intuitively.

From the viewpoint of cognitive modeling, a disadvantage of PFNETs in the present state of development is that we have no way of knowing the features upon which similarity judgments are made. Thus the semantic content of links is not easily discernible (but see Cooke, Chapter 16, this volume). The empirical data we have collected, however, should be viewed as similarity estimates having components which may be unknown; but the use of such data seems important in bridging the gap between the more standard semantic networks (in which the researcher labels links according to his preferences or beliefs at the time) and a more objective representation of the knowledge of interest. For domains in which objective measures of distance are available, PFNETs provide unique representations of underlying structure not obtainable from any other scaling method.

Definitions and Alternatives

In this section we present definitions to provide the proper foundation for the generation procedures and theorems that follow. A PFNET has n nodes, denoted $N_1, N_2,..., N_n$ (or $N_a, N_b,...$). A *link* is an association between a pair of nodes which can be either undirected or directed. A *directed link* is called an *arc*, and an *undirected link* is called an *edge*. In this chapter we will deal mainly with networks having undirected links, or edges, but some of the definitions are more general, and a few examples of directed networks will be given to illustrate this generality. In this spirit, links are labeled e_{ij}, for the edge between N_i and N_j (or for the arc from N_i to N_j). N_i and N_j are *end nodes* of the link e_{ij}. The distance from node N_i to N_j (along the link e_{ij}) is the weight w_{ij}, and these weights are often written in matrix form as an $n \times n$ matrix W. The elements of W can be considered as distances between nodes along the direct paths between every pair of nodes. The distances are often considered as dissimilarities, and W is called either the *adjacency* or *weight* matrix. We assume that $w_{ii} = 0$ and $w_{ij} > 0$ for $1 \le i, j \le n$ where $i \ne j$. If this matrix is symmetric, then a PFNET derived from it is an undirected network. Typically the distance measures (weights) for each pair of entities (nodes) are found either empirically, from similarity estimates by human subjects, or analytically, using some appropriate measure of set intersection and set union, or some distance metric between entities.

A path from node N_a to node N_e, passing through nodes N_b, N_c, and N_d, is denoted by P_{abcde} (if the intermediate nodes are important) or P_{ae} otherwise. The former presumes the existence of edges e_{ab}, e_{bc}, e_{cd}, and e_{de} (either undirected or with appropriate directions), whereas P_{ae} presumes the existence of some unspecified set of edges (or arcs) connecting N_a and N_e. The weight of a path P is denoted $W(P_{ae})$, and the function $W(P)$ is determined by the r-metric and the weights w_{ij}.

The triangle inequality is incorporated into our generation procedure by means of the q parameter.

Definition 1

A network is q-triangular if and only if all possible triangle inequalities involving paths with $m \le q$ links are satisfied, using links and weights in the graph and the r-metric chosen. An example is the triangle inequality

$$w_{ae} \le \left(w_{ab}^r + w_{bc}^r + \ldots + w_{de}^r \right)^{1/r}$$

which is a constraint on the weights of two alternate paths between nodes N_a and N_e. For a graph with n nodes, there can be at most $n-1$ edges in any path in which there is no *cycle*. Thus the q parameter is at most $n-1$. Geodetic distances in the network are unchanged if edges which would violate triangle inequalities are omitted.

Definition 2

The (geodetic) network distance d_{ij} between nodes N_i and N_j is computed as a function of all path weights $W(P_{ij})$, for all paths P_{ij} which connect nodes N_i and N_j as

$$D_{ij} = \text{MIN} \left(W(P_{ij1}), W(P_{ij2}), \ldots, W(P_{ijm}) \right)$$

1 Properties of Pathfinder Networks

That is, the distance between two nodes is the weight of the smallest path between those nodes, with all path weights calculated using the (same) appropriate r-metric.

The r-metric and the q parameter provide the elements needed to assure that the networks generated from a particular set of proximity data possess the metric properties discussed in Hakimi and Yau (1964), with the following provisions:

1. The distance from a node to itself is assumed to be zero.
2. The data matrix must be symmetric so that the PFNET is undirected; then the distance between any pair of nodes is independent of direction.
3. The triangle inequality is satisfied for all paths having as many as q edges. To assure that no triangle inequalities whatsoever are violated, q can be set to the number of nodes less one.

For situations in which these metric axioms are satisfied, the concept of distance along a path is the same as the weight of that path. Because the r-metric can take on values from one through infinity, and the q parameter can take on values from one through the number of nodes less one, many different PFNETs can be constructed from a given set of proximity data. However, different values of r and q can result in the generation of the same (*isomorphic*) PFNETs. Frequently, important information from a given set of proximity data can be obtained from different PFNETs, constructed using different values of r and q. Thus it is often not essential that particular choices for r and q be made, to the exclusion of other values. Furthermore, it is sometimes desirable (in cognitive modeling, for example) to violate the metric axioms presented above (also in Hakimi & Yau, 1964; and in Tversky & Gati, 1982). The possibility of constructing directed PFNETs from asymmetric proximity data and (independently) of varying the q parameter provide ways of violating these axioms which correspond to observations about human performance (see, for example, Ortony, 1979; Tversky, 1977; and Tversky & Gati, 1982). Modeling traffic flow on one-way streets provides another example in which asymmetric data are relevant.

Definition 3

A *PFNET(r, q)* is a septuple (*N, E, W, LLR, LMR, r, q*) in which:

N is the set of nodes (concepts), denoted N_i;

E is a square matrix representing names of links in the *complete graph* (i.e., e_{ij} is the name of the link connecting nodes N_i and N_j);

W is the square weight matrix, and its entries are the weights associated with the links in the corresponding positions of the E matrix. The weights on the main diagonal are assumed to be zero, and the remaining weights are assumed to be finite and nonnegative. Thus w_{ij} is the weight of link e_{ij};

LLR, the link-labeling rule, is the procedure used to determine a label for each link, according to some classification scheme;

LMR, the link-membership rule, is the procedure used to determine whether or not each element of the E matrix is added to the PFNET(r, q);

r is the value of the r-metric, and $1 \le r \le \infty$;

q is the value of the q parameter, and $q \in \{1, 2, ..., n-1\}$, where n is the number of nodes.

Definition 4

The link-membership rule (LMR) for PFNETs (either directed or undirected) is given by the following procedure:

1. Define a network consisting of all nodes (concepts) N_i, but no links;
2. Order all elements e_{ij} of the E matrix in some nondecreasing order of their associated weights w_{ij};
3. Consider each e_{ij}, and include e_{ij} in the PFNET(r, q), if and only if e_{ij} provides a path from N_i to N_j which has a weight at least as small as the weight of any other path having no more than q links, using the r-metric to compute the weights of multiple-link paths.

This definition is useful primarily in establishing the concepts associated with Pathfinder networks; computationally efficient algorithms for generating PFNETs will be given in the next section. As an example of the LMR, consider the weight matrix:

$$W = \begin{matrix} 0 & 1 & 4 & 5 \\ 2 & 0 & 2 & 4 \\ 1 & 4 & 0 & 1 \\ 5 & 3 & 1 & 0 \end{matrix}$$

for the nodes N_1, N_2, N_3, and N_4. The complete graph is as shown in Figure 1. The arcs are not labeled because we have not yet developed a labeling rule for directed PFNETs. (Labeling edges with some LLR does not affect the edge membership of an undirected PFNET, because the edges are put there by the LMR, which makes no use of edge labels.)

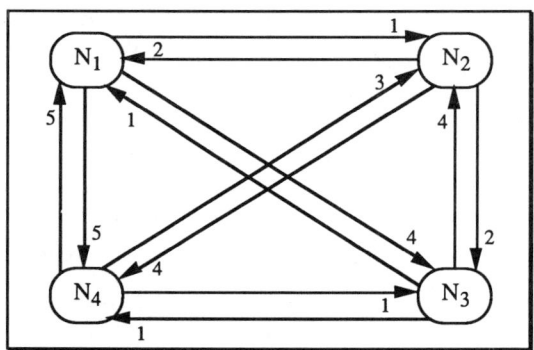

Figure 1. Complete graph for the example.

Let $r = 1$ and $q = 2$. Applying the link membership rule, the PFNET($r = 1, q = 2$) shown in Figure 2 is obtained. Note that e_{14} is in the PFNET because its weight ties with the weight of the path P_{124}, even though the arc e_{24} is not itself in the PFNET; if it were in the PFNET, it would violate the triangle inequality for the alternative path P_{234}. The path P_{1234} has less weight, but is not considered because it has three arcs, and for this example we assumed $q = 2$. The PFNET in Figure 2 is two-triangular, since the q parameter is two.

1 Properties of Pathfinder Networks

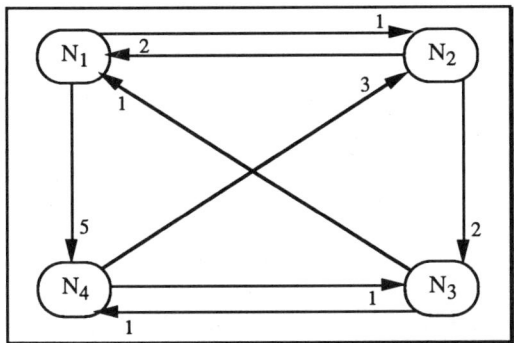

Figure 2. The PFNET($r = 1, q = 2$) corresponding to the complete graph of Figure 1.

Generation Algorithms for PFNETs

A part of our generation procedure, for either directed or undirected PFNETs, requires matrix operations using the weight matrix W. The purpose of these operations is to determine which links providing alternate paths are in the PFNET. For either directed or undirected PFNETs, the matrix operations can be used to determine link membership. These matrix operations find the minimum-weight path(s) having $i \leq q$ links between every pair of nodes, and finally, the minimum-weight path(s) having no more than q edges. The matrix computations are determined partly by the r-metric.

Definition 5
$W^{i+1} = W \oslash W^i$ is computed as follows:

$$w_{jk}^{i+1} = \text{MIN}\left((w_{jm})^r + (w_{mk}^i)^r\right)^{1/r} \text{ for } 1 \leq m \leq n$$

where $w_{jm} \geq 0$ and $w_{mk}^i \geq 0$.

Definition 6
The minimum-distance matrix for paths not exceeding i links is denoted D^i, and its elements are computed as follows:

$$d_{jk}^i = \text{MIN}\left(w_{jk}^1, w_{jk}^2, \ldots, w_{jk}^i\right) \text{ for } j \neq k$$

If the weight matrix W is asymmetric, then the corresponding PFNET(r, q) is directed. The generation procedure we have been using for such data does not label the arcs (a suitable labeling rule is under development). Thus the matrix operations described in definitions 5 and 6 provide the basis of a link membership rule for either directed or undirected PFNETs. Data required are (1) a weight (distance between nodes) matrix W, either symmetric or asymmetric, and (2) values for the r-metric and the q parameter. The following procedure removes links violating triangle inequalities as paths involving more links are considered.

Procedure PATHFINDER PFNET(r, q):
(without link labeling, for either symmetric or asymmetric W matrix)
1. Compute $W^2, D^2, W^3, D^3,..., W^q, D^q$;
2. Comparing elements of D^q and W^1, wherever $d_{ij} = w_{ij}$, then mark e_{ij} as a link in PFNET(r, q).

Actually, W^1 is the weight matrix, and D^1 is identical to it, because it is the distance matrix for paths having one link. W^2 must be computed, however, and provides the minimum cost for paths between each pair of nodes having exactly two links. D^2 provides the lower cost of either one-link or two-link paths for each pair of nodes. Similarly, D^3 provides the lowest cost of paths having one, two, or three links, and finally, D^{n-1} (where n is the number of nodes) would provide the lowest cost of paths having any number of links (without cycles) between every pair of nodes. Thus the second step of the procedure assures that every link e_{ij} in PFNET(r, q) provides a path between nodes N_i and N_j, which has a weight as small as any alternative path having from two to as many as q links. As each W^{i+1} and D^{i+1} are computed from W^i and D^i, links may be removed from the PFNET(r, q), but they cannot be added. Those links which are removed are called *redundant* links, because they do not affect any of the distances between nodes (Schvaneveldt, Dearholt, & Durso, 1988).

This procedure, which is an LMR, can be viewed as a means of uncovering the links responsible for the distance measurements between nodes and omitting all the other links; but if there are multiple paths having the same minimum cost, then links responsible for those ties are included, so that there is no arbitrary aspect to the LMR. As the value of the q parameter increases, links are removed as violations of the triangle inequality are discovered in longer and longer paths by the matrix operations. The distance matrix D^i, for $i < n-1$, may incorporate distances in which links not in the network are utilized as a part of the distance computation between a pair of nodes. This can be so only for cases in which there is a shorter alternative path having more than i links. Therefore, when $q = n-1$, every entry in D^{n-1} is computed using links which are in the network; otherwise, it would mean that some path not in the network is shorter than any path between the two nodes of interest composed of links which are in the network. The procedure guarantees that the weights of the shortest paths having q or fewer links are entered in the D^q matrix at each step.

For situations in which the W matrix must be symmetric (such as the information retrieval problem discussed in the chapter by Fowler and Dearholt (Chapter 12, this volume), or may be assumed to be symmetric, we have developed an LLR which yields important structural information. An LLR based on a node-covering paradigm is included as part of a generation procedure which will be given after a few more definitions.

Definition 7
A set of nodes (of an undirected PFNET) called a *node sublist* is a connected *subgraph* (at a stage of development) of a PFNET, and is denoted by NSL. A family of NSLs partitions the set of nodes, and when an edge joins nodes in two different NSLs, the two NSLs merge to form a single NSL, which consists of all of the nodes in the two original NSLs.

1 Properties of Pathfinder Networks

Definition 8

Let $E(w_i)$ denote the equivalence class of edges having the weight value w_i. Edges in a given equivalence class are assumed to have equal saliency.

The purpose of considering the NSLs defined by increasing equivalence class (weight) values is to establish the most salient edges first. The NSLs are merged as successive equivalence classes and are considered until the entire graph is connected. The NSLs also provide clustering information, as discussed later in this chapter in the section entitled Pathfinder Networks and Hierarchical Clustering.

The following definitions, concerned with edge labeling, distinguish three ways of incorporating nodes into a PFNET during construction.

Definition 9

For an undirected PFNET, a *primary* edge is, for some equivalence class of edges $E(w_i)$, the only path joining a node in some NSL to a single-node NSL. That is, a primary edge provides the only path between a node (NSL having node size one) and some other NSL for a particular equivalence class. A primary edge is labeled PRI. Within a network already constructed, an edge e_{ij} is primary if its weight is smaller than the weight of any other edge *incident* to one or both of the nodes N_i or N_j.

Definition 10

For an undirected PFNET, a *secondary* edge is, for some equivalence class $E(w_i)$, an edge joining node sublists which were distinct before $E(w_i)$ is considered, and in which there are either alternate paths to the end nodes, or the node size of both NSLs exceeds one. A secondary edge is labeled SEC. For primary or secondary edges, no alternative path exists in a smaller equivalence class. The primary edges are seen to provide the lowest cost connections between nodes, and the secondary edges either tie for low cost, or provide paths which connect nodes already connected to other nodes in a smaller (prior) equivalence class. For primary and secondary edges, inclusion into the PFNET is accomplished in stages by forming and merging NSLs much as clusters are merged in a hierarchical clustering scheme.

Definition 11

For an undirected PFNET, a *tertiary* edge is, for some equivalence class $E(w_i)$, a link joining nodes within a single NSL, so that alternate paths existed before the class $E(w_i)$ is considered. A tertiary edge is labeled TER.

For undirected PFNETs, the following more detailed procedure is equivalent to the LMR of Definition 4. This procedure also provides an LLR for the labeling of edges as PRI, SEC, or TER. This LLR has proven helpful in identifying categories and subcategories within the data, and will be related to hierarchical clustering discussed in the section entitled Pathfinder Networks and Hierarchical Clustering later in this chapter. Variations of the procedure have been implemented in APL, LISP, Pascal, and C. The concepts (nodes) N, the weight matrix W, and the values for the r-metric and q parameter are assumed to be given, and W is assumed to be symmetrical. This procedure is called *agglomerative* because it begins with each node or object placed in its own cluster, gradually merging these atomic clusters into larger and larger clusters until all nodes are merged into a single cluster or a *connected graph* (Jain & Dubes, 1988).

Procedure PATHFINDER PFNET(r, q)
(with link labeling, for symmetric matrix W)

1. Define a network consisting of all nodes N_i, but no edges;
2. Partition all edges into equivalence classes $E(w_i)$ according to the values of the weights w_i, and arrange these equivalence classes in increasing order $E(w_i)$, $E(w_j)$, ... according to the weight value of each class;
3. The initial node sublists are the individual nodes themselves, with no edges (equivalent to the *weak clustering* of Johnson, 1967);
4. Construct an *incidence table*, in which the nodes are column headings, and the edges, ordered by equivalence class according to weight values, are row headings. All edges in $E(w_1)$, the equivalence class having the smallest weight value, are listed as the first rows of the table;
5. For each edge e_{ij} in $E(w_1)$, place a check mark in columns N_i and N_j;
6. Any column with exactly one check mark in it identifies a primary edge, which is labeled PRI;
7. All other edges in $E(w_1)$ are labeled SEC;
8. The nodes in each NSL are marked with a symbol designating membership in the appropriate NSL;
9. All edges in $E(w_1)$ are added to the PFNET;

Beginning of Loop:

For each equivalence class $E(w_i)$, taken in increasing order of the weight values, do:

10. If there is only a single NSL (i.e., if all nodes are connected), then all remaining edges which have not been entered into the network as primary or secondary edges are considered as candidates for tertiary edges, and go to Step 15;
11. List the edges in $E(w_i)$ as row headings below the edges in $E(w_{i-1})$, and put check marks in the columns of the end nodes for each edge;
12. Any column with only one check mark throughout, for an edge in $E(w_i)$, identifies a new primary edge which is labeled PRI, and is entered into the network;
13. Any edge in $E(w_i)$ which connects two distinct NSLs, each having more than one node (as determined in $E(w_i)$), is labeled SEC and is entered into the network as a secondary edge (each unlabeled edge is a candidate tertiary edge, connecting nodes within some NSL);
14. The NSLs are relabeled to indicate the merging which has occurred as a result of entering new PRI or SEC edges into the network (note that an NSL can never split—the NSL structure is modified only by PRI or SEC edges merging two NSLs into a new, larger NSL which contains all nodes of both parent NSLs from the prior equivalence class);

End of Loop;

15. Compute W^q and D^q;
16. Wherever w_{ij} (from W^1) = d_{ij} (from D^q) and e_{ij} has not been previously labeled, then label e_{ij} as TER and enter it into the network;

End of Procedure.

1 Properties of Pathfinder Networks

The strategy used for tertiary edges (steps 15 and 16 of the procedure) could have been used to determine link membership for the entire PFNET (as in the earlier procedure for either directed or undirected PFNETs), but the disadvantage of this approach is that the procedure to label edges is less clear. Labeling edges after generation of the PFNET requires the use of a shortest-path algorithm. On the other hand, the procedure described above requires the matrix operations anyway, and thus appears cumbersome. From the procedure, it can easily be determined that every PFNET is connected—when the last NSLs are merged, the graph must be connected.

An example of the construction and labeling of an undirected PFNET is given now, with the symmetric weight matrix:

$$W = \begin{matrix} 0 & 1 & 9 & 5 & 9 & 7 \\ 1 & 0 & 1 & 9 & 9 & 9 \\ 9 & 1 & 0 & 2 & 9 & 9 \\ 5 & 9 & 2 & 0 & 1 & 9 \\ 9 & 9 & 9 & 1 & 0 & 1 \\ 7 & 9 & 9 & 9 & 1 & 0 \end{matrix}$$

For computational simplicity in the example, suppose that $r = 1$ and $q = 2$. Considering only edges in the upper triangular matrix because of symmetry, the equivalence classes are:

$E(1): e_{12}, e_{23}, e_{45}, e_{56}$
$E(2): e_{34}$
$E(5): e_{14}$
$E(7): e_{16}$
$E(9): e_{13}, e_{15}, e_{24}, e_{25}, e_{26}, e_{35}, e_{36}, e_{46}$

The incidence table for $E(1)$ is shown in Table 1.

Table 1. The incidence table for the first equivalence class $E(1)$.

Equivalence Class	Edge	NSL 1			NSL 2			Edge Label
		N_1	N_2	N_3	N_4	N_5	N_6	
$E(1)$	e_{12}	√	√					PRI
$E(1)$	e_{23}		√	√				PRI
$E(1)$	e_{45}				√	√		PRI
$E(1)$	e_{56}					√	√	PRI

Because the columns under N_1, N_3, N_4, and N_6 have only single check marks, the corresponding edges are primary. There are no secondary edges in $E(1)$ for this example. The next step is to label the nodes in the incidence table according to membership in an NSL. For this example, there are two NSLs after $E(1)$ is considered. The PFNET, after Step 9, consists of the edges shown in Figure 3.

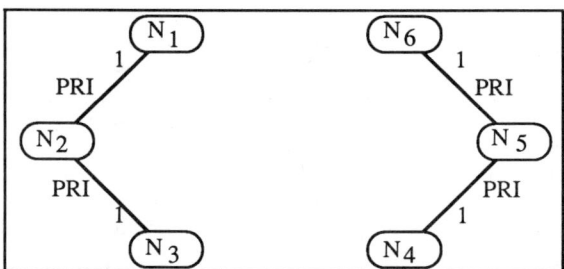

Figure 3. The PFNET after $E(1)$ is considered.

The equivalence class $E(2)$ is considered next, and its (only) edge e_{34} is appended to the edge-node table as the last row. For convenience, the NSL identifiers are placed above the node identifiers. The incidence table is now as shown in Table 2.

Table 2. The incidence table for the first two equivalence classes.

Equivalence Class	Edge	NSL 1						Edge Label
		N_1	N_2	N_3	N_4	N_5	N_6	
$E(1)$	e_{12}	√	√					PRI
	e_{23}		√	√				PRI
	e_{45}				√	√		PRI
	e_{56}					√	√	PRI
$E(2)$	e_{34}			√	√			SEC

Since e_{34} connects nodes in two different NSLs, each having node size greater than one, it is a secondary edge, and it is entered in the PFNET. Because it joins the only two NSLs, none of the candidate edges in the remaining equivalence classes can be either primary or secondary edges. Thus we consider $E(5)$, $E(7)$, and $E(9)$ together (reentering the top of the loop at Step 10 in the procedure), because there is now only one NSL. We next compute $W^2 = W \bigodot W$ as previously described, and

$$W^2 = \begin{matrix} - & 10 & 2 & 10 & 6 & 10 \\ 10 & - & 10 & 3 & 10 & 8 \\ 2 & 10 & - & 10 & 3 & 10 \\ 10 & 3 & 10 & - & 10 & 2 \\ 6 & 10 & 3 & 10 & - & 10 \\ 10 & 8 & 10 & 2 & 10 & - \end{matrix}$$

For $r = 1$, the weight of a path is the sum of the weights of the edges along the path. The entries on the main diagonal of W are not relevant, because they consist of two-edge paths only in a degenerate sense: either (1) no edges are traversed from the indicated node, or (2) an edge is traversed, but then immediately retraced. Similarly, two-edge paths

1 Properties of Pathfinder Networks

involving a zero-valued entry in W are omitted from consideration in off-diagonal elements of W because such paths really consist only of a single edge. Next, D^2 is found by taking the smaller value in each position of W and W^2 and noting from which matrix this smaller distance came.

$$D^2 = \begin{matrix} - & 1 & 2 & 5 & 6 & 7 \\ 1 & - & 1 & 3 & 9 & 8 \\ 2 & 1 & - & 2 & 3 & 9 \\ 5 & 3 & 2 & - & 1 & 2 \\ 6 & 9 & 3 & 1 & - & 1 \\ 7 & 8 & 9 & 2 & 1 & - \end{matrix}$$

Comparing elements of D^2 to elements of W, the entries which are equal are in the locations corresponding to e_{12}, e_{14}, e_{16}, e_{23}, e_{25}, e_{34}, e_{36}, e_{45}, and e_{56}. The edges in this list which have not yet been labeled are labeled TER and are added to the PFNET, shown in final form in Figure 4.

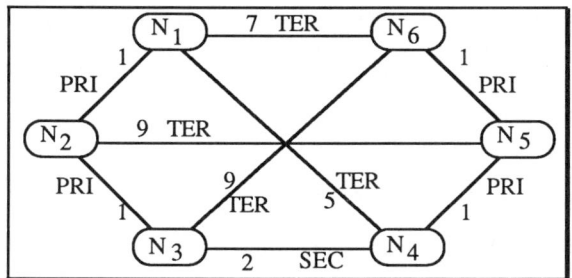

Figure 4. The labeled PFNET($r = 1$, $q = 2$) for the example.

Since no NSLs can be merged, and all edges have been considered, there can be no more edges added, and the procedure terminates.

This network illustrates two aspects of the triangle inequality: No link violates the triangle inequality considering paths of two links ($q = 2$), and some links do violate a triangle inequality considering paths of three or more links. An example of the latter is e_{14}, in which $w_{14} = 5$. An alternate (three-link) path is P_{1234}, which has a weight of 4. Similarly, $w_{16} = 7$ and the path P_{123456} with weight of 6 illustrates the violation of a triangle inequality involving a five-link path.

Fundamental Properties

It is appropriate to begin by showing that the two generation procedures just described add precisely the same edges to the PFNET as the LMR given in Definition 4, for the situation in which the W matrix is symmetrical (the case in which both procedures apply).

Theorem 1

For a given r, q, and symmetric W, either of the PATHFINDER PFNET(r, q) procedures given in the preceding section results in the same edges in the PFNET as does the

LMR given in Definition 4. (Recall that one of the procedures is intended for either asymmetric or symmetric weight matrices, and the other is only for symmetric matrices. An asymmetric weight matrix yields directed networks, and a symmetric weight matrix yields undirected networks, having labels on the edges if the latter procedure is used.)

Proof

We begin the proof by considering the procedure for symmetric W. The edges labeled PRI and SEC, at the time of labeling and entry into the PFNET, represent the only path, or equal and minimum-cost paths, between the nodes involved. Because edges are considered by equivalence classes of increasing weight, no edge to be added later can decrease the cost of the direct path afforded by a single primary or secondary edge. A tertiary edge provides a minimum-cost path alternative to multiple-edge paths of PRI and SEC edges. The matrix operations in Step 15 of the generation procedure for symmetric weight matrices find all edges which provide lowest-cost direct paths, considering all paths having up to q edges. These matrix operations identify all edges in the PFNET(r, q), whether or not the weight matrix W is symmetric, and independently of the edge labeling rule. If the edge labels are not needed, then these matrix operations alone identify all edges providing minimum-cost links between nodes, considering multiple-link paths of q or fewer edges. Thus the Pathfinder procedure used for asymmetric data, and the procedure for symmetric data, both result in the same LMR as given in Definition 4. □

By definition of the LMR, and as described in the procedure, edges providing alternate, equal-cost paths are all entered in the PFNET. This feature eliminates a random or arbitrary choice of edges to represent a particular association, and results in a unique network, given a particular weight matrix W, and values for r and q.

A network of particular interest is PFNET$(r = \infty, q = n-1)$. This PFNET always has the minimum number of edges, because the path-length metric is simply the value of the maximum weight along the path (this is called the *dominant* metric), and no edge in the PFNET can result in the violation of a triangle inequality for any path. The maximum number of edges in a path without a cycle, for a graph having n nodes, is $n-1$. The edges in an undirected PFNET$(r = \infty, q = n-1)$ are all labeled PRI or SEC, because (1) PRI and SEC edges are added to the PFNET based on node-covering properties and are independent of values of r and q, and (2) since TER edges always provide an alternate path to PRI or SEC edges established in an earlier equivalence class, the path length with $r = \infty$ is always less through the PRI or SEC edges. Thus tertiary edges never appear in an undirected PFNET$(r = \infty, q = n-1)$. For a given symmetric weight matrix W, if PFNET(r, q) is generated, then PFNET$(r = \infty, q = n-1)$ can always be found simply by deleting all edges labeled TER in the PFNET(r, q).

Definition 12

A minimum-cost spanning *tree* (mintree) of an undirected PFNET is a connected graph having no cycles in which (1) there is a path between any pair of nodes, and (2) the sum of the weights associated with the edges is a minimum.

This definition is similar to that given in Even (1979, Chapter 2). A PFNET can have several mintrees; a simple example is a PFNET in which each off-diagonal weight in W is one. Then all potential edges are in the first (and only) equivalence class, all edges are secondary edges, and all edges are entered in the PFNET. Clearly there are several mintrees for such a PFNET (which is the complete graph), although this is an extreme case.

1 Properties of Pathfinder Networks

Definition 13
The *minimum-cost network* (MCN) of an undirected PFNET(r, q) is the union of all mintrees of the PFNET.

Theorem 2
For a given symmetric W, r, and q, the MCN of PFNET(r, q) is PFNET($\infty, n-1$).

Proof
Every primary edge is in every mintree, because a primary edge provides the lowest-cost access to at least one of the end nodes responsible for the edge being labeled PRI. Each secondary edge is in some mintree, because a secondary edge provides either (1) a unique, lowest-cost path between two NSLs for which the node size is greater than one, or (2) an equally lowest-cost alternative path between two NSLs, in which the alternatives are provided by (secondary) edges in the same equivalence class. In the first case, the secondary edge is in every mintree; in the second case, alternative mintrees correspond to the secondary edges providing alternative paths in the same equivalence class. A tertiary edge cannot be in any mintree of PFNET(r, q), because such an edge provides an alternative path (covering the same end nodes) to one already existing. Thus the tertiary edges always form cycles, and there is always an alternative path to a tertiary edge which is composed of primary and secondary edges established in earlier equivalence classes. Therefore the PFNET($\infty, n-1$) is independent of the r-metric and the q parameter, and includes all primary and secondary edges (as does the MCN). □

The significance of this result is that the PFNET which has fewest edges, the MCN, is unique, provides all minimum-cost paths between nodes, has no violations of the triangle inequality, and is the sparsest network which possesses these properties. For cognitive modeling, the scaling of data, and for clustering, these properties of economy are significant indeed.

Properties of Inclusion

The networks generated in Procedure PATHFINDER PFNET(r, q) possess properties of inclusion, or nesting, as values of the r-metric and q parameter change.

Definition 14
A PFNET1 is *included in* (or is a *spanning subgraph* of) PFNET2 if and only if:

(1) PFNET1 and PFNET2 have the same set of nodes, and

(2) every link in PFNET1 is also in PFNET2.

The first inclusion property to be discussed is related to the r-metric used in computing path lengths within a PFNET.

Theorem 3
For a weight matrix W, PFNET(r_2, q) is a spanning subgraph of PFNET(r_1, q) if and only if $r_1 \leq r_2$.

Proof
   Since edge labeling is not a consideration, we will show inclusion by using the Pathfinder procedure that uses only the matrix operations. First suppose that $r_1 \leq r_2$, and that e_{ij} is a link in PFNET(r_2, q), then

$$w_{ij}(r_2) = d_{ij}^q(r_2).$$

   We must show that e_{ij} is also in PFNET(r_1, q). The reason that e_{ij} is in PFNET(r_2, q) is that alternative paths having more than one link are at least as costly; these path lengths are computed in $W^2, W^3, ..., W^q$. The following inequality is helpful:

$$\left(\sum_{i=1}^{m}(w_i)^{r_1}\right)^{\frac{1}{r_1}} \geq \left(\sum_{i=1}^{m}(w_i)^{r_2}\right)^{\frac{1}{r_2}} \text{ if and only if } r_1 \leq r_2.$$

   This inequality is well-known in mathematics, and a proof and references are given in Schvaneveldt, Dearholt, and Durso (1988). Thus a given path that is longer than w_{ij} using $r = r_2$ is at least as long using $r = r_1$. Therefore the matrix operations which identify links by computing these alternative path lengths cannot result in a shorter alternative path, and e_{ij} is in PFNET(r_1, q) also. But suppose that the minimum-length paths connecting N_i and N_j in PFNET(r_1, q) and PFNET(r_2, q) are different, because of a single link e_{ij} which is in PFNET(r_1, q) but is not in PFNET(r_2, q). Therefore we know that

$$w_{ij} \leq \left(w_1^{r_1} + w_2^{r_1} + ... + w_m^{r_1}\right)^{\frac{1}{r_1}}$$

and

$$w_{ij} > \left(w_1^{r_2} + w_2^{r_2} + ... + w_m^{r_2}\right)^{\frac{1}{r_2}}$$

   Note that, since $r_1 \leq r_2$, the converse is not possible (because of the previous inequality); that is, a link e_{ij} cannot be in PFNET(r_2, q) without also being in PFNET(r_1, q).
   Now suppose that if e_{ij} is in PFNET(r_2, q), then e_{ij} is also in PFNET(r_1, q). Then any path that exists in PFNET(r_2, q) also exists in PFNET(r_1, q). We must show that $r_1 \leq r_2$. Therefore

$$w_{ij} \leq \left(w_1^{r_2} + w_2^{r_2} + ... + w_m^{r_2}\right)^{\frac{1}{r_2}}$$

implies

$$w_{ij} \leq \left(w_1^{r_1} + w_2^{r_1} + ... + w_m^{r_1}\right)^{\frac{1}{r_1}}$$

over the same path in each network, by hypothesis.

1 Properties of Pathfinder Networks

Therefore

$$\left(w_1^{r_2} + w_2^{r_2} + \ldots + w_m^{r_2}\right)^{\frac{1}{r_2}} \leq \left(w_1^{r_1} + w_2^{r_1} + \ldots + w_m^{r_1}\right)^{\frac{1}{r_1}}$$

which is true provided $r_1 \leq r_2$, as given in the inequality. □

An observation for the symmetric case, in which the edge-labeling procedure is used, is that PFNET(r_1, q) and PFNET(r_2, q) share the same MCN, which is PFNET($\infty, n-1$). Thus PFNET(r_1, q) and PFNET(r_2, q) have the same primary and secondary edges, and may differ only in the tertiary edges, as discussed previously.

The second inclusion property concerns the q parameter, in a manner analogous to the r-metric.

Theorem 4

For a weight matrix W, PFNET(r, q_2) is a spanning subgraph of PFNET(r, q_1) if and only if $q_1 \leq q_2$.

Proof

It is sufficient to observe that, as W^{i+1} is computed from W^i using the given r-metric, links to the resulting PFNET cannot be added, but may be eliminated. This is true because W^i and D^i determine the links satisfying the triangle inequality considering paths of i or fewer links. A potential link e_{jk} not satisfying such a triangle inequality results in an entry

$$d_{jk}^i \leq w_{jk}$$

at step i, so that e_{jk} will not be in PFNET($r, q = i$). □

The inclusion relations can be illustrated by means of a graph in two-dimensional space. As shown in Figure 5, we will use the r-metric for the abscissa axis and the q parameter for the ordinate axis. Then the r and q values used for a particular PFNET are plotted at the intersection of the two lines. The inclusion relations for other PFNETs computed from the same data, but in which different values of r and q are used, are the rectangular areas shown with hatch marks. The PFNET represented by the point (r, q) is a spanning subgraph of every PFNET represented by the r and q values below and to the left, since inclusion is transitive. Similarly, the PFNET represented by the point (r, q) includes every PFNET represented by the r and q values above and to the right. In practice, the PFNET having fewest links (the MCN) is generated with r and q values substantially smaller than the theoretical values of $r = \infty$ and $q = n-1$, which are the values guaranteed to produce the MCN.

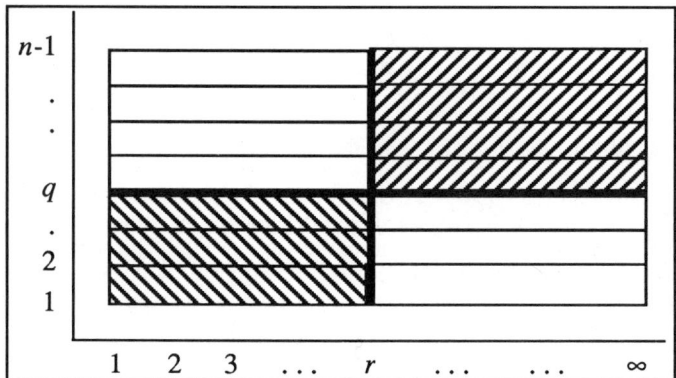

Figure 5. Illustration of the inclusion relations.

Data Transformations and Network Structure

An important issue in data scaling and modeling concerns the effects upon network structure of various types of transformations upon the proximity data. Uncertainty associated with the meaning of data values should be accommodated by appropriate options in the scaling procedure.

Definition 15

Let T be a transformation to be applied to the elements w_{ij} of the weight matrix W, and denote the transformed weight values as $T(w_{ij}) = t_{ij}$. The transformed weight matrix is denoted Wt, and the resulting network is denoted by PFNETt.

The first result is that a multiplicative transformation preserves link structure for both directed and undirected PFNETs, and edge labels are also preserved for undirected, labeled PFNETs.

Theorem 5

Given nodes N_i and a weight matrix W, resulting in PFNET(r, q), and given the transformation,

$$T: t_{ij} = bw_{ij} \text{ where } b > 0,$$

then e_{ij} is in PFNET(r, q) if and only if e_{ij} is in PFNETt(r, q). If e_{ij} is in both networks, and if the networks are undirected and labeled, then it has the same label in both networks.

Proof

We first consider the undirected, labeled case. The transformation T can be viewed as transforming the equivalence classes $E(w_i)$ into equivalence classes $E(bw_i)$. The transformed (corresponding) equivalence classes are hence considered in the same order as the original equivalence classes. Thus the construction of the incidence tables is identical for both PFNET(r, q) and PFNETt(r, q) so the primary and secondary edge structure of the two networks is isomorphic.

1 Properties of Pathfinder Networks

For the tertiary edges, the matrix operations can be expressed by the observation that e_{ij} is added to PFNET$t(r, q)$ if and only if

$$bw_{ij} \leq \left(\sum_{k=1}^{K}(bw_k)^r\right)^{1/r} = b\left(\sum_{k=1}^{K}(w_k)^r\right)^{1/r}$$

where $K \leq q$, for any path connecting N_i and N_j. Both sides can be divided by b to obtain the comparison

$$w_{ij} \leq \left(\sum_{k=1}^{K}(w_k)^r\right)^{1/r}$$

which is the inequality used to determine whether or not e_{ij} is added to PFNET(r, q). Thus e_{ij} is either added to both networks, with the same label in each, or it is not added to either. □

That arc structure is preserved for the directed PFNETs under multiplicative transformations is evident by referring to the matrix operations invoked for tertiary edges, in the undirected case just discussed. The multiplier b factors out, and the decisions for all arcs are the same, with or without the multiplier.

The other type of transformation we consider is a monotonic transformation T.
T: $t_{ij} = f(w_{ij})$ so that,

If $w_{ab} = w_{xy}$, then $t_{ab} = t_{xy}$, and
If $w_{ab} > w_{xy}$, then $t_{ab} > t_{xy}$.

Our second result is that, provided $r = \infty$, a monotonic transformation preserves link structure for both directed and undirected PFNETs, and preserves edge labels for undirected, labeled PFNETs.

Theorem 6
Given nodes N_i and a weight matrix W, and given a monotonic transformation,
$$T: t_{ij} = f(w_{ij}),$$
then e_{ij} is in PFNET$(r = \infty, q)$ if and only if e_{ij} is in PFNET$t(r = \infty, q)$. If e_{ij} is in both networks, and if the networks are undirected and labeled, then it has the same label in both networks.

Proof
We first consider the undirected, labeled case. The transformation T can be viewed as transforming the equivalence classes $E(w_i)$ into equivalence classes $E(f(w_i))$. Furthermore, the order of the transformed equivalence classes remains the same for the generation procedure. Thus the incidence tables are the same for each network, so the primary and secondary edges are also the same for each network. With $r = \infty$, the tertiary edges in the

PFNET are determined by the matrix operations described earlier in the section entitled Generation Algorithms for PFNETs, and the maximum weight on an edge in a path determines the cost of that path. But the transformed maximum weight is the same as the maximum of the transformed weights, since monotonic transformations preserve order. Thus the same tertiary edges are in the PFNET in either the transformed version or the original version. Therefore e_{ij} is added to both networks, with the same label in each, or it is not added to either.

To see that arc structure is preserved for the directed case, we note that a monotonic transformation preserves order. Thus the same decisions are made regarding arc membership on either the original data or the transformed data, provided $r = \infty$. □

In considering properties of data, ordinal data are presumed to be data in which the experimenter is confident that the data values are, at the minimum, in proper order. Further claims on properties of the data, such as the meaning of intervals or of ratios, however, may not be warranted. For these cases, the PFNETs used should be constrained to those in which $r = \infty$. Ratio-scale data are data in which each data value is presumed to be within a multiplicative constant of the "correct" value. For these types of data, any values of r and q can be used in generating a PFNET, for a given data matrix, and the resulting PFNET is independent of such multipliers.

Pathfinder Networks and Hierarchical Clustering

For approximately two decades now, hierarchical clustering has been an important tool in the analysis of proximity data. For certain types of data, it is very appropriate and leads to meaningful clusters, or more precisely, to meaningful families of clusters. The purpose of this section is to explicate the relationships between hierarchical clustering and Pathfinder networks. We will show that Pathfinder networks can provide the same information that is available in the minimum method of Johnson (1967), also called the single-linkage (or single-link) method (Ling, 1972; Sokal & Sneath, 1963). Furthermore, we will show that hierarchical clustering cannot provide the structural information available in Pathfinder networks.

For hierarchical clustering schemes (abbreviated HCS for convenience), distance is regarded as being the primary feature of interest. For Pathfinder networks, however, both structural properties and distance are regarded as important. As shown in the preceding section, the structure of a Pathfinder network is invariant under multiplicative transformations of the weight matrix; and for $r = \infty$, the structure is invariant under monotonic transformations of the weight matrix. Pathfinder networks maintain the original (most salient) proximities (associations) of the data by preserving all minimum-cost paths, thus supporting the representation of these associations explicitly in the network (as shown in Theorem 2, each Pathfinder network contains all mintrees). Thus geodetic paths (minimum distance paths between each pair of nodes) are preserved in Pathfinder networks, and the computation of the distances is a part of the generation procedure (Step 15). But why, if the scaling of data is the objective, should organization be of any interest or consequence when distances are all that is required for clustering? The answer lies in the origin of Pathfinder networks in modeling the organization of human semantic memory and in other applications of these networks which have been explored recently.

We will use *object* (in the same way as Johnson, 1967) to mean either a single entity or node, or a cluster of entities or nodes; the context will be sufficient to establish the precise meaning if it is important. A complication for HCS and Pathfinder is the question of

1 Properties of Pathfinder Networks

measuring the distance between two objects. Should this distance be (1) the smallest distance between some object in one cluster and an object in the other cluster, as in the minimum method; (2) the largest distance between an object in one cluster and an object in the other cluster, as in the maximum or complete-link method; (3) the median distance of the between-cluster distances between objects; or (4) some average distance, perhaps weighted, of the distances between pairs of objects in which one object is in each cluster? Choices (1), (2), and (3) require only ordinal properties of data, while (4) requires ratio properties. The latter is not often assumed for subjective data, but for problems in which objective data are available, then ratio assumptions are frequently appropriate. Because the computations are simple, the minimum method and the maximum method, in the terminology of Johnson (1967), have been examined in great detail for many applications of psychological proximity data.

An assumption made by Johnson (1967, p. 249) is that the off-diagonal distances in the original matrix are all positive and *distinct* (the distances on the main diagonal are all assumed to be zero). This assumption of distinctness of the off-diagonal distances is motivated by his assumption that the data are strictly ordinal, in the sense that the subjects or researchers can differentiate between each pair of entities. Nevertheless, we have found this assumption to be unwarranted—real data we have collected have frequently exhibited multiple entries having the same values. Furthermore, data representing large systems are quite likely to have large numbers of ties. Thus attempts to model proximity data must include the general case in which data can exhibit ties. As illustrated in several of the examples in this chapter, ties in the data are easily modeled in Pathfinder by including them as links in the network (provided they offer paths having smallest weights between node sublists).

We will now review the agglomerative procedure described in Johnson (1967, p. 248) for generating an HCS using the minimum method, although Johnson does not, in that article, begin from a general weight or *adjacency matrix*; instead, he shows only a distance matrix which could have been reduced from any one of a large number of different weight matrices by application of the HCS procedure. He assumes, however, that a weight matrix is available at the beginning of the procedure. In the following statement of Johnson's procedure, the first part of each step is taken verbatim from his paper (1967); the parenthetical comments concluding each step relate the terminology he used to that used for Pathfinder and also include additional comments intended to be helpful.

Johnson's HCS Procedure:

1. Clustering C_0, with value 0, is the *weak clustering*. (It is called the weak clustering because each object (node) is considered to be an individual cluster. The value is simply the distance to other nodes, and corresponds to the equivalence classes discussed in the section Generation Algorithms for PFNETs.)

2. Assume we are given the clustering C_{j-1} with the similarity function d, defined for all objects or clusters in C_{j-1}. Let α_j be a minimal nonzero entry in the matrix. Merge the pair of objects and/or clusters with distance α_j to create C_j, of value α_j. (This step is ambivalent, because "a minimal nonzero entry" seems to imply that there could be ties in the data, but Johnson specifically excludes ties in his discussion of clustering. The merging of two clusters can be considered equivalent to the joining of two NSLs with an edge in the graphical paradigm of Pathfinder. Johnson's similarity function corresponds to the distance matrix of the Pathfinder procedure.)

3. We create a new similarity function for C_j in the following manner: If x and y are clustered in C_j and not in C_{j-1} (i.e., $d(x, y) = \alpha_j$), we define the distance from the cluster $[x, y]$ to any third object or cluster, z, by $d([x, y], z) = \min [d(x, z), d(y, z)]$.

   If x and y are objects and/or clusters in C_{j-1} not clustered in C_j, $d(x, y)$ remains the same. We obtain a new similarity function d for C_j in this way. (The distance between two clusters is defined to be the minimum distance between any pair of elements in which one element is in each cluster. That is, if x and y are two objects—an object can be either a single element or a cluster—at level C_{j-1}, and if $d(x,y) = \alpha_j$ (so that x and y become clustered in C_j), and if z is any other object or cluster at level C_{j-1}, then $d(x,z) = d(y,z)$. Johnson (1967, p. 245) gives a brief proof of this statement, and it is equivalent to the use of infinity for the value of the r-metric, which also implies the use of the *ultrametric inequality* of Johnson.)

4. We now repeat steps 2 and 3 until we finally obtain the *strong clustering*—we are then finished. (The strong clustering is one in which all elements (nodes) are in the same cluster. In the paradigm of Pathfinder, the nodes and edges then form a connected graph.)

Because HCS were not designed to incorporate subtle aspects of structure, it is not surprising that they have limitations in structural matters. Some of these are:

1. If ties in the proximity data are allowed, then the mapping of weight matrices to HCS is not one-to-one; that is, several different weight matrices can yield the same clustering in an HCS (this is also true of PFNETs, but to a lesser extent as will be shown later).

2. Nonhierarchical structural relationships cannot be represented. There are two ways these can occur; first, equal weights can lead to cycles, in the graphical paradigm, and second, clusters can be overlapping, as described in Shepard and Arabie (1979). We will address only the first nonhierarchical situation, as Pathfinder has not yet been sufficiently developed to apply it to the latter.

3. Information regarding the pair of objects or nodes responsible for establishing the distance between two clusters (that is, which pair has the most "salient" relationship) is lost once the clustering is established.

Ties in proximity data used in HCS have been regarded as a problem for many years. Furthermore, there are significant differences in the single-link (minimum) and complete-link (maximum) methods in dealing with ties. The single-link method has a continuity property (Jain & Dubes, 1988, Section 3.2.6) which insures that adding or subtracting small amounts to tied values results in dendograms which merge smoothly into the same dendogram as the added amounts tend to zero, no matter how the ties are broken. The complete-link method, however, does not possess this property, and can yield different dendograms depending upon how the ties are resolved. As mentioned previously, the ultrametric inequality is equivalent to the use of $r = \infty$ in PFNETs; the weight of a path between nodes (objects) not directly linked is the maximum of any of the weights on links making up that path, as computed by the Minkowski metric as r approached infinity. By

1 Properties of Pathfinder Networks

modifying Step 2 of Johnson's procedure, we obtain a version of his minimum method which models ties in the proximity data. This modified step is:

2'. Assume we are given the clustering C_{j-1} with the similarity function d, defined for all objects or clusters in C_{j-1}. Let α_j be a minimal nonzero entry in the matrix. Merge *all pairs of objects and/or clusters* (formerly read: the pair of objects and/or clusters) with distance α_j to create C_j of value α_j.

We will use UHCS to denote the HCS with this step substituted in Johnson's minimum method, because a unique dendogram is generated even if there are ties in the weight matrix (the alternative is to break ties arbitrarily). The next example will illustrate this modified method, and will also show that the mapping from the weight matrix to the UHCS is not one-to-one.

$$W_1 = \begin{matrix} 0 & 1 & 1 & 5 \\ 1 & 0 & 4 & 2 \\ 1 & 4 & 0 & 5 \\ 5 & 2 & 5 & 0 \end{matrix}$$

The UHCS using the minimum method for this matrix is shown in Figure 6, and the PFNET($r = \infty, q = n-1$) for the matrix is shown in Figure 7.

The matrix W_2, shown below, has the same UHCS as does W_1; however, the structural aspects relating to specific associations are not a factor in obtaining the UHCS.

$$W_2 = \begin{matrix} 0 & 1 & 4 & 5 \\ 1 & 0 & 1 & 2 \\ 4 & 1 & 0 & 5 \\ 5 & 2 & 5 & 0 \end{matrix}$$

The PFNET($r = \infty, q = n-1$), however, is different for W_2, as shown in Figure 8.

A bit of combinatorics shows that there are, in fact, 12 different matrices (and 12 different PFNETs) which have the same UHCS as W_1 or W_2 (nodes 1, 2, and 3 can have any one edge missing of the three edges that are possible, or all three present, and the link between this cluster and node 4 can be between node 4 and any one of the other three nodes). If the weight matrix is viewed as a representation of the complete graph, then any change in a weight or weights sufficient to change any

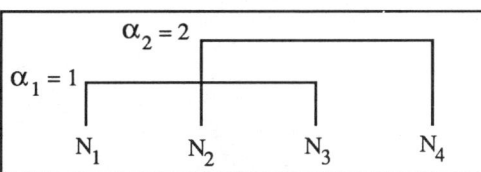

Figure 6. The UHCS for matrices W_1 and W_2.

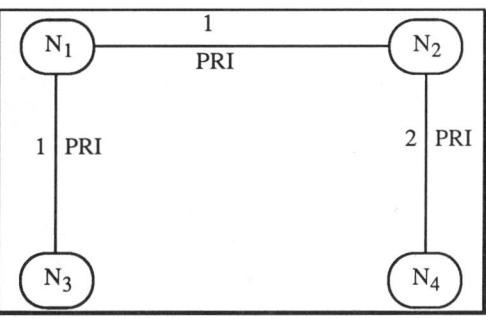

Figure 7. PFNET($r = \infty, q = n-1$) for W_1.

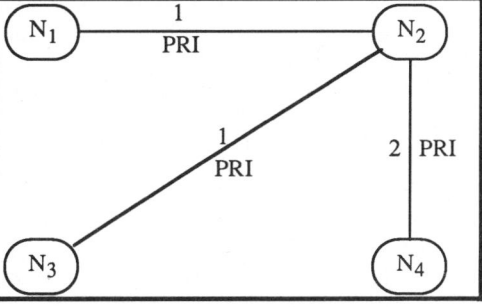

Figure 8. PFNET($r = \infty, q = n-1$) for W_2.

mintree will modify the structure of PFNET(∞, $n-1$), but may not necessarily modify the UHCS. Therefore the primary and secondary edges, considered together, define the same clusters via the node sublists as the UHCS. The labels used in PFNETs (primary, secondary, and tertiary) thus distinguish between edges in ways not possible in HCS or UHCS. Again, if distance is the only concern, then one needs only the UHCS; but if navigating through a data structure by following links is important, or if computing cluster metrics using structural information is necessary, then the structural differences can be important. HCS or UHCS are hierarchical and unambiguous only in the preservation of distances between entities.

While different weight matrices can also yield the same PFNETs, structure (the nodes linked because of salient relationships) depends only upon the minimum-distance paths, which are always preserved in PFNETs. Stated another way, if a given weight is sufficiently large, the corresponding edge will not be in the PFNET. For example, in the matrix W_2 above, suppose that the distance w_{13} (which has a value of 4 in this example) is allowed to vary. Provided that w_{13} has a value greater than 1, then the structure of PFNET($r = \infty$, $q = 3$) remains the same as in Figure 8, since no mintree is changed.

There are other graphical approaches to hierarchical clustering. An example is the family of *threshold graphs*, undirected graphs in which link membership is determined by selecting successively larger thresholds, beginning with the smallest weight value in the weight matrix. For a given threshold, each link having a weight less than or equal to that threshold is in the threshold graph. As the threshold is increased in value, a new graph is generated and the new graph typically has more links (it never has fewer links). Ultimately, the threshold is made large enough to include sufficient links to make the graph connected. This occurs precisely when the threshold value equals the value of α_i which would yield the strong clustering defined in Johnson's procedure. A *proximity graph* is a threshold graph with the links labeled with their corresponding weights. Threshold and proximity graphs are discussed in Jain and Dubes (1988). These graphs simply provide an alternative perspective on the formation of the clusters using the minimum method. A disadvantage, however, is that they have only the context supplied by the global threshold on weight values. In contrast, a PFNET has a more local context provided by the triangle inequality, in which a link with a certain weight value may be included in one part of the network, while another link having the same weight value may not be included in another part of the network. Thus a connected threshold graph can have more links than PFNET ($r = \infty$, $q = n-1$), but cannot have fewer links. However, we will show that the same family of clusters is obtained from the PFNETs.

Before presenting two theorems, we will discuss the relationships between the minimum method UHCS and Pathfinder networks with respect to points 2 and 3 above (modeling ties through cycles in the graphical paradigm, and maintaining the information regarding the pair of entities responsible for establishing the distance between two clusters), in terms of both an example and the procedures involved for the generation of the clustering information and the Pathfinder networks.

We will argue that (1) the information available in a minimum method UHCS is also available in PFNET($r = \infty, q = n-1$), and therefore (2) is also available in any PFNET (r, q) by the inclusion theorems, although some computation may be required, and finally (3) that the converses of (1) and (2) do not hold.

1 Properties of Pathfinder Networks

The aspects in which the minimum method of UHCS and Pathfinder networks are equivalent will be discussed in terms of the following example:

$$W_3 = \begin{matrix} 0 & 1 & 4 & 5 & 6 & 7 \\ 1 & 0 & 4 & 7 & 8 & 6 \\ 4 & 4 & 0 & 2 & 6 & 8 \\ 5 & 7 & 2 & 0 & 5 & 7 \\ 6 & 8 & 6 & 5 & 0 & 3 \\ 7 & 6 & 8 & 7 & 3 & 0 \end{matrix}$$

The incidence table is constructed below, and the membership in node sublists (NSLs) is indicated by the tree growing from the top of the incidence table:

Table 3. The incidence table for the example.

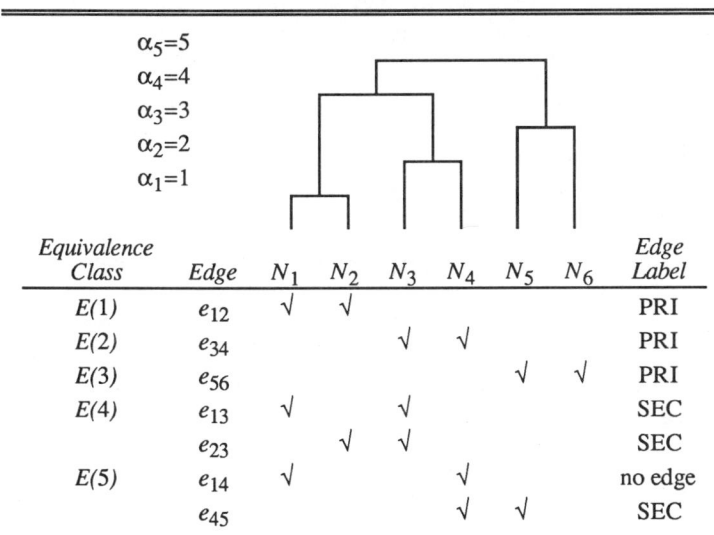

Equivalence Class	Edge	N_1	N_2	N_3	N_4	N_5	N_6	Edge Label
E(1)	e_{12}	√	√					PRI
E(2)	e_{34}			√	√			PRI
E(3)	e_{56}					√	√	PRI
E(4)	e_{13}	√		√				SEC
	e_{23}		√	√				SEC
E(5)	e_{14}	√			√			no edge
	e_{45}				√	√		SEC

There are two observations which can be made on the basis of this example. First, the α_j levels of Johnson (1967) correspond to the equivalence classes $E(w_j)$, which in turn correspond to weights of the Pathfinder algorithm. Second, the clusters formed by the minimum method UHCS are the same as the NSLs of the Pathfinder procedure for symmetric data. A proof showing that the latter claim is general will be offered in Theorem 7.

In the third step of Johnson's HCS procedure, a *similarity function* is updated after each α_j is considered. Usually given in matrix form, this function provides the distances between objects and has dimension equal to the number of objects after merging at each new value of α_j. There is a simple procedure for computing similar distances between nodes in the agglomerative Pathfinder generation procedure illustrated above. This procedure derives a square matrix which we will denote *DM*, and is as follows:

1. Begin with the weight matrix W;
2. Using node membership within NSLs, eliminate from consideration all w_{ij} in W in which N_i and N_j are in the same NSL (this results in an effective distance of zero between nodes within the same NSL);
3. For each pair NSL_a and NSL_b, select the minimum w_{mn} in which $N_m \in NSL_a$, and $N_n \in NSL_b$ to represent the distance between that pair of NSLs (If some NSL has at least two nodes, the dimension of the distance matrix is decreased because some weights are discarded.);
4. The resulting matrix is the distance matrix DM between NSLs.

The results of this reduction of the W matrix to a new distance matrix is a matrix having the same number of rows and columns as the number of NSLs, in which the distance between each pair of NSLs is the minimum of all the weights on edges connecting the pair of NSLs. Like an object in Johnson's paper, an NSL need not have more than one node.

The distance matrix computed in this procedure differs from the matrix D^i computed in Step 15 of the agglomerative procedure, or in Step 1 of the matrix method, because the latter has not examined all paths until $q = n-1$. The former provides the distances between entities considering all paths, and the latter provides the distances between entities in which only paths having i or fewer links are considered. If Johnson had retained intracluster distances within his computations for each new similarity function (Step 3), thus maintaining the same dimension as the original weight function, the matrix resulting after the strong clustering is obtained would be the same as the distance matrix D^{n-1} obtained in the Pathfinder procedure (using r = ∞, of course).

We will now derive the distances between each pair of entities from the PFNET(r = ∞, $q = 5$) for the above example, and then show that the distances are the same for the minimum method UHCS. Indeed, because of the similarities of the procedures, these distances must always be the same.

Addressing the distance question in terms of the previous example, the UHCS distance matrix and the PFNET distance matrix D^5 (defined and discussed in the section entitled Generation Algorithms for PFNETs) both reduce to:

$$D^5 = W_{3r} = \begin{matrix} 0 & 1 & 4 & 4 & 5 & 5 \\ 1 & 0 & 4 & 4 & 5 & 5 \\ 4 & 4 & 0 & 2 & 5 & 5 \\ 4 & 4 & 2 & 0 & 5 & 5 \\ 5 & 5 & 5 & 5 & 0 & 3 \\ 5 & 5 & 5 & 5 & 3 & 0 \end{matrix}$$

The PFNET($r = \infty, q = n-1$) for W_3 is shown in Figure 9. Structurally, it can be seen that the relationships in the PFNET are not hierarchical, in the sense that there is a cycle containing nodes 1, 2, and 3; that is, the NSL formed in E_1, consisting of joining nodes 1 and 2 with an edge having weight 1, is joined with node 3 via both nodes 1 and 2 with edges having weights of 4. Thus nodes 1, 2, and 3 form what is known as a *clique* in graph theory (after $E(4)$ is considered), because each is directly linked to the other. This sort of structural information is not readily available using UHCS. Ties in weight values are evident in Figure 9, in which the edges e_{13} and e_{23} are each in different mintrees, and connect two node sublists with minimum (and equal) costs. Some of the more recent work in clustering has recognized the importance of having a procedure which does not arbitrarily break ties (Shepard & Arabie, 1979, p. 93).

1 Properties of Pathfinder Networks

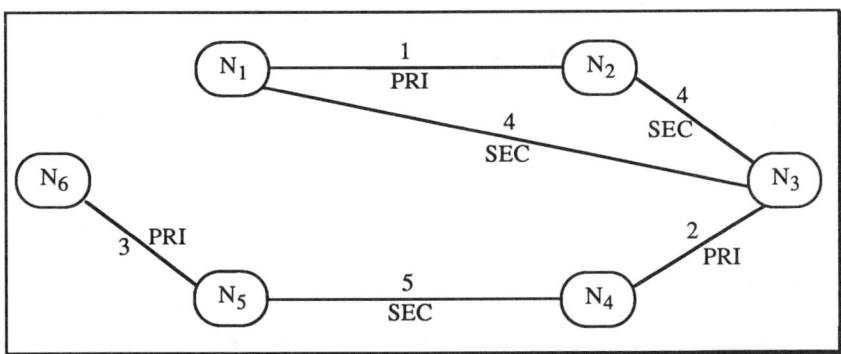

Figure 9. PFNET($r = \infty, q = n-1$) for W_3.

We will now show that the distance matrices for a PFNET($r = \infty, q = n-1$) and the UHCS derived from a common weight matrix W are identical, and the clusters formed are the same at each step in the agglomerative methods.

Theorem 7

Given a symmetric weight matrix W having zero-valued entries on the diagonal and positive values off the diagonal, the clusters formed using the UHCS procedure and the NSLs formed using the agglomerative Pathfinder procedure are isomorphic at each step. Furthermore, the interobject distances obtained from UHCS and PFNET($r = \infty, q = n-1$) are also identical at each step.

Proof

The proof is by induction over equivalence classes for the Pathfinder network and over the α_j levels of the UHCS. Both agglomerative methods begin with the weak clustering, in which each node or object is also a cluster. In the graphical paradigm, the nodes are not linked at this stage; in the clustering scheme, no clusters have been merged. Clearly the distance matrices are now identical to the W matrix, and the clusters are the same for each paradigm. For the induction step, assume that the clustering formed at level α_{j-1}, denoted by C_{j-1}, has the same clusters as the NSLs formed after the equivalence class $E(w_{j-1})$ is considered, using the agglomerative Pathfinder method described in the section entitled Fundamental Properties. We also assume that the interobject distances resulting from both paradigms are identical at this stage.

Now consider the clustering level α_j, which corresponds to a weight value of w_j in the W matrix. In the UHCS procedure, all weights having this value are considered, and objects having this weight value between them are merged, provided that they are not already in the same cluster. In the Pathfinder procedure, the equivalence class of weights having this value is considered, and new NSLs are formed based upon precisely the same criteria as for the UHCS. Therefore the objects in the next clustering, C_j, are isomorphic to the NSLs in the Pathfinder paradigm after $E(w_j)$ is considered.

It remains to show that the interobject distances of the UHCS and of the PFNET($r = \infty$, $q = n-1$) are the same after the merging of clusters which occurs at the clustering level α_j and the merging of NSLs when the equivalence class $E(w_j)$ is considered. Since the level

α_j corresponds to the weight w_j in the ordering of distances given by the original weight matrix W, clusters separated by this distance are then merged in the UHCS, and NSLs separated by links having this weight are also merged in Pathfinder. For the UHCS, suppose that clusters C_a and C_b are merged in this step; then each entity in C_a is at a distance of α_j from each entity in C_b, as in Step 3 of the procedure of Johnson. For PFNET($r = \infty$, $q = n-1$), there are NSL_a and NSL_b corresponding to C_a and C_b (by the inductive hypothesis), and the weight w_j is equivalent to the level α_j. Therefore each node in NSL_a is at a distance of w_j from each node in NSL_b (because the use of $r = \infty$ is equivalent to the use of the ultrametric inequality discussed in Johnson, 1967, p. 245). That is, the weight of a path is equal to the largest weight on any link in the path for $r = \infty$. In UHCS and in Pathfinder, previously determined distances (from smaller-valued equivalence classes or α levels) remain unchanged. Distances larger than w_j may be modified in either UHCS or Pathfinder at this step, because the smallest weight connecting two objects is used to represent the distance between those objects; but they must be modified the same in either paradigm because the weights are the same, and the interobject distances are the same at the preceding step (inductive hypothesis). Therefore the interobject distances obtained in each method are equal. □

Given a UHCS, using the minimum method, can a PFNET(r, q) be derived? In general, the answer is no; the distance matrix obtained from application of the UHCS algorithm contains little information about the structure. For example, consider the information available about two clusters in both UHCS and in Pathfinder. Once the merging of the two clusters has occurred, all entities from one cluster take on the same distance from all the entities in the other cluster. At this point, the only ways to construct networks from the distance matrix are (1) to link each node in one cluster with each node in the other cluster; or (2) to provide some arbitrary link(s) between clusters having the appropriate weight(s). One of these might, of course, correspond to the PFNET(r, q), but the correspondence would surely be incidental in terms of the link structure. In general, the UHCS distance matrix cannot yield enough information to construct a PFNET(r, q) which would be computed from the distance matrix.

Theorem 8
Given W_1 and W_2, suppose that the PFNET($r = \infty$, $q = n-1$) computed from W_1 is isomorphic to the PFNET($r = \infty$, $q = n-1$) computed from W_2. Then the UHCS(W_1) computed from W_1 is also isomorphic to the UHCS(W_2) computed from W_2.

Proof:
From Theorem 7, the clusters and distances of PFNET($r = \infty$, $q = n-1$) computed from W_1 are isomorphic to the clusters and distances computed from UHCS(W_1). Similarly, the clusters and distances of PFNET($r = \infty$, $q = n-1$) computed from W_2 are isomorphic to the clusters and the distances computed from UHCS(W_2). Therefore, the clusters and distances of UHCS(W_1) are isomorphic to those of UHCS(W_2), by transitivity, since the PFNETs are isomorphic. □

The converse of Theorem 8 is not true, however, because a unique PFNET cannot be constructed from a UHCS, as discussed in the examples using W_1 and W_2 and illustrated in Figures 6, 7, and 8 early in this section.

Given a PFNET(r, q), obtaining the minimum method UHCS is simple. If the incidence (edge-node) table has been constructed, then this already contains the UHCS in the merging of the node sublists. If the matrix method was used to determine edge member-

1 Properties of Pathfinder Networks

ship, with labeling done after generation via a shortest-path algorithm, then the UHCS can be computed as follows.

Procedure UHCS given PFNET(r, q):

1. Disregard (remove) all tertiary edges in PFNET(r, q);
2. Compute the distance matrix D^{n-1} using $r = \infty$ over the primary and secondary edges in the PFNET. This can be done using the matrix operations described in the section entitled Generation Algorithms for PFNETs by constructing a weight matrix W^* in which the weights associated with the primary and secondary edges are entered in the appropriate locations, and other off-diagonal entries in W^* are filled with some value greater than any weight associated with a primary or a secondary edge;
3. D^{n-1} is closely related to the similarity function of Johnson, as discussed previously, and provides sufficient information to construct the tree representation of the UHCS if that is desired. Although D^{n-1} is equivalent to the UHCS, it is sometimes desirable to construct a tree to represent the UHCS, as is done in Johnson (1967, p. 243).

As a simple example, suppose that the distance matrix D^5, obtained for the PFNET in Figure 9, has been computed as described above, and the UHCS is desired for this matrix. To construct the tree shown at the top of Table 3 from the matrix D^5, simply apply the UHCS procedure. The significance of this is that the same clusters are obtained whether or not the distance matrix is obtained from a Pathfinder network. Computationally, it is desirable to compute clusters directly, using one procedure or the other. Some of our Pathfinder programs print the NSLs after each equivalence class is considered, so that the UHCS is obtained as the network is generated.

These results (that the minimum method of UHCS is available from any Pathfinder network) are consistent with the intuitive observations of many of those who have worked with Pathfinder; clustering of similar entities is usually obvious from a drawing or display of the network. The importance of graphical display and some of the power of the additional structural information available in Pathfinder beyond that available in UHCS are illustrated in Esposito (Chapter 6, this volume; and 1988). For example, the basic-level categories of Rosch, Mervis, Gray, Johnson, and Boyes-Braem (1976) have a distinctive representation in Pathfinder networks, with the category name in the center of a "wheel," and the exemplars of the category at the end of the "spokes" of the wheel. Thus the *degree* of the node representing the category name (compared with the number of exemplars) is an appropriate measure to investigate whether or not a category can be considered a basic-level category. The degree of a node is also important in a measure proposed by Collins and Loftus (1975), in which the similarity of two concepts depends upon the number of edges shared by those concepts. Edge labels depend only upon ordinal properties of the data (see Theorem 6), so they offer an appealing approach for further measures to investigate. They are, of course, useful in identifying mintrees and deriving the minimum method UHCS, as shown in this chapter.

Conclusions

We have presented a class of graphs, PFNETs, whose structure is determined algorithmically by proximity data and two parameters, the r-metric and q parameter. The proximity data can either be obtained empirically, or in some domains, determined by objective measurements.

It was shown that each Pathfinder network contains all mintrees and is unique, given a particular weight matrix and particular values for the r-metric and the q parameter. Inclusion properties as r and q vary were demonstrated, so that systematic families of PFNETs are generated as r and q vary. PFNET structure becomes sparser (has fewer links) as either r or q increases.

Structure-preserving properties of the PFNETs under monotonic and multiplicative transformations were proven, so that any PFNET can be used with ratio data, and PFNET($r = \infty$, q) can be appropriately used with ordinal data.

Last, it was shown that the information in a minimum method hierarchical clustering scheme is also available in every PFNET, but that there is not sufficient information in a minimum method HCS to construct a unique PFNET. Pathfinder networks retain information concerning the entities responsible for establishing minimum paths (salient associations), unlike HCS, and thus make structural distinctions unavailable in HCS.

Chapter 2

Graphs in the Social and Psychological Sciences: Empirical Contributions of Pathfinder*

Francis T. Durso and Kathy A. Coggins

Graphs and graph theory play important roles in the theories and methods of many sciences. As we have seen (Dearholt & Schvaneveldt, Chapter 1, this volume), a graph is a mathematical formalism and so may be used to represent a wide range of phenomena. Chemical isomers, electrical circuits, Markov chains, statistical mechanics, and network flow in operational research are but a sampling of fields in which graph theory has been useful (Harary, 1969). It is this abstract character that allow graphs to have such general utility. Formally identical graphs could represent the distances from hospitals to homes in a neighborhood, the flow capacity of water through a city, the frequency of information exchange among or within organizations, or the associative network of concepts in human long-term memory. Graph theory allows for the identification of the shortest path between two locations, the identification of the most efficient circuit (i.e., the shortest closed path) in an electrical diagram, the isolation of cliques (i.e., completely connected subgraphs) in a social organization, the prestige (i.e., indegree) of members of a social group, or the hub (i.e., center) of government bureaucracy. In addition, analyses at more microscopic levels allow for determination of particular associations, channels, or influences within a graph.

The value of graphs has not gone unrecognized by social and psychological scientists. In psychology, one does not have to look far for a theory that uses some form of graph as its cornerstone. Clearly, in cognitive psychology, graphs are present in force, from neoassociative networks of memory (Anderson, 1983), through propositional analysis of discourse (van Dijk & Kintsch, 1983), to connectionist networks of cognition (McClelland & Rumelhart, 1986; Rumelhart & McClelland, 1986). Whereas cognitivists have perhaps been the most voracious consumers of graph theory in psychology, other areas of psychology have successfully employed graph theoretic concepts. In social psychology, we find theories that utilize graph structures to account for communication patterns within groups (Shaw, 1981). In fact, social psychology's theories of dissonance (Festinger, 1954) and balance theory (Harary, Norman, & Cartwright, 1965) rest on principles of graphs.

Graph theory also has played a role in psychological studies of infrahuman species. Tolman's (1932) notion of a cognitive map has been pursued in the recent work of Lieblich and Arbib (1982), who argue explicitly that animals learn a graph in the most abstract sense of the term. In fact, Brown (1987) provided empirical evidence that the behavior of rats in a radial arm maze is based on the formal graph constructed of the area and not on more local cues to food (such as the texture of the runway).

---

*The authors thank Deborah Bates-Porter for her help in preparing this chapter and Tom Dayton, Peder Johnson, Wendy Shore, and Mike Betts for their comments on an earlier draft of this chapter. Preparation of this chapter was supported by a grant from the University of Oklahoma.

Other social scientists besides psychologists have also made use of graph-theoretical constructs. For example, sociologists interested in the transfer of commodities (e.g., information, money) from one institution to another have made significant contributions in the application of graph theory to social issues (e.g., Knoke & Wood, 1981). In a related vein, sociologists and psychologists interested in "networks" have helped adapt concepts from the mathematics of graphs to applied domains (e.g., Burt & Minor, 1983). In fact, these efforts by sociologists have been more empirically based than those by psychologists. Network analysis[1] in sociology concerns itself with the collection of data from which networks could be determined and with the ultimate interpretation of those networks. In psychology, graphs are often constructed top-down from the intuitions or theories of the researcher, and thus are rarely empirically derived (cf., Chi & Koeske, 1983; Fillenbaum & Rapoport, 1971; Hutchinson, 1981).

In this chapter, we selectively review research from a number of areas that rely on graphs. Some of these endeavors have utilized Pathfinder, and for these we attempt to highlight the value of this scaling algorithm and report comparisons with other scaling algorithms when appropriate. Other endeavors have not been informed by the Pathfinder algorithm and for these we first review the work and then attempt to illustrate the value of Pathfinder by applying it to a relevant dataset. It is interesting to note that it was often difficult to find data that would allow Pathfinder to demonstrate its full potential. Some data were available that presented a trivial task to Pathfinder: For example, many sociograms are merely matrices of 0's and 1's, for which Pathfinder would simply link the nodes corresponding to the entries of 1 and not link those with a 0 for the cell entry. It seemed clear, at least to us, that many of the issues addressed in this manner could also have been easily addressed with more sensitive scales if there existed a method capable of taking advantage of the increased sensitivity. Pathfinder is one such method.

The output of Pathfinder is a PFNET that can be uniquely specified by two parameters: r and q. The *r parameter* is the Minkowski exponent. With an exponent of infinity, Pathfinder makes only ordinal assumptions about the data. In this chapter, all of the reported PFNETs used an r parameter of infinity. The second parameter, q, is a restriction on the number of edges in a path that Pathfinder will use in deciding if two concepts are already connected. The sparsest PFNET will result when Pathfinder is permitted to consider paths of any length, that is when q is equal to one less than the number of nodes. The most dense graphs result when Pathfinder can only consider a path as consisting of two edges, that is $q = 2$. This PFNET (∞, 2) is identical to solutions produced by NETSCAL (Hutchinson, 1981). Although decisions about the r parameter can be justified on measurement assumptions, the decision concerning q is more difficult. There is, currently, no formal mechanism for choosing among values of q for a given r. This is a situation similar to deciding on the appropriate dimensionality of a multidimensional space. Both when picking q and when picking the dimensionality, several factors, including the illuminating power of the solution, must be considered. In this chapter, we have a bias toward the sparsest solution ($q = n - 1$), especially when the graph is directed. However, when decreasing q provided additional insights, we did not hesitate to report that solution.

We note at the outset that our purpose is to review the contributions of Pathfinder as another tool, although we believe a very valuable tool, for the analysis of several issues that lend themselves to graph-theoretic analysis. In adopting this purpose we often do not do justice to the methods of analysis originally employed in the area, and we often skirt some

---

[1] Much of the sociological work concerns itself with identifying the subgroups within a graph. Subgroups can be identified either by finding the cliques in a graph or by determining what nodes in a graph are structurally equivalent.

2 Empirically Derived Graphs

of the theoretical complexities. These omissions will, of course, be obvious to researchers active in these areas. However, we believe that these researchers will also see the value of Pathfinder for their areas, perhaps more clearly than would have otherwise been the case.

We begin by considering a number of uses of graphs in cognitive domains. In particular, we review how graphs have been used to represent categories, to understand the representation of expertise, to predict details of human memory performance, and to design artifacts more compatible with human information processing.

Following the discussion of graphs in cognitive psychology, we consider the use of graphs in social domains. The graph analysis of these social phenomena, unlike the cognitive phenomena we consider, have not employed Pathfinder. For these areas, we discuss briefly the original analysis and then attempt to apply Pathfinder to the issue. Our application of Pathfinder to these areas should be viewed as, at best, a demonstration of how the algorithm could be of some assistance. We begin with a study of information and money exchange among a number of institutions in Indianapolis. We follow this with an analysis of the friendship graphs of a class of 8th graders. We end this section by considering how graphs have been utilized to understand small group communication dynamics and speculate on how Pathfinder could prove a valuable aid.

Cognitive Graphs

Knowledge Structures

The first work with Pathfinder was with an eye toward determining the types of structures that Pathfinder could, or would tend to, produce. This early work focused on the representation of knowledge. This choice followed naturally from the interests of Pathfinder's developers and also proved a fortunate choice in that clear differences among the PFNETs were observed that fit nicely with past research and theory.

Natural concepts. The first use of Pathfinder was presented at the meetings of the Psychonomic Society (Schvaneveldt & Durso, 1981), and involved a PFNET of 25 natural concepts. Several theorists posit an associative network in semantic memory, but they typically rely on intuition to construct the network (e.g., Collins & Loftus, 1975). Schvaneveldt and Durso reasoned that a minimal requirement for Pathfinder would be to capture many of the intuitive relations one would expect to hold among the concepts. The concepts (see Figure 1) included some that stood in a superordinate-subordinate relation, whereas others were related at the same level in the hierarchy. Some concepts, intuitively, had relations with several concepts across the network, whereas others had more specific relations. That first effort was very encouraging. Pathfinder reduced the 300 (25 items taken two at a time) pairwise similarity ratings considerably. The sparsest graph, PFNET(∞, 24), contained 25 links; even the most dense graph, PFNET(1, 2) was a reduction to 119 links. PFNET(∞, 2) had 32 links and appears in Figure 1. The connections certainly did not violate intuitions and in fact revealed a number of interesting relations.

For example, Figure 1 shows that some nodes played a restricted role in the graph (e.g., *hooves*), whereas others enter into categorical and property relations (e.g., *green*). Further, typical members of a category tended to connect directly to the category, whereas atypical members tended to connect indirectly with their superordinate category. The category *mammal* is especially interesting, in part because of its history in semantic memory research (see Rips, Shoben, & Smith, 1973; Smith, Shoben, & Rips, 1974). The suspicion of some, that *mammal* is not psychologically a natural category, seems to receive some support here. In fact, when biology graduate students rated the same items, *mammal* played the more central role that the scientific taxonomy predicts (Schvaneveldt, Durso, & Dearholt, 1989). Overall, the graph was simple and the relations were interesting and consistent with intuitions.

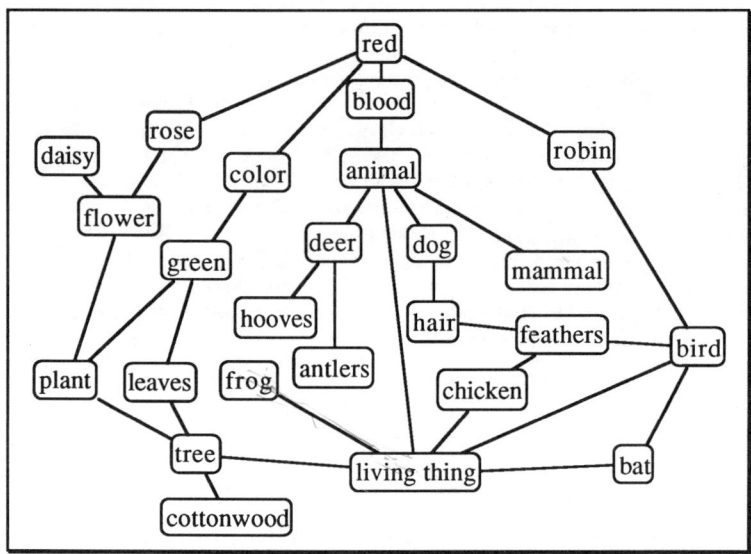

Figure 1. The most dense PFNET assuming ordinal data ($r = \infty$) for the 25 natural concepts used by Schvaneveldt and Durso (1981).

Basic-level categories. Following this initial success, we investigated a number of cognitive structures that we thought might produce interesting structures and at the same time extend the earlier work of others. One area derived from Rosch's contention (e.g., Rosch, Mervis, Gray, Johnson, & Boyes-Braem, 1976) that all Aristotelian natural categories were not created equal (at least not psychologically). Rosch and her colleagues have supplied a good deal of evidence to support the idea that some categories, such as basic-level categories, are psychologically special. For basic-level categories it is the world (not the language, cf., Whorf, 1956) that articulates thought.

Schvaneveldt et al. (1989; see also Hutchinson, 1981) used existing association norms (i.e., Cohen, Bousfield, & Whitmarsh, 1957; Marshall & Cofer, 1970) to establish matrices of associations for the six categories used by Rosch. According to Rosch, *fish, bird*, and *tree* are basic-level categories, whereas *clothes, fruit*, and *musical instruments* are not. In addition to this theoretical motivation, the association matrices were of interest to the development of Pathfinder because they were asymmetric: *Bird* is more likely to be a response to *thrush* than *thrush* is to the stimulus *bird*. Traditionally, when the judgment a_{ij} differed from a_{ji}, the scaling algorithm assumed that both were measures of the same underlying distribution and that the difference was due to noise. Recognition of the psychological reality of asymmetries (e.g., Tversky, 1977), however, makes it clear that an algorithm that could handle asymmetries in a meaningful way would be a useful step in understanding how categories are structured.

Figure 2 presents the sparsest PFNETs (∞, $n-1$) for Rosch's six categories. The "starness" of the graphs is apparent for the basic-level categories. Starness is easily

2 Empirically Derived Graphs

quantified by dividing the degree of each category node by the total number of arcs in the graph. Table 1 shows the correspondence between this index and Rosch's original classifications. This starness index also suggested that the category *flower* could have been treated as a basic-level category by Rosch, but that *professions* could not.

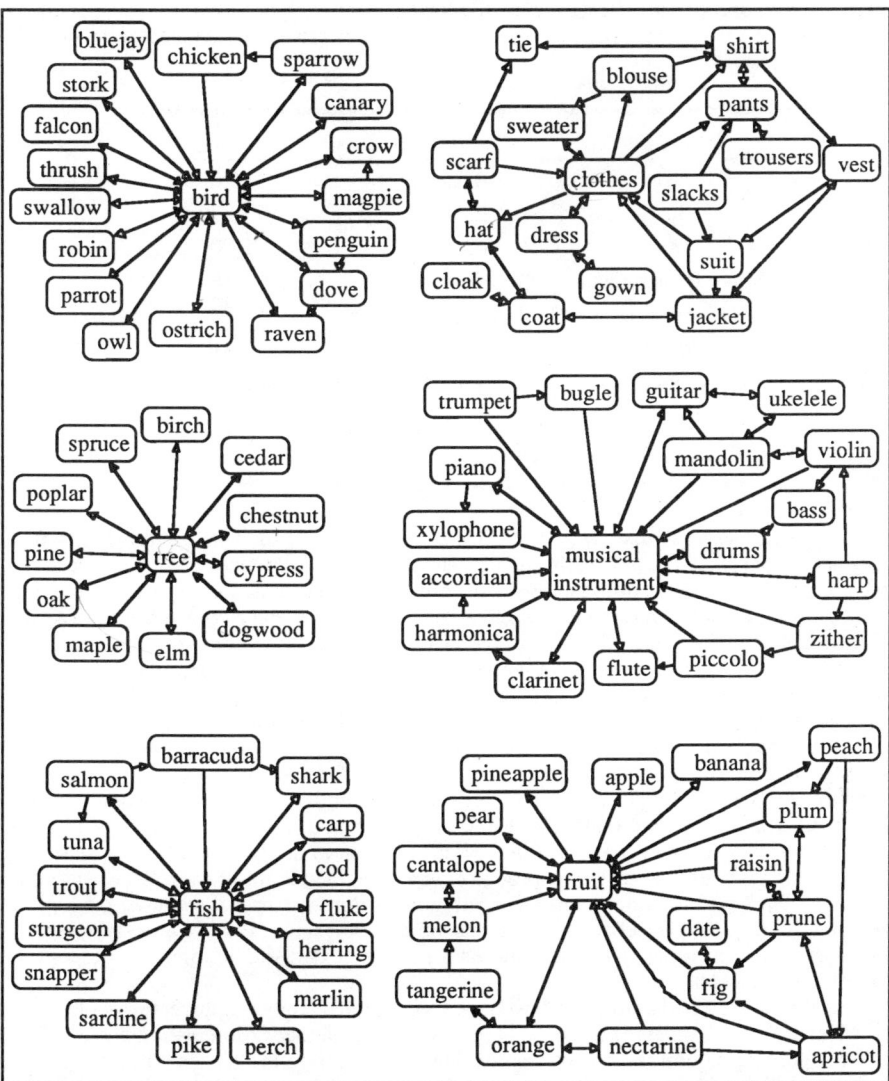

Figure 2. The sparsest directed PFNETs of three basic-level categories (left) and three superordinate-level categories (right).

We should note that applying a spatial algorithm to these data failed to distinguish between the basic-level and the superordinate categories. Although category labels tended to fall at the center of a two-dimensional space, this was true of all categories. Perhaps the discrete nature of language and categories makes a graph-theoretic model like Pathfinder particularly appropriate. However, as the next section suggests, Pathfinder appears to contribute even in domains where spatial algorithms have proven quite successful.

Table 1. Starness index for eight natural categories.

Category	Classification	Starness
Fish	Basic[a,b]	.90
Bird	Basic[a,b]	.89
Tree	Basic[a,b]	1.00
Musical Instrument	Superordinate[a,b]	.56
Fruit	Superordinate[a,b]	.50
Clothes	Superordinate[a,b]	.31
Flower	Basic[b]	1.00
Professions	Superordinate[b]	.50

[a]Rosch's classification [b]starness classification

Concepts with underlying dimensions. Pathfinder, like other algorithms that produce graphs, is likely to have its greatest success in representing discrete concepts. We were interested in how Pathfinder would perform when the stimuli varied along underlying dimensions. With this in mind we looked at (a) judgments of color borrowed from Ekman (1954) and used by Shepard (1962) to reproduce the Newton Color Circle with multidimensional scaling; (b) judgments of words signifying length of time (e.g., *second, minute*) that we collected from undergraduates and intuitively should fall on a single underlying dimension; and (c) judgments of restaurant-script concepts (Maxwell, 1983).

As Figures 3 and 4 attest, Pathfinder revealed interesting structures even though a spatial algorithm might have been the a priori procedure of choice. For the Ekman data, the sparsest PFNET (∞, 13) mirrored the physical wavelengths, but the PFNET (∞, 2) solution produced the Newton Color Circle. It is interesting to note that when Shepard first presented his Multidimensional Scaling (MDS) solution to the Ekman data, he connected the terms to highlight the structure, producing the graph that Pathfinder produces algorithmically. The length-of-time terms fell neatly onto a simple path that captured the logical relations among the concepts. Finding this simple dimension was not a trivial exercise given that a one-dimensional MDS solution for the same data did not preserve the logical ordering of concepts.

Finally, the temporal dimension presumed to underlie scripts is apparent in the PFNET of Maxwell's (1983) data (Figure 5), augmented by a number of interesting cycles. The cycle involving paying the bill was particularly appealing to the students in our classes: Perhaps it allowed for the possibility of not leaving a tip (although it probably represents a difference between "paying the cashier" and more formal eating establishments).

2 Empirically Derived Graphs

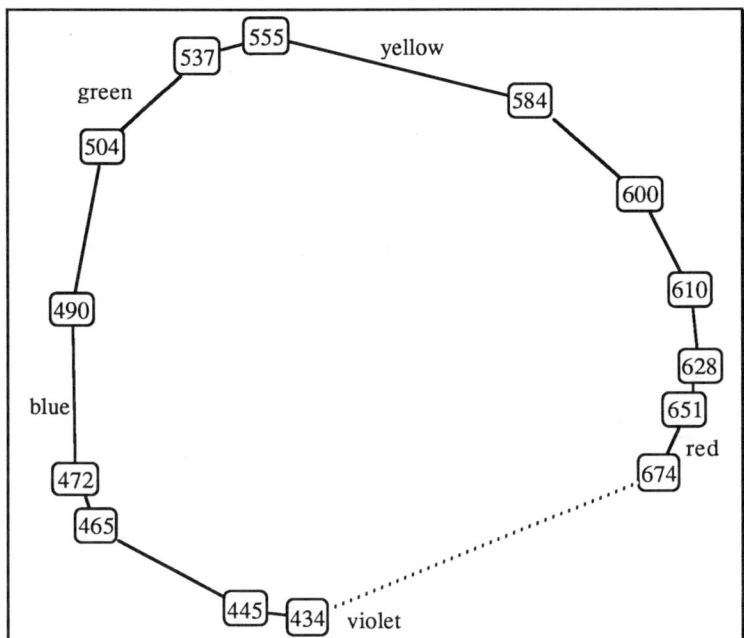

Figure 3. A PFNET capturing the Newton Color Circle. The solid links are from the sparsest PFNET and the dotted link is the single link added by the most dense ordinal PFNET.

These initial efforts demonstrated the viability of Pathfinder. What was discovered across a number of knowledge domains fit intuitions and conformed to previous theory and data. These initial PFNETs continue to be of particular interest because they highlight differences among the types of graphs that could emerge from the Pathfinder algorithm: a single cycle for the color data, a single path for the time-duration data, star-patterns for the basic-level data, as well as more general graphs for the natural concepts and the script data.

These PFNETs highlight one of the assets of Pathfinder. That is, looking at the graph in its entirety, rather than at only subgraphs, can often give additional insights into the domain under scrutiny. For example, the sociometric study of graphs is often restricted to the discovery of substructures like cliques, but consideration of the "big picture," the Gestalt of the graph, has been difficult. This does not imply that a study of subgraphs and other graph-theoretic summaries are without value. However, without the use of a device like Pathfinder to reduce the data to a tractable graph, the researchers do not have the freedom to look at both the overall graph structure and at the subgraphs, but instead are restricted to analysis of only the more manageable substructures.

Finally, Pathfinder can supply information that multidimensional scaling does not. The failure of MDS to capture the logical relations among the time-length terms and the failure of MDS to distinguish between basic-level and superordinate-level categories suggests that Pathfinder can, at least, complement spatial analyses. In some cases, Pathfinder provides information akin to that of a nearest neighbor analysis (Tversky & Hutchinson, 1986).

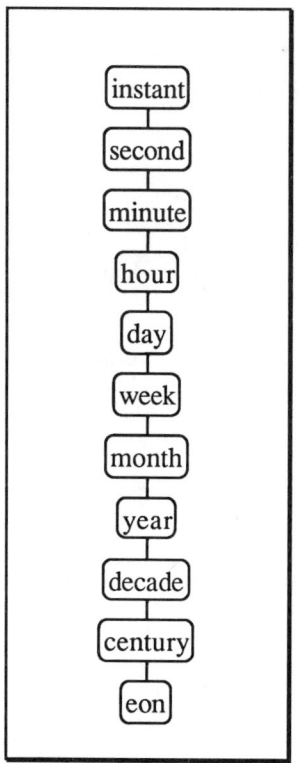 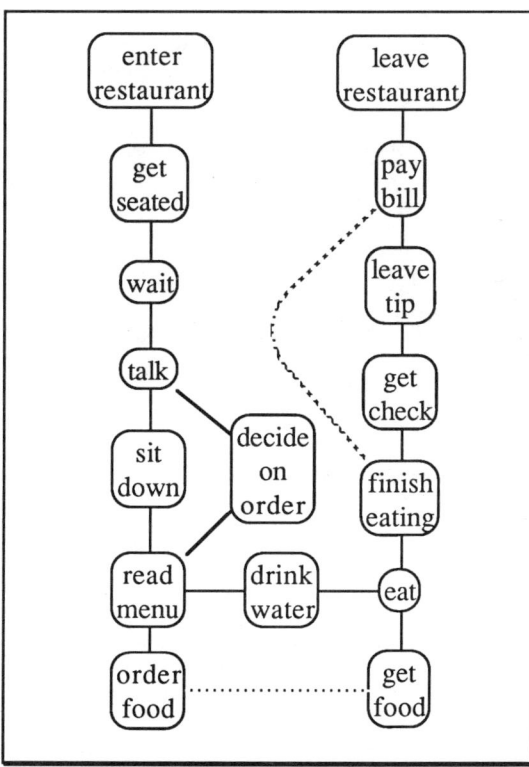

Figure 4. The sparsest PFNET for the time-duration terms.

Figure 5. A PFNET for terms from a restaurant script.

Expertise

Differences among knowledge structures have been implied to underlie a number of phenomena in cognitive psychology. This implication has been strongest in explanations of expertise. From the seminal work of Chase and Simon (1973), through the numerous studies it inspired (e.g., McKeithen, Reitman, Rueter, & Hirtle, 1981), to the recent work of Chi and associates (e.g., Chi & Koeske, 1983), the underlying structure of knowledge has been the theoretical focus.

For example, Chi and Koeske (1983) explicitly constructed graphs from the protocol of a 4-1/2-year-old child with an avid interest in dinosaurs. Although Chi and Koeske did not directly compare experts and novices, they did construct separate graphs for the better-known dinosaurs and for the lesser-known dinosaurs. The nodes associated with the better-known dinosaurs had higher degrees (i.e., more *dinosaur-dinosaur* connections) and stronger linkages than did the lesser-known dinosaurs. Finally, the graph of better-known dinosaurs was apparently more cohesive in that the subgroups defined in that graph were characterized by stronger within-group paths and weaker between-group paths than were the subgroups of the "novice" graph.

2 Empirically Derived Graphs 39

Several studies lead to the conclusion that experts understand at a deeper, more abstract level than do novices (e.g., Adelson, 1981; Chi, Feltovich, & Glaser, 1981). This, however, does not require that the knowledge of experts and novices differ structurally. For example, experts have more domain-specific knowledge than do novices, but beyond this the knowledge may be organized or configured in the same way. A more compelling demonstration of the role of structural differences would be to demonstrate that a common set of concepts are interrelated differently for experts and novices.

Some research has attempted to discern directly if structural differences can be associated with superior knowledge. For example, McKeithen et al. (1981) showed that multitrial-free recall of ALGOL-W reserved words by expert programmers led to structures apparently more tree-like than was the knowledge of less expert programmers.

This literature suggested that Pathfinder could help provide evidence for the role of structure in our understanding of expertise. With Pathfinder it is relatively easy to present a common set of items to experts and novices, obtain judgments of relatedness, and then construct graphs for the two groups.

Schvaneveldt, Durso, Goldsmith, Breen, Cooke, Tucker, and DeMaio (1985) selected 30 concepts from each of two air-combat situations: split-plane maneuvers (air-to-air) and strafe maneuvers (air-to-ground). All possible pairs of these concepts (435 from each domain) were judged (on a scale of 0 to 9) by members of the U.S. Air Force or members of the Air National Guard. The Air Force pilots differed in their level of expertise (e.g., flight time) with some being undergraduate pilot trainees and others serving as their instructors. The Guard Pilots were all expert. PFNETs were constructed for each group and for each pilot. Schvaneveldt et al. (1985) presented a number of different analyses, but here we discuss two that we feel highlight the value of Pathfinder particularly well.

One analysis investigated the extent to which an individual pilot could be classified as an expert or a novice based on his or her cognitive representation.[2] The analysis began by constructing graphs for each pilot. These graphs for individuals were dense. This *density* highlights the consequence of ensuring a unique graph: When there are several ties in the data (as when graphing the data of individuals using a limited scale), Pathfinder will not arbitrarily discriminate between two (or more) possible edges of the same weight, but will instead include both (or all). Ties in the data are, of course, rarely a problem when summary data are submitted to Pathfinder. Further, it could be reduced in the construction of individual data, perhaps by using magnitude estimation procedures (Stevens, 1975) rather than more traditional Likert scales.

The density of the graphs notwithstanding, Schvaneveldt et al. computed three types of patterns for each individual: graph patterns indicating the presence or absence of a link, MDS patterns of the distance in k-dimensional space between each pair of concepts, and the original empirical ratings. The questions became: Does the adjacency information provided by Pathfinder discriminate between experts and novices? And if so, is this discrimination superior to that which could be accomplished by pilots' relatedness judgments?

Nilsson's (1965) pattern recognition algorithm was used to define two prototypes (e.g., undergraduates vs. instructors, instructors vs. guard pilots) based on a subset of the matrices. The algorithm then classified a pilot who was not used to constructing the prototype by indicating to which prototype the "unknown" pilot belonged. Schvaneveldt et al. (1985) repeated the procedure until every pilot served as the to-be-classified pilot; the percent correct classifications were then calculated for the population of decisions. Figure 6 shows that discrimination was quite successful using Pathfinder. Further, discrimination

---

[2]Cooke and Schvaneveldt (1988) had the same intent when they analyzed the PFNETs of expert, intermediate, novice, and naive programmers.

was superior using the Pathfinder network compared with the original rating data. Despite a possible ceiling effect, it is interesting to note that MDS classification was at least as accurate as Pathfinder and was superior in some cases.

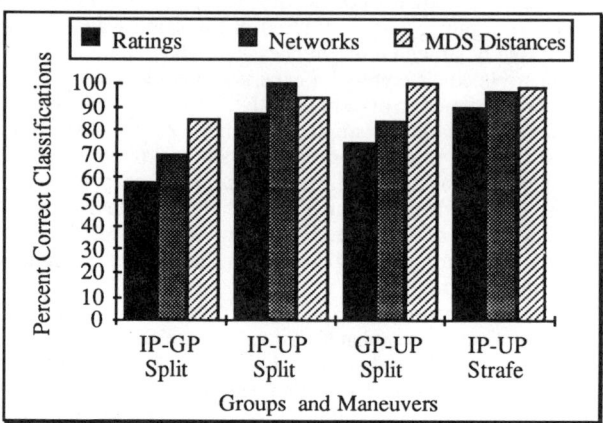

Figure 6. Comparative classification success of Pathfinder, MDS, and the original ratings in a study of fighter pilots. IP = Instructor Pilot, UP = Undergraduate Pilot Trainee, GP = Air National Guard Pilot. From "Measuring the structure of expertise" by Schvaneveldt et al., 1985, *International Journal of Man-Machine Studies 23*, p. 717. Copyright 1985 by Academic Press Inc. (London) Limited. Adapted with permission.

The second analysis of interest focused on an attempt to establish the relationships critical to expertise. Schvaneveldt et al. reasoned that if a link was present in the graph of one group of experts, but not in the graph of the other group of experts, then that connection must not be necessary to the cognitive structure of expert fighter pilots. The graph of the links that were shared by the two groups of experts (the "right stuff," see Figure 7) constructed by Schvaneveldt et al., and its comparison with the novice graph, did receive some validation. Concepts that were particularly poorly understood by the undergraduate pilots (i.e., those with few connections in common with the experts) were isolated and then used to classify individuals as described above. This set of only 10 "misunderstood" concepts perfected the novice-expert classifications: 100% of the novices and experts were correctly classified.

In summary, there is some evidence that experts can be distinguished from novices based on their cognitive structures. Classifications based on Pathfinder were superior to classifications based on the rating data suggesting that Pathfinder was successful at uncovering the latent structure inherent in the empirical ratings. Thus, a comparison of experts with novices supplies some validation of the psychological utility of Pathfinder.

To the extent that Pathfinder can capture important structural aspects of the expert's knowledge, it presents an interesting methodology that could be used to assist in solving important applied problems that rely on an understanding of human expertise. Cooke and McDonald (1987) and Schvaneveldt and Goldsmith (1985) have both pursued the implications of Pathfinder for artificial expertise: The former have focused on knowledge elicitation for use in expert systems and the latter have used empirically derived graphs as a basis for ACES, an air-to-air combat simulation.

2 Empirically Derived Graphs 41

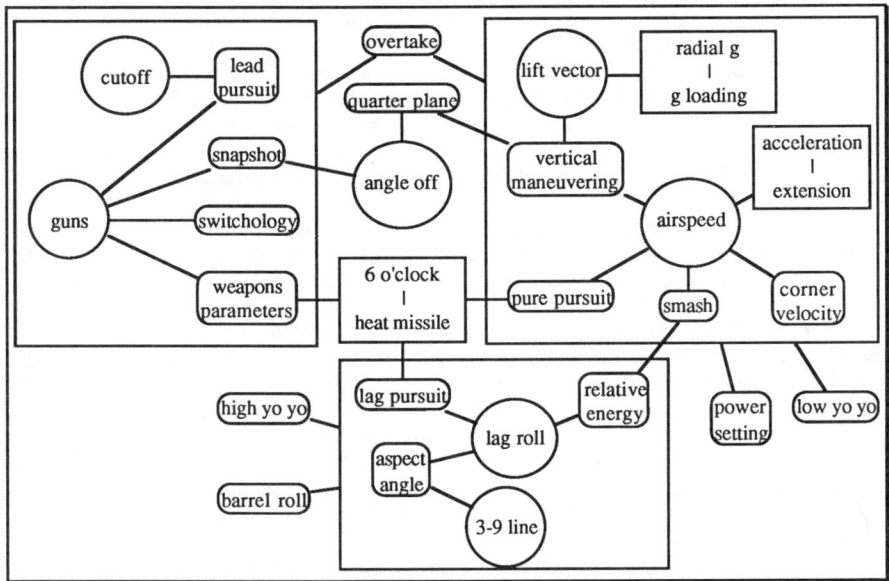

Figure 7. A graph of those links shared by Instructor Pilots and National Guard Pilots. The circle nodes constitute the minimal dominating set of concepts. From "Measuring the structure of expertise" by Schvaneveldt et al., 1985, *International Journal of Man-Machine Studies 23*. Adapted with permission from Academic Press Inc. (London) and the authors.

Memory

We now turn to a study that supplies further validation of Pathfinder, this time by taking advantage of the fact that knowledge structures have an impact on memory and recall. The differences in structure that apparently exist in our underlying knowledge should also be revealed in how subjects recall events. This contention has a long history in the study of human memory. Indices of underlying structure (e.g., meaningfulness) were used extensively by verbal learning theorists; a number of learning and cognitive theorists have shown relations between organization and recall; and most recently researchers have explored how various scaling solutions predict recall. For example, the work of Chi and Koeske (1983) mentioned earlier included a demonstration that recall was superior for the dinosaurs that came from the more cohesive graph.

Analyses of recall have also been conducted using more formal scaling procedures including hierarchical cluster analysis and multidimensional scaling. Friendly (1977) and others (e.g., Caramazza, Hersch, & Torgenson, 1976) have explored this approach extensively; the interested reader is referred to Friendly (1979) for a very readable review of a myriad of procedures. Of interest to our review of Pathfinder is that multidimensional scaling solutions apparently predict the organization in recall quite well.

Cooke, Durso, and Schvaneveldt (1986) compared Pathfinder with MDS-derived structures on their ability to predict the recall of a list of words. Subjects studied lists of words chosen from the natural concept experiment (Schvaneveldt & Durso, 1981) that we discussed earlier. The lists were organized according to different scaling solutions. A list could be tightly organized according to Pathfinder but not according to MDS (Pathfinder list); it could be organized according to MDS but unorganized according to Pathfinder (MDS list); or a list could be unorganized according to either scaling procedure. The Pathfinder list was constructed by selecting items that were linked in the graph but distant in MDS; the MDS list comprised a sequence of items that were close in multidimensional space but not linked in the graph; and the unorganized lists comprised items that were neither linked nor close.

The results of the serial learning tasks were quite dramatic. Subjects who studied the Pathfinder lists learned more quickly than those who studied the MDS lists. In fact, on some indices, learning the MDS list was as poor as learning an unorganized list. Despite these dramatic effects, it could be argued that serial learning was a task especially well suited to Pathfinder. Unlike MDS, which focuses on a global configuration, Pathfinder tends to focus on more local relations. The highly related concepts are especially important to the Pathfinder algorithm. In a task such as free recall, the advantage of Pathfinder might be lost.

Cooke et al. presented the items in a random order and then tried to predict the recall order. Predictor variables were either based on Pathfinder, MDS, or the unscaled empirical ratings. The predicted variable was the average interitem distance in the recall protocols.

Cooke et al. found, as have others (Caramazza et al., 1976), that MDS predicted recall order. They also found that Pathfinder and the original ratings predicted recall order. More interesting than these simple correlations were the partial correlations that separated the contribution of the scaling procedure from the predictive power inherent in the ratings. If the purpose is to predict recall order, what is gained by transforming the data using either MDS or Pathfinder?

When the contribution of the original ratings were partialled out of the correlation, Pathfinder, but not MDS, proved independently predictive of recall order. These findings have two important implications. First, the reports that MDS could predict recall order may have been based on the predictive power of the proximity judgments, with MDS adding little. Second, and more important, is that Pathfinder revealed some latent structure useful for predicting recall order beyond that which could be linearly predicted from the ratings.

Other investigations of Pathfinder in episodic memory tasks appear in this volume. Branaghan (Chapter 8) presents work with paired-associates and Goldsmith and Johnson (Chapter 17) investigate memory organization as a function of classroom experience.

Human-Machine Interaction

In addition to being able to predict learning and recall, some recent work suggests that Pathfinder-derived structures may be a useful tool for the design of human-machine interfaces. The main objectives of this type of research are to make the devices easier to use and quicker to learn, thus making it possible for people to perform tasks that they might not otherwise be able to accomplish. As McDonald and Schvaneveldt (1988) reminded us, the main problem in this area is that the interfaces are generally based on the perspectives of the designer, which are not necessarily congruent with the *user's* perspectives.

Guidelines and standards are often established without adequate research or a theoretical basis. One possibility is that the mental model of the user should be examined and used in developing the interfaces. Several researchers have used Pathfinder to aid in uncovering the mental model.

Roske-Hofstrand and Paap (1986a) were the first to take this tack by extending the Pathfinder procedure to the important applied concern of human-machine interaction. They tapped the user's cognitive organization to develop a menu-driven system for the control-

2 Empirically Derived Graphs

display unit (CDU) within a simulator. Their objective was to ensure that a new component within an automated cockpit would be easy to use by making it "compatible with the content and organization of the pilot's existing cognitive structure" (p. 1302).

To obtain the conceptual organization of the panels, they had four pilots rate the similarity of each chunk of information associated with 34 panels of the CDU. Using Pathfinder, they developed three menus. One had high redundancy in that there was more than one route to a specific goal. The price of this was an increased menu size. Using the same basic graph, they also developed a menu that eliminated most of the redundancy and one that had no redundancy. A fourth menu was based on the intuitions of a design team.

To test these menus, they had four subjects work with each prototype. The subjects were given a set of 34 scenarios and questions that were to be answered using one of the four prototype menus. They found that subjects using the highest redundancy prototype had the lowest failure rate and were the fastest to solve the questions. The prototype of the design team was at the opposite extreme in both failure rate and speed.

McDonald and Schvaneveldt (1988) used Pathfinder and cluster analysis (Johnson, 1967) in an effort to capture information about the human-UNIX interface. They had 15 experienced UNIX users go through 219 documented functions printed on cards and sort familiar functions into piles based on relatedness. To prevent hierarchical filtering, the subjects were encouraged to make duplicate cards if they felt that a command belonged in more than one pile. From this sort, a conditional probability matrix was constructed. Both hierarchical cluster and Pathfinder analyses were performed on this matrix. Pathfinder provided more information than the hierarchical cluster analysis. For example, although cluster analysis placed the UNIX commands, *pc, pi, pix,* and *px* in one cluster, the PFNET revealed the commands formed a clique; but cluster analysis had these four in a large cluster of six commands. In short, the cluster solution was derivable from the PFNET, but not vice versa.

In addition to structure, the design of an interface requires some abstraction of the underlying categories to allow the structure to be implemented in a Von Neumann architecture. (It is worthy of note that a connectionist architecture may allow implementation of Pathfinder networks without abstraction of the underlying categories, and we have begun investigating this possibility.) Category labels for such clusters are one such abstraction. Thus (as discussed by both Cooke & McDonald, 1987; and McDonald & Schvaneveldt, 1988), 4 of the 15 experienced UNIX users rated 83 clusters of two or more UNIX commands for goodness on a 5-point scale and provided names for all but the very "bad" clusters. This procedure also allowed the researchers to distinguish between artifactual clusters and real conceptual clusters. Most ratings were 4's and 5's suggesting that the cluster analysis did permit meaningful abstractions.

In a related endeavor, McDonald and Schvaneveldt (1988) investigated task sequences by looking at the co-occurrences of commands given by nine experienced UNIX users during a session with the operating system. By capturing a graph of probable-next-commands, context-sensitive help could take advantage of the previous sequence of commands. For example, McDonald and Schvaneveldt found only one arc leading to the command *kill* in UNIX, suggesting that only one command frequently precedes it. If an interface had this type of information, it could easily "anticipate" likely next commands and offer assistance. In fact, it is interesting to speculate on how such a system could be custom designed to the user's level of experience. As the user learns more of the system, a monitoring interface might keep track of command sequences and modify the PFNET, and thus the assistance offered.

The uses of Pathfinder discussed in this section are nascent, but they promise to facilitate human-machine interaction by adapting the machine to match our underlying knowledge networks. Context sensitive help, generic operating system advice, and artificially intelligent adaptive computer interfaces may all benefit from this work.

Social Graphs

In principle, Pathfinder could be a useful scaling procedure in a number of social sciences. To date, however, its use has been exclusively in the cognitive sciences. Unlike our review of that work, the current section is more a call for further work than it is a review. Although we consider some of the graph-theoretic literature, our discussion of Pathfinder is based on PFNETs constructed to illustrate the potential of this scaling procedure. Admittedly, in this way we are able to sidestep many of the issues, theoretical and methodological, that have consumed the efforts of many an insightful scholar. What we do accomplish, however, is a demonstration that applications of Pathfinder are likely to bear fruit in the social sciences.

Graphs have clearly become a central concern to sociologists and social psychologists. The methodology has been addressed explicitly in *Network Analysis* (Knoke & Kuklinski, 1982) and *Applied Network Analysis* (Burt & Minor, 1983), and a number of journals routinely report graph-theoretic treatments of sociological issues (e.g., *Social Networks*). Much of this work is concerned with how to collect data for network analysis. We will not be concerned with this here except to note that much of the data collected has been less quantitative than it would need to be if Pathfinder informed the work. For example, social exchange analyses often begin with simple 0 and 1 sociograms, in which a 1 indicates that there is some exchange and a 0 indicates that there is not. With Pathfinder, more sensitive measures (e.g., how much money is exchanged) could be handled easily. To date, however, even when more sensitive measures are collected, the data are often considered without any scaling procedure, thus allowing measurement error to exert strong influences, or the data are reduced to a simpler form (e.g., 0/1 sociograms).

Interorganization Exchange

In our consideration of the sociological literature, the interaction of large groups (e.g., institutions) struck us as an important question to which one could successfully apply Pathfinder. Graph-theoretic constructs had been employed in this work; in fact, Knoke and Kuklinski's monograph *Network Analysis*, used exchange among organizations as a vehicle for illustrating applications of graph theory.

Sociologists have identified a number of methods for analyzing network data. We borrow the data from Knoke and Kuklinski to illustrate those methods and to make a comparison to Pathfinder. Those data were a subset of the complete study reported in Knoke and Wood (1981); we return to the complete dataset later.

Knoke and Kuklinski began with two 0/1 sociograms of 10 organizations: one for information exchange and one for money exchange. These sociograms were collapsed in different ways depending on whether structural equivalence was determined by continuous distance procedures or by discrete distance procedures (blockmodel procedures; White, Boorman, & Breiger, 1976). The differences between these procedures need not concern us here. However, both procedures attempt to define structurally equivalent subgroups: "two objects a and b of a set C are *structurally equivalent* if, for any given relation R and any object x of C, aRx if and only if bRx, and xRa if and only if xRb" (Knoke & Kuklinski, 1982, p. 59). In other words, if a and b are identical in their relations to all

2 Empirically Derived Graphs

other objects, then the objects *a* and *b* are equivalent. For both procedures, this equivalence is established by using hierarchical cluster procedures, such as Johnson's (1967) cluster analysis (for continuous distance) or CONCOR (for discrete distance).

The continuous distance procedure and the blockmodel procedure each yielded four clusters. The clusters appear in Table 2.

Table 2. Clustering of 10 organizations[a] in Indianapolis according to two different procedures (Knoke & Kuklinski, 1982).

Continuous Distance	*Blockmodel*
(WRO, WEST)	(WRO, WEST)
(COMM, MAYO)	(COMM, MAYO)
(COUN, INDU, NEWS, EDUC, WELF)	(COUN, INDU, NEWS)
(UWAY)	(EDUC, UWAY, WELF)

[a]WRO (Women's Rights Organization); WEST (West End Organization); COMM (Chamber of Commerce); MAYO (Mayor's Office); COUN (City/County Council); INDU (Local Industry); NEWS (Star-News); EDUC (Education); WELF (Welfare); UWAY (United Way).

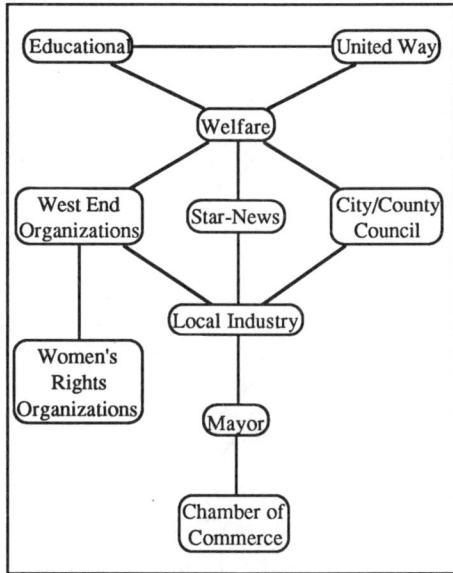

Figure 8. The sparsest PFNET for Knoke and Kuklinski's illustrative data taken from a sample of organizations in Indianapolis.

To create a PFNET for these organizations, a matrix of social role distances from Knoke and Kuklinski (1982, p. 63) was submitted to Pathfinder. The sparsest PFNET appears in Figure 8. Several facets of the graph are of interest. First, there is one clique (EDUCation, United WAY, WELFare) similar to the result of the blockmodel. Second, there are three cycles involving City/County COUNcil, INDUStries, NEWS, WELFare, and West End Organizations in different ways. Finally, Women's Rights Organization and the Chamber of COMMerce-MAYOr's office connection are relatively isolated from the rest of the graph.

We think the PFNET agrees well with the structural equivalence analyses. It also helps show how the blocks hang together in a more Gestalt way. For example, the fact that WELF is a *cutpoint* helps explain why it is clustered with EDUC and UWAY in the discrete distance analysis and why it is clustered with COUN, INDU, and NEWS in the continuous distance output. Although it is true that there may be few cliques if a strict graph-theoretic definition is used, the detection of circuits or cycles seems to supply some information about the substructure of the graph.

We now turn briefly to the original Knoke and Wood (1981) dataset from which the above organizations were sampled. We focus here on their analysis of perceived influence among seven blocks of organizations. Knoke and Wood asked organizations to indicate those organizations that had "policies or programs which your organization has tried to influence." We computed an index of the interconnections between blocks: For $i \neq j$, we computed the proportion of organizations within a block that had connections to organizations in other blocks; or, for $i = j$, this index reflected the proportion of organizations within a block that had connections, that is the intrablock connectedness. These data were submitted to Pathfinder (see Figure 9).

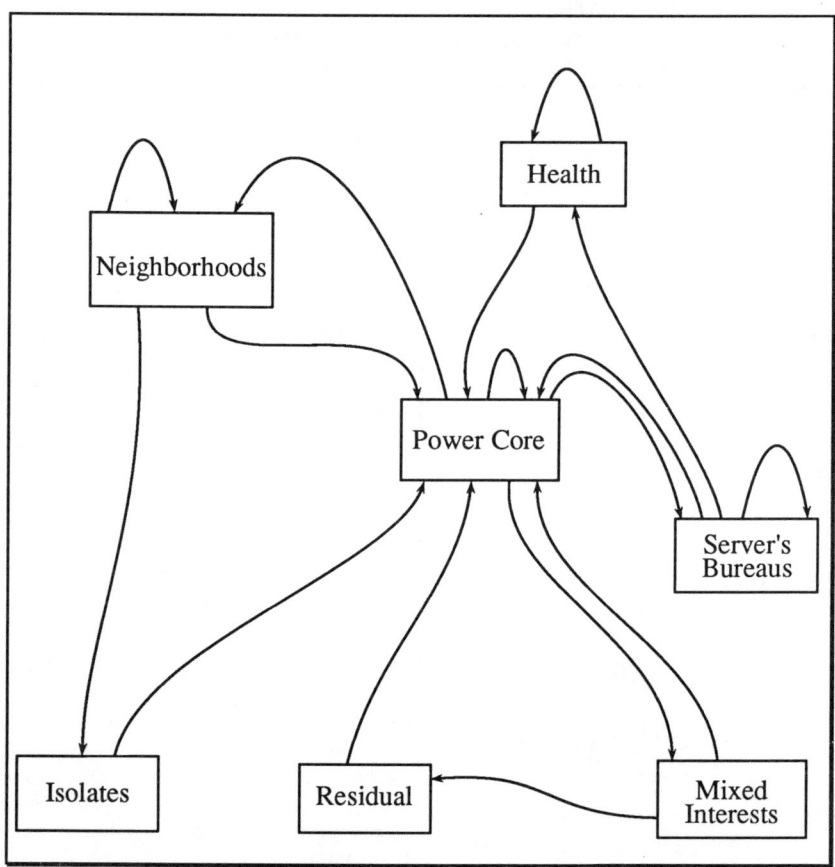

Figure 9. The directed PFNET for the seven blocks discerned by Knoke and Wood (1981). The graph is based on the data in Table 2.

2 Empirically Derived Graphs

The resultant PFNET revealed a number of features consistent with Knoke and Wood. The power core block, which should have been at the center of the graph, was represented by the node with the highest degree. Further, every node had an arc terminating at the power core. We calculated the degree of each node and then correlated these values with an independent assessment of influence collected by Knoke and Wood. The correlation was .75 between node-degree and perceived influence.

Blocks representing a tightly knit collection of organizations have a *loop* in the PFNET (i.e., an arc leaving and entering the same node). This is interesting in that Knoke and Wood point out that neither the neighborhood block nor the isolate block were influential, but the blocks differed in terms of the intrablock connections. Consistent with this, the neighborhood block had a loop in the PFNET but the isolate block did not. This indicates that the organizations making up the neighborhood block showed substantial interrelations, whereas the isolate members showed little connection among themselves as well as little connection with the rest of the graph.

Friendship

Another area of social graphs that emerges from the literature is the study of friendships. A typical method of collecting data about friends is to ask participants to nominate one or more individuals (Coie, Dodge, & Coppotelli, 1982; Peery, 1979; Wright, Giammarino, & Parad, 1986). This nomination procedure strikes us as a less than optimal way of gathering information from which to construct a graph, but researchers may have settled for this because of the lack of an algorithm for constructing a graph from complex data.

Some researchers in the area (e.g., Asher & Dodge, 1986; Asher, Singleton, Tinsley, & Hymel, 1979; Roistacher, 1974) have, in fact, criticized the nomination procedure. Hallinan (1982) has advocated giving the children free choice in naming as many friends as they desire, in order to avoid "a major source of measurement error in the data" (p. 57). Similarly, Roistacher (1974) chose to use a rating-scale measure to determine cliques within several high schools. For the rating-scale measure, subjects are given a roster with everyone's name and asked to rate their liking of the people. This procedure clearly lends itself to analysis by Pathfinder, more so than the nomination data. The rating-scale measure allows more total friendship choices than the nomination method, essentially allowing the construction of an $M \times M$ asymmetric input matrix for Pathfinder. In addition, the rating-scale procedure has produced results more consistent with what is suspected of friendships. For example, Slavin and Hansell (1983) reported that, for a cooperative learning study, the results based on the rating-scale measure indicated weaker friendships became stronger, whereas an earlier study using the nomination measure had reached the somewhat counterintuitive conclusion that it was the strong friendship that changed.

From either the nomination or rating-scale approach, several measures can be derived; one of the most common measures is popularity. This can be the averaged scaling (Asher & Dodge, 1986; Perry, 1987), the averaged nominations (Asher et al., 1979), or simply the frequency of positive nominations (Peery, 1979). There are also several measures of rejection or dislike, such as the proportion or frequency of negative nominations (Peery, 1979; Wright et al., 1986) and the proportion of the least-liked ratings given under the rating-scale instruction (Dodge & Somberg, 1987; Perry, 1987).

We thought Pathfinder might be useful for identifying children who differ in their friendship graphs. We used Perry's (1987) dissertation data.[3] She had eighth-grade girls rate each other on a 0 to 5 Likert scale, where "1" meant dislike, "5" meant strong liking,

---

[3] We thank Bridgette Perry for supplying us with her data and for discussing the area of friendship with us.

and "0" meant this person was not known well enough to give a rating. Her matrix was transformed to dissimilarity (smaller numbers indicated greater liking) data and submitted to Pathfinder. Of course, the girls differed in the number of arcs terminating on a node (indegree) and in the number of arcs originating from a node (*outdegree*). In this context, the *indegree* is suggestive of popularity. Our indegree index correlated .69 with her measure of popularity and .76 with a measure of social preference.

Researchers (Coie et al., 1982; Dodge & Somberg, 1987; Peery, 1979) introduced a combination of the above measures to determine social preference, and classified children into one of four categories: popular children who are liked and relatively rarely disliked; controversial children who are both liked and disliked; rejected children who are disliked; and neglected children who are neither liked nor disliked. The rejected and neglected classifications could help target children with social deficits (e.g., Dodge & Somberg, 1987; Hansell & Karweit, 1983; Perry, 1987) for an intervention procedure.

We created four-fold classifications of Perry's students using measures based on the literature. We also created the same four-fold table based on Pathfinder outputs. We first created a PFNET as described above. Nodes with a high indegree could represent students who are popular *or* controversial. Nodes with a low indegree could represent students who are neglected *or* rejected. To distinguish these groups further, another PFNET was constructed; this PFNET was based on the data before being transformed to dissimilarity. Thus, with these input data, nodes with high indegree would be students who are rejected *or* controversial and those with low indegree would be neglected *or* popular. Combining these two PFNETs classifies each student into one of four categories.

Table 3. Classification[a] of 8th graders used by Perry.

Rating Classification	Pathfinder Classification			
	Controversial	Neglected	Popular	Rejected
Controversial	0	0	0	4
Neglected	2	11	7	19
Popular	12	2	19	4
Rejected	1	2	1	0

[a]Rating classification was based on the data and followed procedures typically found in the literature. Pathfinder classification was determined by median splits on the indegrees from a PFNET where liked individuals were connected and from a PFNET where disliked individuals were connected.

As Table 3 indicates, when classification using measures from the literature was compared with classification based on Pathfinder, the agreement was at best fair. Only 38% of the students were classified in the same way by both procedures. The largest disagreements resulted from the tendency for Pathfinder to classify students as rejected rather than neglected, and to classify students as controversial rather than popular. Unfortunately, independent measures are not available to support one of the classification schemes over the other. The value of Pathfinder in assisting in categorizing students as controversial or rejected awaits further research.

Further, such classification research need not be restricted to the above table. One possibility is that if outdegree measures from Pathfinder are considered, friendliness can

2 Empirically Derived Graphs

become another classification variable. For example, those girls who are very friendly, but who are not liked as much in return, may have difficulties in relating to others either by not expressing themselves or because they lack the ability to empathize with others. As with neglected children, those falling in this unrequited group may have sociocognitive characteristics that suggest particular interventions.

We find it interesting that there has not been much work trying to investigate the interaction of friendliness and popularity, although the two factors have been considered separately (e.g., Hallinan, 1982). Of course, the nomination procedure precludes such efforts because it restricts the number of friends nominated.

Communication networks

We conclude this section on social graphs by indulging in pure speculation. The study of communication patterns has made considerable use of graphs, but has not taken advantage of methods for inducing the structure of the graphs. Although much of the work on small group behavior makes reference to differences in underlying communication structure, no empirical proof of the assumed differences has been attempted (Lawson, 1964; Leavitt, 1951; Shaw, 1954b, 1964). Rather, the empirical work on communication structure is characterized by restricting the structure of the communication a priori.

The communication pattern of a five-person group falls into 1 of 12 configurations (Shaw, 1981). These configurations (see Figure 10) range from a completely connected graph (or a comcon) where interchange is unrestricted, to more constrained configurations where, for example, four members can speak directly only to a central member and thus are forced to communicate indirectly to others (a wheel).

Leavitt (1951) has imposed these patterns on groups of people and has found that the structure affects performance on the task. For example, it seems that simple tasks (e.g., information gathering, Shaw, 1954a) are performed well when the imposed pattern has a center (e.g., wheel, Y), whereas performance in complex tasks is better when more strongly connected patterns are imposed, such as a completely connected pattern (i.e., a clique).

These controlled experiments put researchers in a unique theoretical position. There is evidence of the superiority of some structures over others, and the problem is now simply to determine if those are the patterns that emerge in communication situations where the experimenter has not restricted the pattern. For example, some work that has compared ad hoc with established groups has found that the established groups perform better (e.g., Hall & Williams, 1966). The explanation of this superiority has included the presumption that the established group used a more effective communication structure, presumably a comcon. Established groups do not, however, always outperform ad hoc groups (Ford, Nemiroff, & Pasmore, 1977; Hall & Williams, 1970). The inconsistency may be due to established groups sometimes adopting a completely connected pattern and sometimes not.

With Pathfinder, it would be possible to test this speculation. Subjects could be allowed to communicate without restriction. Telephones could be used to allow the experimenter to collect the necessary data about who talked to whom and for how long. These data could be submitted to Pathfinder, and the resultant PFNETs compared to each other and to objective measures of task performance.

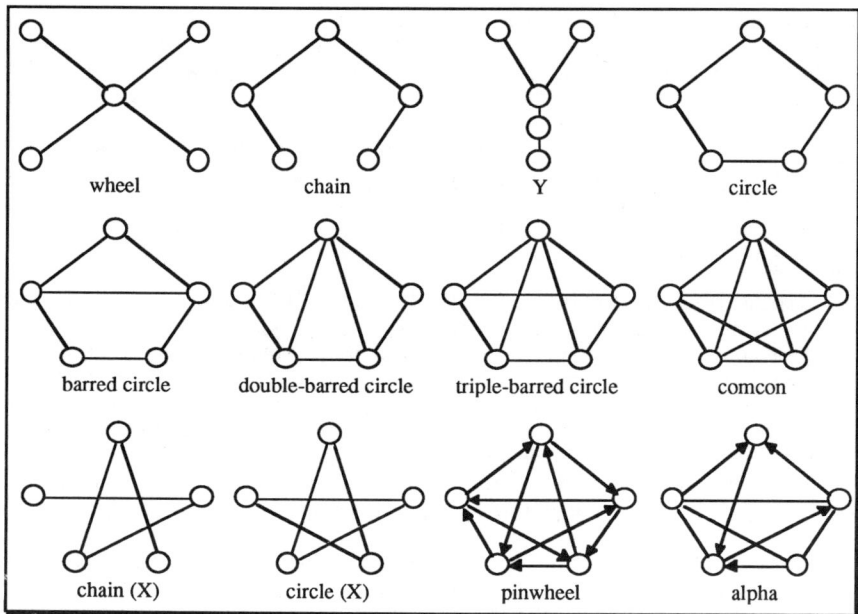

Figure 10. The possible configurations of five people in a communication graph. From Shaw (1964). Communication networks. In L. Berkowitz (Ed.), *Advances in experimental social psychology* (Vol. 1, pp. 111-147). New York: Academic Press. Adapted with permission from Academic Press and the author.

Conclusions

Our primary purpose in this chapter was to demonstrate the potential of Pathfinder and its resultant PFNETs. Pathfinder had been applied with some success to a number of issues in cognitive psychology, and we supplied some evidence that it would be of value to researchers interested in social phenomena.

PFNETs were useful in a number of ways. In some cases, simply obtaining a particular configuration was of interest. In other cases, comparisons of graphs as a whole was of prime concern, as in the classification of individuals or in the comparison of basic- and superordinate-level categories. Analysis at a more microscopic level was also informative in some cases, as when shared links were combined to define expertise or when specific cliques and cycles were identified in the social exchange data. Finally, in some cases, graph-theoretic indices were correlated with other measures to show important relations.

In the studies that compared Pathfinder with other scaling procedures (usually MDS), Pathfinder did not fare badly. It was superior in predicting serial learning and free recall but was surpassed by MDS in some classifications of pilots. It is interesting to speculate on the differences between these two scaling methods. We suspect that Pathfinder will do well in situations where the closest relations are of prime importance, but that MDS will prove superior when the more distant relations bear on the task. It may be that MDS will be better for one class of tasks, and Pathfinder for another, or it may be that the two

2 Empirically Derived Graphs

procedures should be used in concert at each opportunity. In either case, it seems reasonable to us to use both routinely in order to reach some conclusion.

In our review of the literature, we were struck by just how much power is provided Pathfinder by its roots in graph theory. Clearly, researchers have only begun to scratch the surface of this power. More sophisticated uses of measures of centrality, dominance, and path length are only beginning to be explored.

It also became clear that PFNETs with a Minkowski exponent of infinity were of sufficient utility to question the use of other exponents, at least for psychological and sociological data. This is encouraging because these types of data rarely meet more than ordinal assumptions. On the other hand, we chose different q values in different situations. Directed graphs were considerably complex, even with the sparsest PFNET. Undirected graphs seemed to add a number of interesting connections when q was reduced to 2 (Hutchinson, 1981). Although as an exploratory tool, PFNETs of several q's (and perhaps several r's) could be considered, in confirmatory studies researchers will have to consider these parameters carefully. We believe that Pathfinder will prove especially useful as a confirmatory tool, if rational bases for selections of q are devised.

In addition to the fact that most PFNETs reported here had an r parameter of infinity and a q parameter of $n-1$, it was also the case that the PFNETs were treated as graphs rather than networks. We were surprised by the number of important applications that allowed us to consider only the graph properties of the PFNETs. We did not discuss any work that required the weights of the edges in the PFNET. Thus, although a PFNET can produce networks (i.e., graphs with weighted edges), the work reviewed here did not need or did not take advantage of this additional information.

Although there are issues that remain to be considered in the development of Pathfinder, it is nevertheless true that the algorithm provides a powerful tool for applications that have a graph-theoretic connection; and as we have shown there are several such connections.

Chapter 3

Fuzzy PFNETs: Coping with Variability in Proximity Data

Chris Esposito

Many of the quantities that cognitive scientists try to measure are more evasive than the physical constants measured by other sciences. The subjective similarity or distance measurements that are used to generate a Pathfinder network are no exception. As soon as one has decided to obtain similarity data from different individuals, one must also decide how to deal with any individual differences in these data. Several possibilities present themselves. If variability across subjects is high, it may be more useful to treat the set of subjects as several groups instead of one. Another approach is to treat each individual's data separately, generate individual solutions (Pathfinder networks, for example), and then compare the individual solutions in an attempt to synthesize a group solution. Another approach is to combine the individual datasets before the scaling method is applied; averaging the similarity ratings across subjects for each pair of items is a common step in preparing data prior to using Pathfinder. The issues of whether to use intersubject variability or not, and how to use it, must also be addressed.

The work to be presented in this chapter focuses on combining individual data matrices into a composite matrix and using the variability between the individual matrices in generating the network. We also introduce a new generation parameter, z, and discuss how it's value affects the generated network. To begin, we discuss some problems that arise from an interaction between the edge membership rule and how intersubject rating variability is used. Then we present some changes to the Pathfinder algorithms that add "missing" edges and often improve the statistical fit between network solutions and the data matrix they were derived from. Finally, some empirical evidence is presented to support the claim that the statistical version of Pathfinder produces networks that fit the original data better than networks generated by the nonstatistical version of the algorithm.

FUZZYPF—An Interval-Based Version of Pathfinder

When gathering subjective similarity data from several subjects, the standard procedure for getting a representative composite matrix is to compute the average of the corresponding elements in the individual subject matrices (Schvaneveldt, Durso, & Dearholt, 1985). Another standard procedure is that the average is the only composite measure that is used; intersubject variation for every pair is ignored. This has both advantages and disadvantages. One of the claimed advantages is that for any given pair of items, the average retains what is common among the individual ratings while filtering out relatively unimportant individual differences. If the variation across subjects for a given pair is fairly small, this advantage is probably real.

However, if there is substantial disagreement between subjects over the rating for a particular item pair, then the amount of variability increases sharply. This raises two problems. First, given that the goal of the averaging is to get a representative composite value, it is no longer clear that a point value (rather than an interval) can genuinely represent widely disparate ratings. The second problem flows from how the edge membership rule compares path and edge lengths. The basic decision that occurs in generating a network is that if a path length is less than the edge length, the edge is *not* added. The problem is that *any difference*, no matter how small, is currently treated as if it were significant. As pointed out in Roske-Hofstrand and Paap (Chapter 4, this volume), many subjects will assign a slightly different score even to the same pair over time, so there appears to be an irreducible amount of variability in the score-assigning process. Failing to properly take this variability into account in network generation can result in edges being added to or omitted from the network based on differences between edge and path lengths that are sufficiently small as to not be significant. As previously mentioned, this can also reduce the fit between the network and the original data.

The proposed solution to these problems is to replace the point values for edge and path lengths with interval values constructed in such a way that both distance estimate intervals contain the "real" or "most likely" values for edge or path lengths. The basic decision then becomes one of comparing intervals to see if they overlap; if they do not and the path-length interval is below the edge-length interval, the difference between them is sufficiently large that it justifies not adding the edge. What follows below is a conceptual framework for developing a revised version of Pathfinder that uses intervals instead of point values in its decision making.

The basic model adopted is that for each edge (or pair of items) ij there is an independent random variable RV_{ij} and that the set of rating values RV_{ij} takes on across subjects constitutes a sample from the population of subjects of that type (e.g., the rating data for a command pair across a sample of UNIX experts can be used to estimate the "true" rating for that pair across the entire population of UNIX experts). The length of a path from node i to node j is also a random variable P_{ij} whose distribution is a function of the distributions of its constituent edge random variables.

As we remarked above, one view of using Pathfinder is that it consists of a sequence of decisions to be made, both by the user and the program. Several of these decisions are critically relevant here. First, how do individual ratings for a pair get combined into a composite? Second, how do individual edge lengths get combined into a path length? Third, how do we compare edge and path lengths? The statistical versions of these questions are given below, and it is those we shall attempt to answer.

(1) What is the distribution for each random variable?

(2) What is the appropriate measure of central tendency for the random variable? In other words, how do we compute its expected value?

(3) How do we determine the distribution for a path length given the distributions for its edge lengths?

(4) Given path and edge distributions, how do we compare them?

Before we begin to answer these questions it should be stated that the answers to questions 2–4 will be much easier to come by if we can demonstrate either that the random variables are each normally distributed, or that the normal distribution is a good approximation. The reason for this is that normal random variables are generally more well-behaved and easier to manipulate than those with other distributions. For example, Freund (1971) points out that normal random variables are closed under addition, minimum, and

3 Fuzzy PFNETs

maximum (the three basic operations in Pathfinder algorithms), while random variables with other distributions, such as exponential or geometric, are not.

Let us begin to answer these questions by taking another look at the basic operation of judging the similarity of two items. Without loss of generality, we can assume that these judgments are done on a scale of 1 to 10, with 1 meaning *not related* and 10 meaning *very related*. The ratings distribution across all pairs for any particular subject is generally bimodal, with one group of ratings at the high end and another group at the low end. This suggests that, at one level, the decision being made is simply *related* versus *unrelated*; the first outcome will be labeled *successes*, and the second outcome will be labeled *failures*. We can also divide the rating scale in such a way as to reflect this. Choose some position on the rating scale and let all scores less than the scale midpoint represent the *unrelated* outcome, with all scores greater than or equal to the midpoint as the *related* outcome. Without loss of generality, we can assume that for the moment the chosen position is the scale midpoint. Once the rating process has been recast in this way, a single rating of a single pair has all the characteristics of a Bernoulli trial: there are two possible outcomes, success (*related*) and failure (*unrelated*). If we repeat this procedure for a pair across n subjects then we have all the ingredients of a binomial distribution.

We are now in a position to answer the first question. As most statistics texts point out (for example, see Mendenhall, McClave, & Ramey, 1977), the Central Limit Theorem states that under certain conditions the normal distribution is a good approximation to the binomial distribution. The three parameters in a binomial distribution are n, p, and q, where p is the probability of success, $q = 1-p$ is the probability of failure, and n is the number of trials. Since we use one trial per subject, this is also the number of subjects. The two basic descriptors in a normal distribution are μ and σ, the mean and standard deviation. To determine whether the normal approximation will be adequate, calculate $\mu = np$ and $\sigma = \sqrt{npq}$. If the inequality, $0 \leq \mu \pm \sigma \leq n$, holds then the approximation will be reasonably good (Mendenhall, McClave, & Ramey, 1977). Assume, for a moment, that $p = q = 0.5$. We shall determine the minimum value of n for the normal approximation to be acceptable. Taking the left half of the inequality first, we have $0 \leq \mu - 2\sigma$. Substituting our values $\mu = 0.5n$ and $\sigma = \sqrt{0.25n}$ and simplifying, we get $0.4 \leq n$. So the left-hand inequality requires that the number of subjects is at least four. Taking the right half of the inequality next, we have $\mu + 2\sigma \leq n$. Substituting the same values for μ and σ and simplifying, we get $4 \leq n$. The right-hand inequality also requires that the number of subjects is at least four. For $p = 0.5$, as long as four or more subjects are used, the normal distribution is an acceptable approximation.

A closer examination of these bimodal distributions for a subjects ratings across pairs is likely to reveal that the region of the scale actually used for *related* judgments is much smaller than the entire top half. On our example scale that goes from 1 to 10, even if we require that *related* means a score of at least 9 (which leads to a p of 0.2), solving the above inequality yields a requirement that n be at least 16, which is still a very modest requirement.

Given that the assumption of a normal distribution is reasonable even for small numbers of subjects, we can now answer the other three questions posed earlier. If x is a normal random variable, then the appropriate measure of central tendency is the mean, or the average of the sample values. Therefore, averaging individual ratings for a pair to get a mean value is a defensible procedure. Since we are going to use sample means in calculating edge lengths and path lengths, the appropriate measure of variability is the equally familiar standard error of the sample mean, $\sigma_{\bar{x}}$, the formula for which can be found in any statistics text. As we shall see, an addition to the algorithm will be to calculate a matrix containing pairwise standard errors of the mean in addition to the means we are already computing.

As defined in Schvaneveldt, Durso, and Dearholt (1987), the length of a path is the rth root of the sum of the weights for the edges in the path, each raised to the rth power, or the maximum of these edge weights when $r = \infty$. As pointed out earlier, the distribution of a sum of normally distributed random variables is also normal, as is the maximum and minimum. The mean (or expected value) of a sum of normal random variables is the sum of the means. Also, the variance of such a sum is the sum of the variances. Unfortunately, this result is exactly true only when the random variables in the sum are not raised to powers, that is, when $r = 1$ or $r = \infty$. The square of a normal random variable has a chi-square distribution, rather than a normal one (Freund, 1971). However, the Central Limit Theorem states that if a sample of n observations (with sample mean $\bar{x}$) is drawn from a population with an arbitrary distribution, finite mean μ and standard deviation σ, then as n increases the sample mean $\bar{x}$ will be increasingly normally distributed with mean μ and standard deviation $\sigma/\sqrt{n}$. As a result, apart from some minor modifications to be described shortly, the path-length calculation does not change significantly.

At this point, let us review the decisions that have been made. Individual ratings are averaged to obtain composite edge lengths. Both edges and paths are represented by normally distributed random variables, with the latter as the sum of the former. The only issue left to resolve is how to compare an edge and a path and determine if one is shorter than another. The initial impetus for this revision of Pathfinder along statistical lines was that using point values to represent edge and path lengths led to some problems (e.g., the omission of edges), so we have replaced these point values by distributions. However, normal distributions extend infinitely in both directions, so a simple measure such as overlapping distributions goes too far in the opposite direction. The solution adopted in this work is to take the central portions of the distributions (the mean ± some user-specified amount on either side) and compare those intervals. This has the advantage of capturing the most likely values for edge and path lengths while being flexible enough to accommodate differing amounts of variability across different datasets and applications. An algorithm that incorporates all of these features is described below.

The data collection procedure for a single individual is unchanged. Given the set of individual distance matrices, we create two more matrices. The first is the average matrix W, where W_{ij} is the average of the corresponding entries in the subject matrices. The second is the variation matrix V, where V_{ij} is the value of one standard error of the mean for that pair and sample size.

FUZZYPF accepts these two data matrices as input. In addition to the q and r parameters, the user must specify the value of a third parameter z, which determines how much variability the program will use in making edge membership decisions. The parameter is called z because it determines how many standard errors the bounds on the edge weight intervals will be from their respective means, that is, their z scores. In order to create intervals of the desired size, the value zv_{ij} is added to and subtracted from w_{ij} in order to create the upper and lower bounds w_{ij_u} and w_{ij_l} for that edge weight, so that as z increases, the intervals widen. The upper bound on the path length between nodes i and j (denoted P_{ij_u}) is the sum of the upper bounds on its constituent edge weights. Conversely, the lower bound on path length P_{ij_l} is the sum of the lower bounds on its edge weights. The revised edge membership rule is that an edge is added to the network if the edge weight interval is less than or overlaps with the path-length interval. A necessary and sufficient condition for this to occur is $w_{ij_l} \leq P_{ij_u}$.

3 Fuzzy PFNETs

A somewhat surprising result is how little the network generation algorithms need to be changed in order to accommodate the modifications just described. The one unfortunate result is that the amount of space required has gone up. Two additional $n \times n$ matrices are required, one to hold the lower bounds on the edge weights and one to hold the variation matrix. The matrix that is used to hold the current minimum path lengths starts out holding the upper bounds on edge lengths and is updated using these values as each algorithm proceeds. The most significant change in using Pathfinder is that the user must now choose a value for the new z *parameter*. The next section examines some of the issues surrounding this choice.

Choosing a z Value

The issues to be explored in this section deal with choosing a z value for the statistical version of Pathfinder that was just described. It should be fairly evident from the description of FUZZYPF that as z increases the intervals widen, and so it is increasingly likely that they will overlap and the edge will be added. As Figure 1 indicates, this results in increasingly dense networks as z increases. This raises the all too familiar question of what parameter value to choose. Some related questions are (1) how much improvement in network fit does FUZZYPF provide over nonstatistical versions, and (2) how does this improvement vary as a function of z?

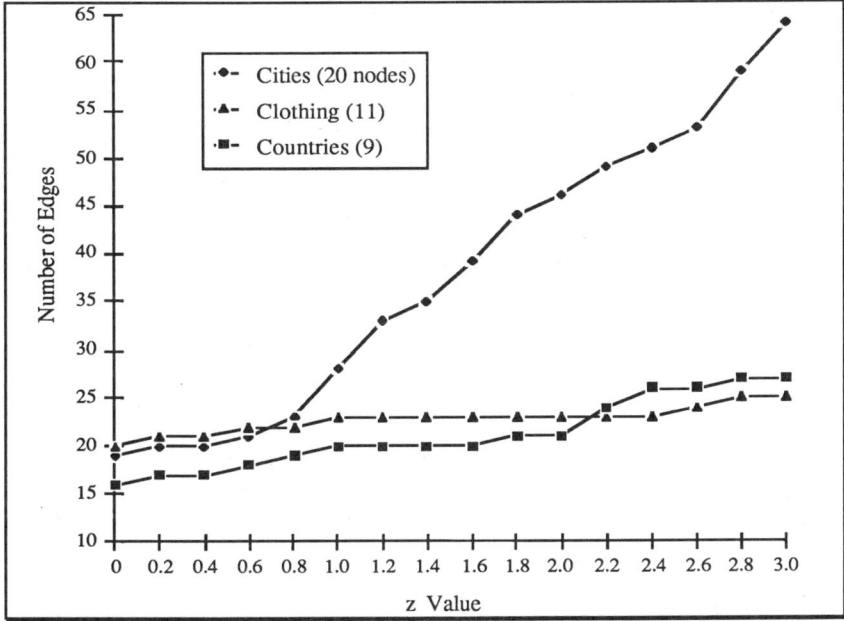

Figure 1. Number of edges as a function of z value. PFNETs($r = \infty$, $q = n-1$).

A useful way of thinking about these questions is to look at the set of edges not present in a network generated by a nonstatistical version of Pathfinder. These can be divided roughly into two types. Type 1 edges are those that clearly don't belong in the network because they are significantly longer than the shortest alternate paths. Type 2 edges are those that are longer than the shortest alternate paths by an insignificant amount. Since the length differences are insignificant, the edge length and the path length should be considered as actually tied and the edge should be added, which often also improves the fit between the network solution and the original data. The goal is to find the z value that results in adding all of the Type 2 edges but none of the Type 1 edges.

The attempts to answer these questions are admittedly more exploratory than final. The approach used was to examine the fit between networks and original data matrices across several domains and a wide variety of z values. The datasets and measures of fit used here are also part of a larger study on the relationship between generation parameters, distance measures, and measures of fit between networks and the original distance matrices. For more details, see Esposito (Chapter 6, this volume). A brief summary of the relevant information is presented here.

The three domains used in this study were cities in New Mexico (20 terms, 12 subjects), items of clothing (11 terms, 20 subjects), and countries (9 terms, 9 subjects). For each domain, a set of networks was generated with $q = n-1$, $r = \infty$, and z varying from 0.0 to 3.0 in 0.1 increments. For each network we derived a distance matrix using the graph-theoretic definition of distance. Since the r value used in network generation entailed making only ordinal assumptions about the data, Spearman's ρ statistic was used to measure the fit between the derived and the original distance matrices (ρ also makes only ordinal assumptions).

Figure 2 presents a graph of fit as a function of z value for the three domains. Notice first that for this set of graphs, $r = \infty$ and $q = n-1$. The leftmost data point (network) is at $z = 0$. Since variability is effectively ignored at this z value, for each domain this network represents the PFNET($r = \infty$, $q = n-1$) as it would be calculated by any of the "regular" versions of Pathfinder and will serve as a reference standard to compare to networks generated with other values of z.

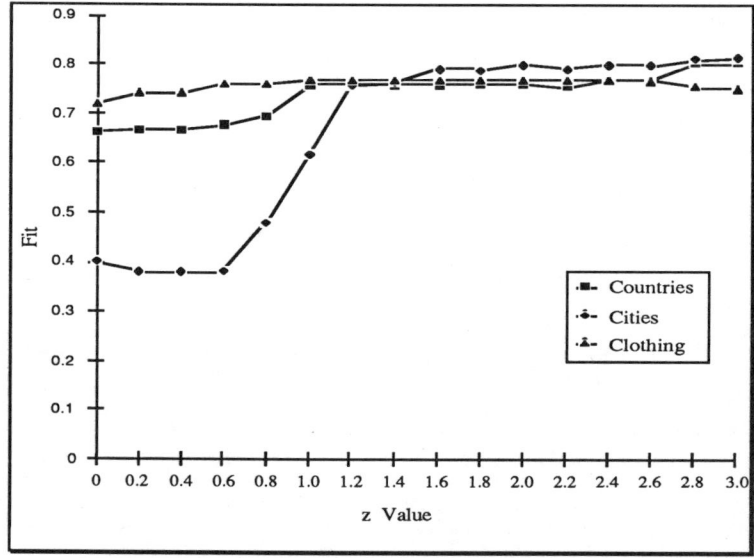

Figure 2. Network fit as a function of z value. PFNETs($r = \infty$, $q = n-1$).

3 Fuzzy PFNETs

As z increases, the edge-based fit curve for each dataset rises above where it was when $z = 0$, which strongly suggests that FUZZYPF produces networks that fit the original data better than networks produced by the regular version. In each of these datasets there is a fairly sharp rise in fit, followed by a leveling off, and then by a decline in fit as z gets rather large. The decline is easy to explain. As z increases, the networks get progressively denser. Since we are using a measure of fit that is based on the number of edges that separate two nodes, denser networks mean that everything is closer together; at some sufficiently large value of z, every pair of nodes will be nearly equidistant and so these distances will not fit very well with the original data matrix.

If we use the technique of looking for "elbows" in a fit curve, then all three of these curves have them, although different amounts of variability in the datasets means that they occur at different z scores. However, Table 1 shows that the same rising and leveling off in the number of edges added is also present. For the three domains used, the elbow in the edge curve (Figure 1) occurs at roughly the same z score that it occurs at in the fit curve (Figure 2). One can therefore make a case that a correspondence between the z value at which an elbow occurs in the fit curve and in the edges-added curve indicates that this is the z value required to create Type 2 edges in the network for that dataset and choice of q and r.

Table 1. Number of edges in various domains as a function of z value (values in parentheses are numbers of nodes).

z value	Domain				
	bank(20)	fruit(11)	cities(20)	clothing(11)	countries(9)
0.00	20	20	19	20	16
0.20	23	20	20	21	17
0.40	23	20	20	21	17
0.60	24	20	21	22	18
0.80	26	22	23	22	19
1.00	26	22	28	23	19
1.20	27	22	33	23	20
1.40	27	22	35	23	20
1.60	30	22	39	23	20
1.80	31	22	44	23	21
2.00	32	22	46	23	23
Complete	190	110	190	110	72
	Undirected	Directed	Undirected	Directed	Directed

Conclusions

In this chapter we presented a revised version of Pathfinder in order to deal with some statistical problems. The problems stem from the fact that there is often variability in subjective data, and in the old algorithm this variability was ignored for lack of a principled way of dealing with it. This often led to the omission of edges from the network because they were slightly longer than the shortest alternate path, even though the difference in length was due to the random fluctuations in length rather than some statistically significant difference in length. The structure of the generated network was therefore affected in unpredictable ways that also often adversely affected the fit between the network scaling solution and the original data.

The new algorithm, FUZZYPF, explicitly incorporates a user-specified measure of variability by replacing the point values for edge and path length with intervals of user-specified width. The new criterion for edge membership in a Pathfinder network is that the edge is included if its length interval overlaps with the path-length interval. This new algorithm requires no more time than the old one, but does require two additional matrices, one to hold the standard error matrix and the other to hold edge weight lower bounds.

As a test of FUZZYPF, note that when $z = 0$, all variability is ignored and this version produces networks identical to the nonstatistical version. In order to compare the networks produced by FUZZYPF with those produced by a regular version of Pathfinder, we took five domains and computed the PFNET($r = \infty$, $q = n-1$) (with $z = 0$) for each one. We then let the z value range between 0.1 and 3.0 in increments of 0.1 and computed the fit (Spearman's ρ) between distance matrices derived from the networks and the original data. In every domain, the fit value rose and then fell as z increased, supporting the contention that for the "right" value of z, FUZZYPF produced better networks than the regular version of Pathfinder.

Despite the successful results reported above, it should not be concluded that the issue of how to deal with variability in subjective data when doing data scaling has been resolved in any final sense. Several related issues and approaches that were not addressed in this work are worth exploring. As mentioned at the beginning of this chapter, if between-subject variability is very high, a single network solution will not represent the group very well. On the other hand, if variability is very low, then using the averaged distance matrix by itself is probably sufficient. FUZZYPF offers the greatest advantage when the variability is moderate, so a more precise understanding of the size of this "moderate" range would be very useful. On a related note, different pairs of items will have differing amounts of variability in their similarity ratings, so some idea of what parts of the network are stable (less variable) and what parts are more variable would also be useful. A more general issue is the role of variability in data scaling in general. Some measure of how sensitive a clustering is to perturbations in the original distance data would be very useful. In addition, a version of multidimensional scaling based on interval rather than point values would likely produce some interesting results.

Chapter 4

Discriminating Between Degrees of Low or High Similarity: Implications for Scaling Techniques Using Semantic Judgments[*]

Renate J. Roske-Hofstrand and Kenneth R. Paap

Several scaling techniques (Pathfinder, multidimensional scaling [MDS], hierarchical clustering) require as input a proximity matrix that specifies the distance between each pair of concepts in the domain of interest. Many applications derive these distance estimates from similarity or relatedness judgments obtained with the method of paired comparisons. For example, ratings have been used to obtain distance estimates between: the display panels of the interface to a flight management system (Roske-Hofstrand & Paap, 1986a), items on a fast food cash register (McDonald, Dayton, & McDonald, 1988), and concepts in an experimental methods course (Goldsmith & Johnson, Chapter 17, this volume). In this method all possible pairs of concepts are presented, one pair at a time, and judges are asked to rate the degree of similarity or relatedness on some scale. This chapter focuses on some issues concerning the sensitivity and reliability of distance estimates derived from ratings.

The primary issue concerns the type of measurement scale inherent in a proximity matrix derived from ratings. On the basis of our own introspections and the debriefing of many raters, we began with the working hypothesis that judges are able to make very fine discriminations between pairs of concepts that are obviously similar, but that it is much more difficult to judge the degree of dissimilarity for a pair of concepts that have *weak* or *secondary* dimensions of similarity.

For example, the reader is invited to rate the similarity of *plate* and *bowl*. Many primary or important dimensions of comparison come readily to mind and most of these comparisons yield good matches, for example, they have the same round shapes (but different depths), they function as holders for food (but different types of food), they are likely to be made from the same material and come in matched sets, and so on. Is the *plate-bowl* pair more or less similar than the pair *plate-cup* or *plate-fork*? To us, these types of judgments seem easy and a feeling of degree of similarity pops up fairly quickly and automatically. This initial feeling can usually be supported by the type of deliberate analysis illustrated above for the dimensions of shape, function, and material.

In contrast, rate the similarity between *washing machine* and *rocking horse*. Sharing our introspections again, if we sense any fast and automatic evaluation to this pair of concepts, it is the feeling that they are completely dissimilar. However, we can search for weak or very abstract dimensions in which they have something in common. For example, the size and structure are such that a child could sit on either one (but is not likely to sit on the former) and movement is an important part of the function of both items. Given agreement that the pair has some small degree of similarity, how does its similarity compare to

---

[*]Portions of this study were presented at the 27th Annual Meeting of the Psychonomic Society, November, 1986, New Orleans, LA. We thank Roger Schvaneveldt and Jim McDonald for valuable discussions of this work and Dean Berry and Judy Northrup for collecting much of the data.

sink and *rocking horse*? Our intuitions are that it is much more difficult to discriminate between pairs of concepts with low similarity than pairs of high similarity.

If the working hypothesis is correct, what implications might this have for the way the judges use the rating scale? Two related behaviors seem likely. Assume that judges are asked to rate the similarity of pairs on a scale of 0 to 9 with lower values indicating greater similarity. Given that raters experience many degrees of high similarity they should choose to distribute their responses over several of the values at the similar end of the scale. That is, if the subjective experience is that some pairs are extremely similar, some very similar, some quite similar, and others only moderately similar, then a separate scale value is likely to be used to reflect each of these different states. On the other hand, if degrees of dissimilarity (or extremely low similarity) are difficult to discern, then raters might restrict their responses on the dissimilar end of the scale to only one or two extreme values. This is one prediction we tested in the experiment reported later in this chapter.

If judges are more sensitive to similarity differences among more similar pairs than less similar pairs, then they may also show greater consistency in their ratings. One approach to examining this possibility would be to assess the reliability of judges when they are asked to rate the same set of concepts a second time. This would be a good direct test of the consistency hypothesis, however, we used a somewhat less direct attack that has some other desirable characteristics that will be discussed following a brief presentation of our method.

Suppose an individual has rated the similarity of all pairs in a given domain and is then invited back to the laboratory a few days later. In this second session subjects are presented with two pairs on every trial, such as, *plate-bowl* and *cup-fork*, and asked whether one pair was more similar than the other. Assume for the moment that this judge had, in the first phase of the experiment, rated the first pair as highly similar, say a "1," and the second pair somewhat less similar, say a "2." If the same assessment of similarity occurs during the second task, then the judge should answer "yes," that *plate-bowl* is the more similar pair. However, if the experience of similarity has some instability and *plate-bowl* is assessed to have just a little less similarity during this second session, then the judge may say "no," they seem to have the same degree of similarity. Thus, consistency is reflected in the likelihood that such pairs of pairs are still perceived to have different degrees of similarity.

It should be intuitive that as we present pairs of pairs with greater discrepancy in the original rating that the likelihood of a judge to fail to experience them as different should decline despite some amount of instability. For example, if the pair *plate-bowl* is presented together with a pair that was originally rated as considerably less similar, such as, *bed-dresser*, then *plate-bowl* should still be experienced as more similar, even if, due to instability, it has shifted a bit in the direction of less similarity. The procedure described in more detail in the method section, computes an index of consistency by asking how far down the similarity scale must one go before judges consistently experience a difference in degree of similarity.

The underlying logic is similar to that first used by Weber (1846) in determining the just-noticeable-difference (jnd) between a standard and comparison stimulus at any point along some physical dimension. For example, if we start with the highest pitched tone a listener can hear, how far down the frequency scale would we have to go in order for the listener to consistently report that the two tones differ in pitch? This analogy helps show that our consistency index can also be thought of as a measure of sensitivity. The more sensitive a listener is to frequency differences, then the smaller the difference required to produce consistent "different" responses. Similarly, the more sensitive a judge is to

4 Similarity Discrimination 63

similarity differences, the smaller the rating-scale difference required to produce consistent "different" responses.

The attentive reader is sure to observe that the psychophysicist who computes a jnd obtains a measure of the subject's sensitivity relative to some physical scale, but that the jnd we compute is relative to the subjective similarity scale used by our judges in the first phase of the experiment. This is true, but psychological sensitivity to rating scale differences can have significant implications for knowledge representations derived from ratings. One such scenario is outlined next.

Suppose we reverse the procedure described earlier and form pairs of pairs by selecting a standard that was originally given the lowest similarity rating. This standard is compared to other pairs which differ by one, two, or three points on the rating scale. How large a discrepancy between the two pairs is required in order for the judges to consistently report that the standard and comparison differ in degree of similarity? We hypothesize that judges are less sensitive to their own rating scale differences when presented with pairs of low-similarity pairs.

If the difference between an extremely similar pair (rating scale = 0) and a very similar pair (scale = 1) is consistently experienced then that one-unit difference in distance in the proximity matrix is psychologically real. In contrast, if the difference between a pair given the lowest similarity rating (scale = 9) and a pair given the next lowest rating (scale = 8) cannot be consistently discriminated, then this one-unit difference is mostly noise. Some scaling techniques will be much more susceptible to this noise than others. For example, MDS tries to consider all pairs of distances simultaneously. In doing so, it gives considerable weight to the longer distances associated with the dissimilar end of the rating scale.

However, Pathfinder is more protected from this noise, since the presence of links is determined, almost exclusively, by differences among highly similar pairs. Dearholt and Schvaneveldt (Chapter 1, this volume) on the foundations of Pathfinder can be consulted for a more complete description of why this is true, but briefly stated the reliance on distances between highly similar pairs comes from Pathfinder's link membership rule. A direct link between two nodes is maintained if, and only if, the weight of that link is shorter than any alternative path that connects those two nodes. Thus, critical decisions determining the structure of the Pathfinder network usually turn on the difference between one short path and another. Links in a Pathfinder network with substantial weights (long distances) are relatively rare and tend to occur as bridges between distinctive subnets or as trails to hermit nodes a great distance from their nearest neighbor.

In summary, our working hypothesis is that judges in a rating task are more sensitive to differences among pairs of highly similar concepts than to differences among pairs that tend to be quite dissimilar. This hypothesis, in turn, leads to the prediction that judges will tend to restrict their responses on the dissimilar end of the scale to only one or two extreme values. It was further suggested that rating scale differences at the high-similarity end of the scale should be more discriminable than equivalent differences drawn from the low-similarity end. Confirmation of the latter prediction would suggest that knowledge representations derived from Pathfinder's scaling of rating data may be less susceptible to bias than those derived from MDS.

Experiment 1

Method

Subjects. Fifty-seven students participated in order to fulfill a requirement of their introductory psychology class.

Procedure for Phase 1 Similarity Ratings. Stimulus presentation was controlled by a Terak 8510 microcomputer. The first part of the displayed instructions were as follows: "The next part of this experiment involves the assignment of similarity ratings to pairs of objects that might be found in a house. Your task will be to judge the similarity or relatedness of several pairs of objects."

The remainder of the instructions described the rating scale and the procedure for making a response:

> In the actual task you will be shown a horizontal rating scale above each pair of terms. The numbers "0" through "9" will be displayed at regular intervals below the line and you are to choose one of these numbers to reflect your judgment. Use "0" for those items that you feel are highly dissimilar and respond with a "9" for items that appear highly similar. The numbers between "0" and "9" should be used to represent degrees of similarity with higher numbers representing greater similarity. Upon responding a vertical bar will move directly above the number you pressed. A second vertical bar will remain at the right end of the scale. The vertical bars also represent similarity. When they are closer together the concepts are more similar.

Two aspects of the instructions deserve discussion. First, the beginning part of the instructions refer to both "similarity" and "relatedness," but the remainder of the instructions refer only to similarity. Furthermore, the rating scale always displayed the phrase "low similarity" at the left end and "high similarity" at the right end. The choice between similarity and relatedness may be important. Some thoughtful individuals have argued that the two are not synonymous and that concepts can be highly related, but not particularly similar. For example, one might judge *can* and *can opener* to be highly related, but not very similar. We doubt that this subtlety was appreciated by most of our subjects, but the effect of similarity versus relatedness remains an empirical issue. If subjects asked the experimenter how to judge similarity, they were told that it was up to them and that there were many different features or attributes that could be considered for any pair of objects.

A second important point to note is that high similarity was represented by larger numbers on the rating scale. This seems more natural for subjects. However, in order to transform these similarity ratings into estimates of distance it was necessary to subtract each rating from nine. Since the distance transformation is required in order to use the ratings as input to a scaling algorithm, all subsequent discussion will refer to the transformed scores. That is, a reference to a 0-pair is a pair of objects that received the highest rating of similarity (a 9 on the similarity scale) and, accordingly, has the shortest psychological distance.

The rating scale with the vertical bar markers described in the instructions was presented in the top half of the screen. The two objects comprising any pair were centered in the bottom half of the screen with one object directly above the other. The object on top was determined randomly. Subjects were presented with all of the 496 possible pairs of

4 Similarity Discrimination 65

items from their stimulus set of 32 objects. Pairs were presented one at a time, in a different random order for each subject.

Materials for Phase 1 Similarity Ratings. The domain consisted of 64 objects that might be found in a house. The 64 objects were randomly assigned to 4 subsets of 16. Each subject was presented with the items from two of the subsets. Thus, each subject rated the 496 pairs formed from all combinations of 32 objects. Six groups of subjects were required in order to obtain ratings for all pairs in the complete set of 64. The number of subjects in each group ranged from 8 to 11.

Procedure for Phase 2 Discrimination Judgments. One week following their initial ratings, the subjects returned for Phase 2 of the experiment. Stimuli were displayed on a Datamedia 3510 terminal under the control of an APL function that was selecting materials according to the scheme described in the next section. On each trial subjects were presented with a pair of pairs, such as:

1. *spoon-fork*
2. *pillow-cushion*

and were given the prompt: "Is one pair of objects more related than the other?" If the subject pressed "Y" for yes, he was then asked to indicate which of the pairs was more related by pressing either the 1 key or the 2 key.

In retrospect, it may have been better to ask if one pair of objects was more similar than the other since the initial rating task emphasized similarity more than relatedness. It is therefore conceivable that some failures to accurately predict *discrimination in relatedness* in the second phase on the basis of *ratings of similarity* could be attributed to individual pairs having different degrees of perceived similarity and relatedness.

Materials for Phase 2 Discrimination Judgments. To avoid confusion, the following terminology will be adopted. A trial in Phase 1 consisted of the presentation of a pair of objects that the subject rated on the 10-point scale. This pair of objects will be referred to simply as a *pair*. A number in front of a pair indicates its similarity rating (transformed to distance), for example, a 0-pair is a pair of objects perceived to be highly similar. A trial in Phase 2 consisted of the presentation of a pair of pairs (POP) as illustrated above. This type of stimulus will be referred to as a *POP*. The two numbers in front of a POP indicate the similarity rating of each of its constituent pairs, for example, a 0-0 POP is a pair of pairs both of which were assigned the highest degree of similarity.

POPs were selected for each individual subject on the basis of his or her previous ratings of 496 pairs of objects. Four pairs were randomly selected from the set of all pairs given the highest similarity rating (rating value = 0). Five pairs were randomly selected from each of the set of pairs assigned rating values of 1, 2, and 3. Six 0-0 POPs were created from all possible pairings of the four 0-pairs. The four 0-pairs were next crossed with the five 1-pairs, yielding 20 different POPs with a rating-scale difference of 1. The four 0-pairs were similarly crossed with the five 2-pairs and the five 3-pairs. Thus, there were 66 potential POPs drawn from the related end of the rating scale. A full complement of 66 POPs could not be formed for every subject, since not everyone rated a sufficient number of pairs at each of the critical values. For example, if a subject generated only four 1-pairs (rather than 5 or more) then he would be presented with only sixteen 0-1 POPs (rather than 20). Of the 51 subjects who returned and completed phase 2, 29 were presented with the full complement of 66 POPs formed from the related end of the rating scale.

A similar pattern of selections was used at the unrelated end, resulting in full complements of six 9-9 POPs, and 20 each of 9-8 POPs, 9-7 POPs, and 9-6 POPs. Thirty-five subjects received the full complement of POPs at the unrelated end of the rating scale. The set of POPs selected for each subject on the basis of their own earlier ratings were presented in random order. On any given trial the pair appearing on top and designated as 1 (one) for purposes of responding was determined randomly.

Results & Discussion

Phase 1. Figure 1 shows the distribution of ratings responses (transformed to distance). It is heavily skewed toward judgments of low similarity. As predicted, many pairs (45.1%) are experienced as having an undifferentiated degree of very low similarity.

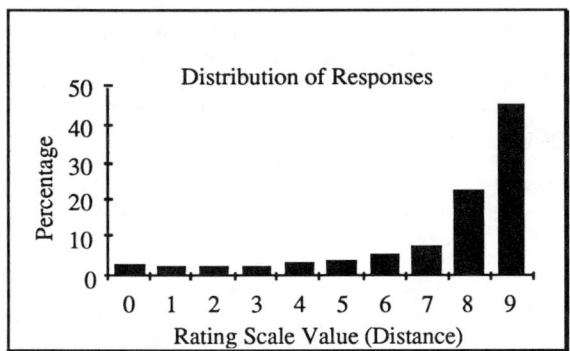

Figure 1. Percentage of each rating-scale response. Low values indicate high degrees of similarity (small psychological distances).

This outcome would be even more pronounced if it were not for an unanticipated characteristic of the response-mapping. Recall that the lowest-similarity value of 9 in Figure 1 is a distance transformation of the actual rating response of 0. Seventeen subjects elected to never use the 0 because "it was out of order on the keyboard." These subjects assigned a 1 to those pairs they judged to be least similar. If this strategy is taken into account by summing the total number of responses in the most extreme category actually used by each subject, the percentage of responses grows to 64.6%. Thus, it appears that almost two-thirds of all pairs are experienced as simply *dissimilar* and assigned the same low value on the rating scale.

There does appear to be a fairly even distribution of responses across the high-similarity values. The percentage of responses across the four values of highest similarity are 2.6, 2.6, 2.5, and 3.2. Although these values cover only 11% of the complete proximity matrix, they roughly correspond to where all the action is during the generation of a *Pathfinder graph*. Path-length differences of 1 or 2 units among these pairs will heavily influence the link structure. Since Pathfinder treats these differences as important, it is informative to determine if people are sensitive to these differences. The results of Phase 2 are, of course, relevant to this question.

Phase 2. The purpose of the second phase was to see how well an individual's original rating could predict his ability to discriminate the similarity of one pair relative to another. POPs composed of two pairs with nearly identical ratings (e.g., 0-1 POPs or 9-8 POPs) should be difficult to discriminate and should receive a relatively low percentage of "different" responses. The percentage of "different" responses should increase as the size of the rating-scale difference increases, for example, the greater similarity of a 0-pair as compared to a 3-pair in a 0-3 POP should be more evident and lead to a higher percentage of "different" responses.

4 Similarity Discrimination

Inspection of Figure 2 shows that the percentage of "different" responses in Phase 2 increases linearly as a function of the size of the rating-scale difference obtained in Phase 1. This is true regardless of whether the standard is the value with lowest or highest similarity. However, the function for differences occurring at the low-similarity end of the scale falls completely below that for the high-similarity end!

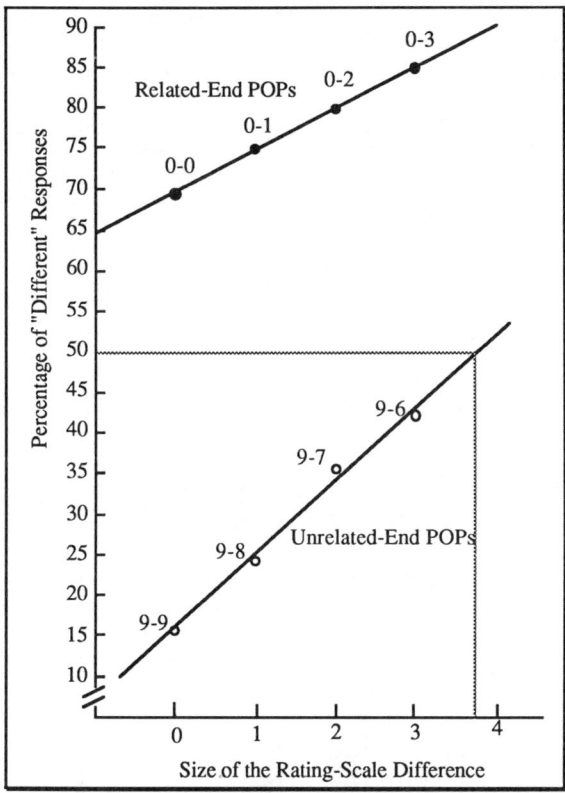

Figure 2. Percentage of "different" responses as a function of the size of the rating-scale difference obtained from the same individual.

The sensitivity index, based on an analogy to the computation of a jnd in classical psychophysics, can be defined as that rating-scale difference required to produce 50% "different" responses. Before examining these results it may be helpful to note that the choice of the 50% criterion cannot be guided by traditional use of the method of constant stimuli. This method usually entails the use of comparison stimuli that are both above and below the standard of interest (Baird & Noma, 1978) and response alternatives of "above" and "below." In this traditional application of the method, 50% "above" responses defines the point of subjective equality, and jnds are usually defined as the stimulus difference corresponding to 25% and 75%. This procedure was not followed in the current study because we were specifically interested in the size of the jnd at the two extreme ends of the rating scale and, in fact, predicted less sensitivity at the low-similarity end. Having selected the 0-pairs and 9-pairs as standards for theoretical reasons, we could not have comparison stimuli more similar than a 0-pair or less similar than a 9-pair. The criterion of 50% "different" responses is midway between chance and perfect performance under the assumption that subjects would generate 100% "different" responses under conditions of complete discriminability and 0% in the total absence of any sensitivity.

Although, a priori, a criterion of 50% "different" responses seemed reasonable this cutoff cannot be readily applied or interpreted to the data obtained at the related end of the rating scale. Since about 69% of the responses to the 0-0 POPs were "different," we can only infer that the index based on the 50% criterion would be less than one rating-scale unit. Apparently subjects have a tendency to run into the upper end of the rating scale and are capable of making very fine discriminations among highly related items.

The results are quite different at the unrelated end. Even with a rating-scale difference of 3, the function has not reached the 50% cutoff. Linear extrapolation suggests that the size of the index at the low-similarity end of the rating scale is about 3.75 units. Apparently subjects have a difficult time judging the relative degree of similarity between pairs that are only marginally related. Not only do they assign a sizable majority to the lowest-similarity class to begin with, but they also fail to consistently appreciate the difference between this large group of unrelated pairs and those pairs they originally rated as having a little more similarity.

When subjects in Phase 2 indicate that one pair is more related than another, they do not always pick the pair they previously rated as more similar. Figure 3 shows the percentage of "different" responses that were followed by a selection opposite from that predicted on the basis of the similarity ratings. The percentage of reversals is fairly high for the POPs differing by only one unit, but systematically declines with the rating-scale difference. Reversals are indicative of considerable instability in judgments of similarity.

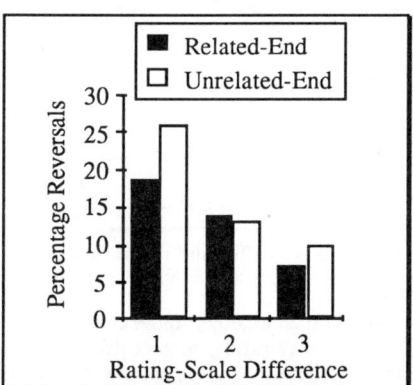

Figure 3. Percentage of "different" responses in which the pair rated less similar during Phase 1 was selected as the more related pair in Phase 2.

It would be informative to determine if the reversals were fairly random or systematic. Rating mistakes in Phase 1 should lead to a systematic pattern of reversals in Phase 2. The following analysis specifically focuses on 0-pairs that may have been overrated and 9-pairs

4 Similarity Discrimination

that may have been underrated. In Phase 2 subjects judged each 0-pair against three other 0-pairs. Based on the original ratings, these should be judged as equally related and receive "same" responses. However, as shown in Figure 2, two-thirds of the 0-0 POPs were judged as "different."

The "different" responses can be used to derive an index of strength for each 0-pair by counting the number of times an individual 0-pair dominates the other three 0-pairs in the corresponding 0-0 POPs. For any given 0-pair, this strength index can range from 0 to 3, with 3 indicating greatest strength. The 0-pairs with the least strength (index = 0) should be most susceptible to reversals. For each 0-pair we determined the number of reversals it produced in all of the 0-1, 0-2, and 0-3 POPs to which it belonged. The strength index was also calculated for each 9-pair based on the number of times it dominated the three other 9-pairs in a subject's 9-9 POPs. In contrast to the 0-pairs, it is the strong 9-pairs that should lead to a greater number of reversals.

Figure 4 shows the mean number of reversals associated with the strength index. Reversals decrease with the strength of a 0-pair and increase with the strength of a 9-pair. It is evident that it is extremely rare for the best 0-pairs to be perceived as less related than a pair previously assigned a lower similarity rating, but that there are some weak 0-pairs that are involved in many reversals. The leading cause of reversals is a strong (index = 3) 9-pair.

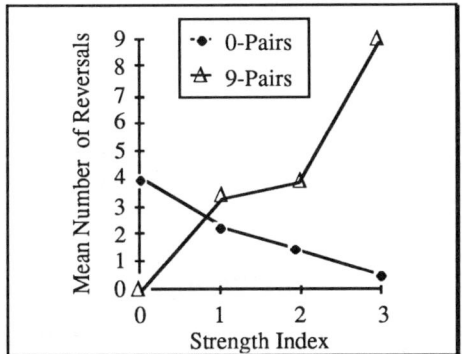

Figure 4. Mean number of reversals (out of 15 possible) as a function of the strength index of the 0-pairs and the 9-pairs.

The left side of Table 1 shows some examples of weak 0-pairs that generated a large number of reversals, while the right side shows examples of strong 9-pairs that produced many reversals. Both the index and the number of reversals for each pair are with respect to some particular individual. Accordingly, given the rules for selecting POPs for each subject the maximum possible number of reversals is 15. The systematic relationship between the strength index and the number of reversals is consistent with the view that some pairs suffer a meaningful shift in similarity from one test session to the next, but that judgments made during Phase 2 are internally consistent. For example, the subject that rated *blender* and *can opener* as completely dissimilar in Phase 1, is clearly treating this pair as having a fair amount of relatedness in Phase 2 and, accordingly, it is selected as the more related pair in 11 of the 15 opportunities in which it was tested with pairs originally rated as 1, 2, or 3 scale units more similar.

Table 1. Examples of weak 0-pairs and strong 9-pairs taken from individual ratings.

Weak 0-Pairs	Index	Reversals	Strong 9-Pairs	Index	Reversals
desk - night table	0	9	toy chest - mirror	2	9
radio - record	0	4	spoon - record	3	11
book - record	0	7	can opener - blender	3	10
towel rack - hanger	0	4	toy chest - bowl	3	10

Most of the pairs associated with high reversal rates are probably simply mistakes, in the sense that if the judge had it to do over again that pair would be given a different similarity rating. However, some could be caused by the switch in focus from similarity in Phase 1 to relatedness in Phase 2. The *blender-can opener* pair could be viewed as such an example, if this subject judges the pair to be reasonably related, but not particularly similar. Note, that this potential cause does not apply as readily to reversals at the related end. The problem with the 0-pair *book-record* for some particular subject was that this pair is experienced in Phase 2 as having significantly less relatedness than its rating would indicate. The hypothesis that the pair is judged to be highly related, but not so similar, would predict the opposite pattern.

In summary, subjects are more likely to respond "same" to POPs whose original ratings were actually different, if the POP is drawn from the low-similarity end of the rating scale. This is consistent with the hypothesis that it is difficult to discriminate between low levels of similarity. It was also determined that the leading cause of reversals is strong 9-pairs that are judged to be more related than the three other 9-pairs selected for testing. Thus, the low-similarity end of the rating scale is plagued by less sensitivity and greater likelihood of systematic error.

Experiment 2

When ratings are obtained using the method of paired comparisons, scaling solutions can be derived for individual subjects or for an entire group of subjects who have comparable levels of domain expertise. There are potential tradeoffs for either theoretical or applied work. Only individual networks can reflect idiosyncratic differences in conceptual organization between subjects. On the other hand, as revealed above, individual ratings may be prone to *mistakes* and some of the error is likely to be corrected by averaging ratings across subjects. To investigate this issue the second phase of Experiment 1 was replicated with a new set of subjects. In this case the size of the rating-scale difference for each POP was based on the first group's average ratings.

Method

Subjects. Twenty-six judges from the same subject pool participated in the second experiment.

Materials. The POPs for Experiment 2 were selected using the following procedure. The six groups that participated in the first experiment rated a total of 2,016 pairs. These correspond to all possible pairs formed from the complete set of 64 household objects. The mean rating for each pair was rank ordered and the number of pairs within a half unit of

4 Similarity Discrimination

each rating-scale value was determined. For example, if the mean for the pair *teddy bear-pillow* was 3.62, the frequency for the rating-scale value 4 was incremented since it was within the range of 3.5 to 4.5. The distribution of pairs within the specified range is shown in the first row of Table 2. Because only two pairs have means less than 0.5, the most related pairs were selected from those within a half unit of a rating-scale value of 1. Nine pairs were selected from the pool of 14 in an effort to both minimize the variance and come as close as possible to a mean of 1.00. The selected means ranged from 0.71 to 1.30, with a mean of 1.00 and a standard deviation of 0.23. Similar procedures were used to select nine pairs with mean ratings close to 2, 3, 4, 6, 7, 8, and 9. These means are shown in the second row of Table 2.

Table 2. Distribution of mean-rating values from Experiment 1 and means of pairs used in Experiment 2.

	Rating-Scale Value									
	0	1	2	3	4	5	6	7	8	9
Number of pairs within a half unit	2	14	29	31	64	101	186	380	911	290
Mean of nine pairs selected for Exp. 2		1.00	2.02	3.00	3.99		6.00	7.00	8.00	9.00

The nine pairs selected for each of the mean rating-scale values were randomly divided into two sets. Set A consisted of five 1-pairs (most similar), five 9-pairs (least similar), and four pairs for each of the intermediate rating-scale values. This permitted the formation of 10 different POPs for rating-scale differences of 0 (i.e., 1-1 POPs and 9-9 POPs), and 20 different POPs for rating scale differences of 1, 2, and 3. Fourteen subjects were presented with the 140 POPs formed from the Set A materials. Set B consisted of four 1-pairs, four 9-pairs, and five pairs for each of the intermediate values. This permitted the formation of 6 different 1-1 POPs and 9-9 POPs, and 20 different POPs for each of the non-zero rating-scale differences. Twelve subjects were presented with the 132 POPs formed from the Set B materials.

Procedure. The subject's task was the same as that used in Phase 2 of Experiment 1. On each trial the subject was presented with a POP and asked to indicate whether the two pairs were the same or different in terms of their degree of relatedness. Whenever subjects responded "different," they were then required to indicate which pair was more related.

Results & Discussion

The same pattern of results was obtained with each set of materials and the analyses described below are collapsed across both sets. Figure 5 shows the percentage of "different" responses as a function of the size of the rating-scale difference for Experiment 2. The results for individual-based (Figure 2) and group-based (Figure 5) stimuli are very similar at the related end. Both would yield a cognitive jnd value less than zero. At the unrelated end, the group-based pairs yield greater discriminability. The 50% cutoff yields a sensitivity index for the group ratings of about 1.75 units compared to the 3.75 when the POPs were formed on the basis of an individual's own ratings.

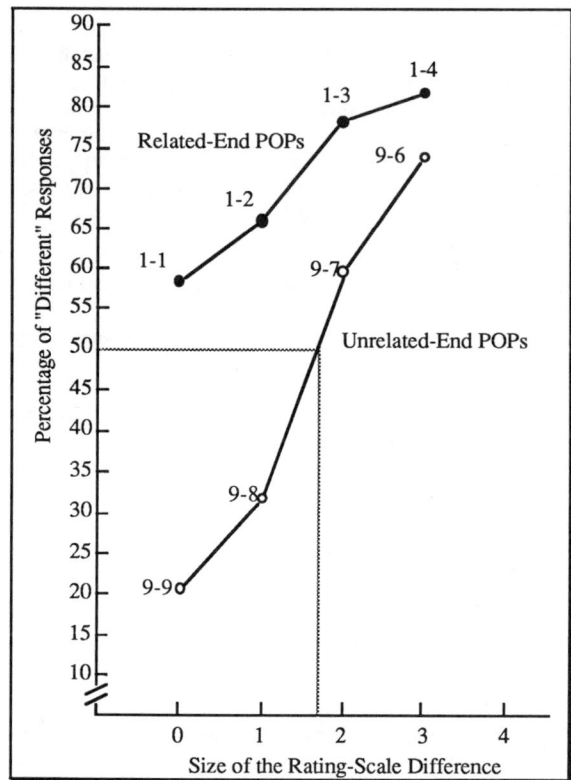

Figure 5. Percentage of "different" responses as a function of the size of the rating-scale difference for the mean ratings from Experiment 1.

Figure 6 shows that the percentage of reversals is associated with the various rating-scale differences. A reversal occurs when the subject responds "different" and then selects as more related, the pair that was rated less similar by the subjects in Experiment 1. The group-based reversals, like the individual-based reversals shown in Figure 3, systematically decrease with the size of the rating-scale difference.

In comparing the percentage of reversals computed by reference to an individual's ratings (Figure 3) to those based on group averages (Figure 6), it appears that the percentage of group-based reversals is somewhat higher at the related end and somewhat lower at the unrelated end. In terms of the tradeoff discussed earlier, this trend suggests that subjects are not in complete agreement when making the fine discriminations among related pairs. However, ratings based on group averages will remove some of the error associated with the difficult discriminations at the unrelated end of the rating scale.

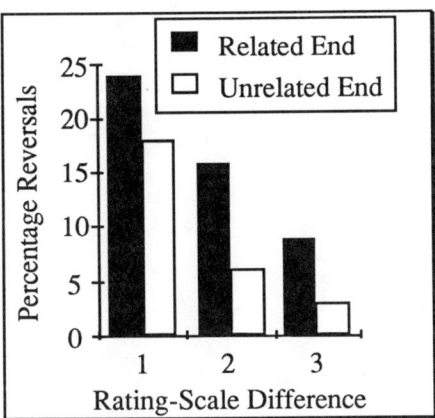

Figure 6. Percentage of "different" responses in which subjects selected the pair rated less similar in Experiment 1 as the more related pair of the POP.

Conclusions

A number of implications follow from these analyses. First, it appears that judges make very fine distinctions among pairs of highly related concepts, but that discriminability suffers at the unrelated end. This is precisely the environment where Pathfinder thrives, since the presence of links is determined, almost exclusively, by differences among highly related pairs. In contrast, scaling techniques like MDS try to consider all distances simultaneously. In doing so, they may be more susceptible to the lack of sensitivity associated with the unrelated end of the rating scale. A second implication is that the conceptual models derived from a homogeneous group of subjects may provide as accurate a model as representations based on an individual's own data. This seems reasonable given that discriminations between degrees of psychological distance could be predicted as well, or better, by the means of an entire group compared to the predictions based on each individual's own ratings. This claim would benefit, of course, from empirical support in which performance was measured as a function of group-based versus individual-based organizations.

Chapter 5

Assessing Structural Similarity of Graphs*

Timothy E. Goldsmith and Daniel M. Davenport

In this chapter, we discuss some measures of the similarity of two graphs. Our work was initially motivated by a need to measure the similarity between Pathfinder networks (Schvaneveldt, Durso, & Dearholt, 1985). The problem of how to compare two different representations, however, exists more generally in the fields of scaling and modeling. A central assumption of our work is that representations have structural properties and comparisons of these representations ought to reflect these structural properties. The aim of this current work, then, is to identify a method of assessing graph similarity that is sensitive to structural information.

We begin by describing a particular view of structure and similarity and its implications for comparing graphs. Next, we identify two basic properties of graphs, paths and neighborhoods, and show how each of these properties can be used as a basis for defining graph similarity. We then describe several related measures for assessing graph similarity, discuss some of their properties, and report results of an initial comparison of the measures. Finally, we offer some generalizations and extensions of the measures.

Similarity and Structure

The basic problem we wish to address is how to measure the similarity of two graphs. More specifically, we would like to define a function that maps any two graphs onto a real number that reflects the graphs' similarity. The set of such functions is very large because similarity itself is not well-defined. Graph similarity is somewhat akin to making human judgments of similarity. Such judgments are inherently subjective because perceived similarity may depend on a multitude of factors including those characteristics of the objects that are psychologically salient to the perceiver and the beliefs the perceiver has about the purpose of the judgment. Similar concerns arise in defining measures of graph similarity.

Consider, for example, the graphs in Figure 1. Graph A is a simple binary tree with seven nodes, and graphs B and C are deviations of A; B differs from A in three edges, whereas C differs in just one edge. Which graph, B or C, is more similar to A? There is, of course, no absolutely right answer, and in fact, we will show shortly that either B or C can be viewed as more similar to A.

Notice in Figure 1 that we have arranged the nodes of graphs B and C in the same spatial layout as in A. We assume that the graphs we compare are always composed of a common set of labeled nodes, and so switching node labels to assess similarity is disallowed. This assumption is realistic for those applications of graph theory where the nodes

---

*This work, performed in part at Sandia National Laboratories, was supported by the U.S. Department of Energy under contract No. DE-AC04-76DP00789.

of a graph correspond to some specific set of objects under study and therefore transposing nodes would be meaningless. However, we do assume that the edges of the graphs are unlabeled. This assumption is also reasonable for many applications of graph theory. (We discuss at the end of the chapter the case of labeled edges.) By graph similarity, then, we mean the similarity of the patterns of edges that define how two graphs with common node sets are linked. We take as axiomatic that it is the structure of these edge patterns that we wish to measure.

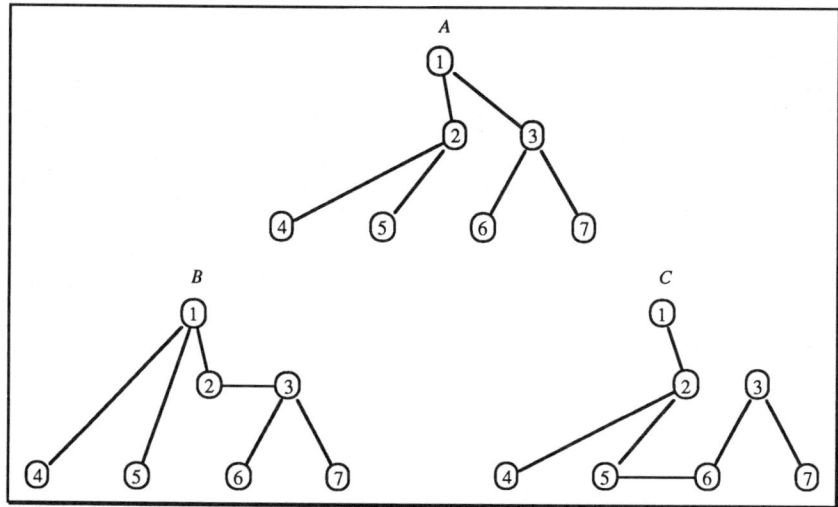

Figure 1. Original Graph *A* and distortions *B* and *C*.

By structure we mean the organization of an object's constituent parts as viewed from the perspective of the whole object. Fundamental to structure is the idea of a relation. An object is defined by the primitive relations on the parts that make it up. An object's structure, in contrast, exists at a level of higher-order relations, or relations on relations. A family tree, for example, is defined by the primitive family relations *son of* and *daughter of*. A higher-order relation on a family tree is the relation *descendant of*. This relation, by our definition, reveals more of the structure of the family tree than either *son of* or *daughter of*. Higher-order relations necessarily appear at a global rather than primitive level. In this regard, structure is an emergent property.

Another way that structure can be viewed is as a collection of subobjects of an object rather than higher-order relations on an object. In this sense structure is viewed as an entity (specifically, a collection of subobjects). Although this distinction in the term may seem subtle, we believe it is important for defining structural similarity. If structure is viewed as an entity, then structural similarity of objects should be assessable by identifying the objects' structural subobjects. On the other hand, if structure is viewed as a property, a measure of structural similarity should compare the objects' structural properties.

Consider first structure as an entity. To define graph similarity under this view, we might begin by identifying specific subentities to serve as the basis for comparing graphs. Subgraphs, such as cycles, stars, and cliques, which are already well-defined and

5 Graph Similarity

intuitively represent graph structures, could be used to define an index of similarity by counting the common substructures of two graphs. Graham (1987) describes an approach of Ulam's using a technique based on this idea for measuring graph similarity. Ulam's method is to partition the edges of two graphs into pairwise isomorphic subgraphs; the smaller the number of subgraphs needed to decompose the original graphs, the more similar the graphs. Two graphs with the same number of edges, no matter how dissimilar, can always be partitioned into sets of identical subgraphs by letting each subgraph be a single edge. However, graphs that are more structurally similar will decompose into fewer identical subgraphs, with the extreme case being graphs that are isomorphic, for which *no* partitioning is needed. Ulam's measure of similarity seems appropriate for abstract graphs, however it is problematic for defining similarity for graphs as representations. First, his method assumes that the graphs are unlabeled, whereas the applications of graph theory we consider usually deal with labeled graphs, and second, the task of finding the minimum Ulam decomposition for two graphs is a very difficult computational problem in itself for which there is no known tractable solution.

Consider next the view of structure as a property. Under this perspective, graph similarity might be assessed by comparing two graphs' higher-order property relations, such as distance between two nodes or the constituents of a neighborhood about a node. This is the approach that we adopt, and in the next section we employ these two higher-order relations of graphs to define graph similarity.

Graph Properties and Similarity

First, we begin with a few basic definitions of graphs and their properties. We limit our discussion to undirected graphs, without loops, and with a common set of labeled nodes. An edge relation is a binary relation defined on pairs of nodes and is the most primitive relation of a graph. All higher-order graph properties are derived from the edge relation. One such property is a path. The distance between two nodes, v and v', is the minimum path length for all paths between v and v' provided such a path exists. Because a path is a relation defined on the edge relation, it is a higher-order relation. A graph can be completely described by either its edge relations (i.e., adjacency matrix) or its path lengths (i.e., distance matrix). In the case of undirected graphs, without loops, each of these matrices can be reduced to a vector consisting of only the upper triangular cell entries, and so a graph with n nodes may be represented by a vector of $(n^2-n)/2$ distances.

A second higher-order graph property is a neighborhood. A *neighborhood* about some node, v, is defined to be the set of nodes that are within distance one from v, excluding v itself. Referring back to Figure 1, the neighborhoods for Node 1 in graphs A, B, and C are $\{2, 3\}$, $\{2, 4, 5\}$, and $\{2\}$, respectively. By excluding v from the set, which diverges from the normal definition of a neighborhood, we can simplify our definitions of graph similarity. Notice that a neighborhood is also a relation on the edge relation, and thus a higher-order relation. So, both path length and neighborhood content are, by our definition, structural properties of graphs. Further, these structural properties are sufficiently general such that every graph may be described in terms of them. Next we show how both of these properties can be used to define graph similarity.

First, two graphs may be compared by computing the correlation coefficient of the the two graphs' distance vectors. A correlation coefficient assesses the shared pool of variability between two sets of numbers, standardized by a measure of the total pool of variability. The idea of forming a ratio of shared attributes to total attributes seems intuitively appealing for measuring similarity, and in fact comparisons of Pathfinder networks with this approach have proved meaningful (e.g., Schvaneveldt, Durso, Goldsmith, Breen, Cooke, Tucker, & DeMaio, 1985).

Turning next to neighborhoods, two graphs may be compared by assessing the similarity of their neighborhoods for corresponding nodes. As with the correlation coefficient, we would like to measure the degree of shared elements relative to some total pool of elements. This may be accomplished with sets by examining their intersection and union. More specifically, an index of similarity for a common node in two graphs is the cardinality of the intersection of the node neighborhoods divided by the cardinality of the union of the neighborhoods. One measure of overall graph similarity is the mean of these n values. This measure will vary from zero to one with higher values indicating greater similarity.

We now have two ways of defining graph similarity, one comparing path distances with a correlation coefficient and the other assessing neighborhood regions with simple set operations. The next logical question is whether these two measures actually differ in their assessments of graph similarity. Consider again the graphs in Figure 1 and the question of which graph, B or C, is more similar to A. We now have a means for answering this question, and the answer is that B is closer to A in terms of path lengths, but C is closer to A in terms of neighborhoods. The correlation of path lengths between A and B is .79 and between A and C is .42, whereas the neighborhood similarities between A and B is .43 and between A and C is .74. So we come to exactly opposite conclusions about the graph's relative similarity with these two measures. Although there are undoubtedly cases where both approaches would agree, we believe that, in general, path lengths and neighborhoods offer qualitatively different ways of assessing structural similarity of graphs. In the following section, we describe some closely related similarity measures and their properties. (A more formal treatment of these definitions and their properties is given in Appendix A.) Following this we describe the results of a study comparing these various measures.

Definitions and Properties of Graph Similarity Measures

In this section we describe a class of related graph similarity measures. Let $C_i(A, B)$ be the similarity between graphs A and B with a common labeled node set as measured by C_i for $i = 1$ to 8. The first four measures are based on neighborhoods and the second four are based on path lengths. The measures C_1 and C_8 are simply the neighborhood and path length measures, respectively, described above.

Two other measures, C_2 and C_3, are similar to C_1 and measure the average similarity of neighborhoods. In each case, the cardinality of the intersection of the neighborhoods is divided by some number which normalizes the index. C_2 and C_3 differ from C_1 only in their normalizations. C_4 is the number of edges that match between A and B divided by the number of possible matches (i.e., the number of node pairs). It is also one minus the mean absolute difference between entries in the adjacency vectors of A and B.

C_5 is the mean of the ratios of smallest to largest values for corresponding entries in the two graphs' distance vectors. C_6 is one minus the mean absolute difference between entries in the graphs' distance vectors normalized by the sum of these distances. Finally, C_7 is the correlation coefficient of the graphs' adjacency vectors. The interested reader may refer to Appendix A for formal definitions.

We next evaluate the above measures with respect to several desirable properties for a graph similarity measure. First, the measure should be independent of the size of the node set or the density of the graph. Some of the C_i's appear to meet this criterion better than

5 Graph Similarity

others, a point that comes up later. C_1 through C_6 are normalized over the range [0, 1] where 0 is *least similar* and 1 is *most similar*. C_7 and C_8 are normalized over the range [-1, 1] where -1 is *least similar* and 1 is *most similar*. Each measure takes on its maximum value for identical graphs. C_1 through C_4 will take on their minimum value when comparing complementary graphs. Finally, algorithms computing the various C_i's are easily coded in standard computer languages which run in $O(n^2)$ to $O(n^3)$ time in the number of nodes.

Comparison of the Measures

We next compare and contrast the various measures by applying them to a common set of graphs. Consider again the graphs in Figure 1. The similarities between each pair of graphs are given in Table 1 for all eight measures. Realize that the various measures are not directly comparable (except C_7 and C_8) because each one occurs on a unique scale. Even C_1 through C_3, which employ similar set-theoretic definitions, are scaled differently because of different normalizations. Therefore, only relative differences of their values are meaningful.

Table 1. The similarity between each pair of graphs A, B, C, in Figure 1 as measured by C_1 through C_8 and rank orders in parentheses from most (1) to least (3) similar.

Similarity Measure	C(A, B)	C(A, C)	C(B, C)
C_1	0.43 (2)	0.74 (1)	0.39 (3)
C_2	0.50 (2)	0.83 (1)	0.48 (3)
C_3	0.50 (3)	0.87 (1)	0.52 (2)
C_4	0.71 (2.5)	0.91 (1)	0.71 (2.5)
C_5	0.79 (1)	0.78 (2)	0.71 (3)
C_6	0.87 (1)	0.85 (2)	0.81 (3)
C_7	0.30 (2.5)	0.77 (1)	0.30 (2.5)
C_8	0.79 (1)	0.42 (3)	0.45 (2)

Notice first that C_1 through C_4 and C_7 all agree that graphs A and C are most similar, whereas C_5, C_6, and C_8 show that A and B are most similar. However, the rank orders of similarity are not identical for these five measures. Notice also that C_5 and C_6 show an identical pattern of ranks but C_8 has a different pattern. Hence, with these simple graphs there appear to be important similarities and differences in what the measures are assessing.

We turn next to a similar analysis of more complex graphs. The graphs are Pathfinder solutions to relatedness ratings of 30 course-relevant concepts given by 20 students and one instructor. For each of the 21 datasets, four classes of graphs varying in *graph density* were derived. Each graph was then compared with every other graph (210 comparisons) in its class using all eight measures. Graph-theoretic distances were used by all of the measures. The resulting similarity values for each measure were then correlated with every other measure. C_1 through C_3 correlated very highly across all of the graphs, as did C_5

and C_6. For this reason we do not report the results of C_2, C_3, and C_6. The resulting correlation matrices for each of the four sets of graphs are shown in Tables 2 through 5.

Notice first, as we also saw in Table 1, that C_1 correlates highly with C_7, and this holds for both sparse and dense graphs. Apparently the method of comparing (set-theoretic functions or correlation) edge relation information has little effect on the resulting similarity indices. C_4 correlates highly with C_1 and C_7 for sparser graphs, but less so for denser graphs. Recall that C_4 and C_7 compare edge relations with mean absolute difference and correlation, respectively. We speculate that the normalization of C_4 is not as good as C_7's or C_1's, and this weakness becomes especially apparent with denser graphs.

Table 2. Correlations on results of C's applied to 30 node graphs (mean graph density = 0.067).

	C_4	C_5	C_7	C_8
C_1	0.97	0.59	0.97	0.46
C_4		0.62	1.00	0.51
C_5			0.62	0.73
C_7				0.50

Table 3. Correlations on results of C's applied to 30 node graphs (mean graph density = 0.124).

	C_4	C_5	C_7	C_8
C_1	0.93	0.84	0.98	0.73
C_4		0.81	0.97	0.78
C_5			0.85	0.84
C_7				0.75

Table 4. Correlations on results of C's applied to 30 node graphs (mean graph density = 0.188).

	C_4	C_5	C_7	C_8
C_1	0.49	0.76	0.96	0.83
C_4		0.17	0.66	0.60
C_5			0.67	0.61
C_7				0.88

5 Graph Similarity

Table 5. Correlations on results of C's applied to 30 node graphs (mean graph density = 0.278).

	C_4	C_5	C_7	C_8
C_1	0.66	0.83	0.94	0.86
C_4		0.46	0.86	0.79
C_5			0.74	0.69
C_7				0.92

C_8 correlates most highly with C_5 for sparser, but lowest with denser graphs. Also, C_8's correlations with C_1 and C_7 steadily increase with increasing graph density. A likely reason for this is because as graph density increases, the difference between the adjacency and distance vectors decreases, until for completely connected graphs the two are identical. In summary, although certain of the measures appear quite similar, overall there are some important and systematic differences. Additional analyses employing other types of graphs and graphs from other applications are needed before more general conclusions can be reached.

The above analyses simply examined relative agreement of the various measures. Evidence that one measure is actually "better" than another for assessing graph similarity would require some external index of similarity. As reported elsewhere in this volume (Goldsmith & Johnson, Chapter 17), there is evidence that the similarity between a student's PFNET of course concepts and the instructor's PFNET predicts final course grades better for similarity measured by C_1 than by C_8. This finding was interpreted as support for the idea that a neighborhood comparison of graphs is more sensitive to configural (i.e., structural) information, and it is this type of information that is important in knowledge structures.

Generalizations and Extensions of the Measures

In this section, we briefly describe some generalizations and extensions of the neighborhood measures. First, there may be occasions when we want to compare two unlabeled graphs. For example, we might want to find a *best fit* of a test graph with some target graph by permuting the nodes of the test graph. Brute force methods are useless here due to the explosive growth of the number of permutations as node size increases. Further, the problem does not seem to lend itself to a linear programming approach. Instead, we could employ a particular C_i as an optimization function in a simulated annealing process (Kirkpatrick, Gelatt, & Vecchi, 1983). Here, we would attempt to find that permutation of nodes for the test graph giving a minimum value for C when compared to the target graph. When the two graphs are not isomorphic, the annealing technique yields a *best fit* as defined by the particular C being used. Initial results based on a visual comparison of the target graph with permuted test graph are promising.

Finally, with some simple modifications we may extend the definitions of C to work for edge- and node-weighted graphs, as well as edge- and node-colored graphs. This will allow C to compare, for example, two chemical molecules by representing each with a node-colored, edge-weighted graph, whose nodes represent individual atoms and whose edge weights represent bond valence, or perhaps, bond energy. Interest in assessing similarity of molecular structures has recently led chemists to define graph measures of similarity. For example, Herndon (1988) has developed a technique for computing a quantitative index of similarity between molecular graphs that requires first translating each graph into a string of symbols and then employing string comparison methods to determine the original graphs' structural similarity. In Appendix B we describe in more detail some modifications of a neighborhood measure of C that may be useful in comparing graph representations of molecules.

Conclusions

We began by assuming that a measure of graph similarity should assess structural information. We attempted to argue that structure is best viewed as a property of graphs rather than as an entity. We then defined structural property as a higher-order relation and identified two distinct types of structural properties in graphs: paths and neighborhoods. Several related measures of graph similarity employing either path lengths or neighborhoods were defined and some of their properties noted. An initial analysis of these measures indicated that use of path lengths and neighborhoods to determine similarity assess different characteristics of graphs.

What lies behind these differences? Path distances describe how far away nodes are located; neighborhoods describe which nodes are linked. Path distances employ node pairs as the unit of comparison; neighborhoods use nodes. The path distance approach first converts the information contained in a path to a real number (the path length) and then compares corresponding path lengths between graphs; the neighborhood approach first compares corresponding neighborhoods with set-theoretic measures and then converts this result to a real number. Whether one approach is better than the other of course ultimately depends on its functional utility within a particular application. If the phenomena being represented by graphs is inherently described by distance information, then path distances will likely be better. However, if what is being represented is more accurately reflected by associations within neighborhoods, then the neighborhood approach should prove better.

5 Graph Similarity

Appendix A

Formal Definitions and Properties of Some Graph Similarity Measures

We identify a class of similarity functions for comparing graphs as $\mathfrak{C}$. We denote a particular function from $\mathfrak{C}$ as C and use subscripts to distinguish different measures. A graph $G = (V, E)$ is a finite set V of n nodes and a set E of edges where E is a subset of $V \times V$ (Carre, 1979). We say two graphs $G_1(V_1, E_1)$ and $G_2(V_2, E_2)$ have a common node set if $V_1 = V_2$. Given two undirected and labeled graphs A and B with common node set V, $C(A, B)$ is the similarity between A and B as measured by C. We define a neighborhood as a region about a particular node in a graph. Let $\delta_G(v,v')$ be the graph distance between nodes v and v'. Define $\alpha_G(v,v')$ to be 1 if $\delta_G(v,v') = 1$ and 0 otherwise. We denote by G_v the set of nodes v' such that $\alpha_G(v,v') = 1$. This set is the neighborhood about v.

We define the following similarity measures. In cases where the denominator of a summand is zero, we employ the convention that if both neighborhoods for an element are empty then the summand for that element is one, but if only one of the neighborhoods is empty, then the summand for that element is zero.

$$C_1(A, B) = \frac{1}{n} \sum_{v \in V} \frac{|A_v \cap B_v|}{|A_v \cup B_v|}$$

$$C_2(A, B) = \frac{2}{n} \sum_{v \in V} \frac{|A_v \cap B_v|}{|A_v| + |B_v|}$$

$$C_3(A, B) = \frac{1}{2n} \sum_{v \in V} |A_v \cap B_v| \left[\frac{1}{|A_v|} + \frac{1}{|B_v|} \right]$$

Notice that if we write C_3 as

$$C_3(A, B) = \frac{1}{2} \left[\frac{1}{n} \sum_{v \in V} \frac{|A_v \cap B_v|}{|A_v|} + \frac{1}{n} \sum_{v \in V} \frac{|A_v \cap B_v|}{|B_v|} \right]$$

we see that it is the average of two other similarity measures. These other measures are interesting in their own right and are reminiscent of conditional probabilities. In assessing similarity between A and B in one case, the measure is sensitive to those edges in A omitted by B, and in the other case it is sensitive to edges in B not found in A. As the average of the two, C_3 captures both cases. There may be applications where one would want to use only one of the individual measures. For example, if we assume that graph A represents a prototype of some sort, then it may be meaningful to assess the similarity of an exemplar (B) to the prototype (A). We would expect a different result when the similarity of A to B is computed.

Next, we define, for two nonnegative real numbers a and b, $a \theta b$ to be one if $a = b$, a/b if $a < b$, and b/a if $b < a$. Also, let $A \oplus B$ be the symmetric difference between sets A and B. Then, C_4 is defined as follows:

$$C_4(A, B) = 1 - \frac{1}{n^2-n} \sum_{v \neq v'} |\alpha_A(v,v') - \alpha_B(v,v')|$$

$$= \frac{1}{n^2-n} \sum_{v \neq v'} \alpha_A(v,v') \, \theta \, \alpha_B(v,v')$$

$$= 1 - \frac{1}{n} \sum_{v \in V} |A_v \oplus B_v|$$

Interestingly, C_4 is also one minus the average of the symmetric differences of the neighborhoods. This may be seen by noting that

$$\sum_{v \neq v'} \alpha_A(v,v') \cdot \alpha_B(v,v') = |A_v \cap B_v|$$

and

$$|\alpha_A(v,v') - \alpha_B(v,v')| = \alpha_A(v,v') + \alpha_B(v,v') - 2(\alpha_A(v,v') \cdot \alpha_B(v,v'))$$

5 Graph Similarity

We now define two other measures C_5 and C_6 which are based on path lengths. Notice that C_5 is similar to C_4 except that the function θ is applied to actual graph distances for C_5, but to edge relations for C_4.

$$C_5(A, B) = \frac{1}{n^2-n} \sum_{v \neq v'} \delta_A(v,v') \; \theta \; \delta_B(v,v')$$

$$C_6(A, B) = 1 - \frac{1}{n^2-n} \sum_{v \neq v'} \frac{|\delta_A(v,v') - \delta_B(v,v')|}{\delta_A(v,v') + \delta_B(v,v')}$$

For similarity measures C_1 through C_6 we may derive distance measures, D, by letting $D = 1-C$. Some of the measures have interesting forms as distances.

$$D_1(A, B) = \frac{1}{n} \sum_{v \in V} \frac{|A_v \oplus B_v|}{|A_v \cup B_v|}$$

$$D_2(A, B) = \frac{1}{n} \sum_{v \in V} \frac{|A_v \oplus B_v|}{|A_v| + |B_v|}$$

$$D_4(A, B) = \frac{1}{n} \sum_{v \in V} |A_v \oplus B_v|$$

Also, from the definitions of C we may discover several interesting properties. Since

$$2 \frac{|A_v| \, |B_v|}{|A_v| + |B_v|} \leq \frac{1}{2}(|A_v| + |B_v|) \leq |A_v \cup B_v| \leq n-1$$

we get $C_1 \leq C_2 \leq C_3$, $C_1 \leq C_4$, and $C_5 \leq C_6$. In general, C_4 is incomparable to C_2 and C_3. Similar relations hold for the distance measures as well. Also, it is possible (but tedious) to show that the distance indices for all of the similarity measures, except C_3, are metrics on the space of subsets of $V \times V$.

We can think of graphs with common node set V as subsets of $V \times V$. As such, the set of all such graphs is a Boolean ring whose multiplicative identity, I, is the completely connected graph and whose zero is the empty graph. In this ring, multiplication is given by intersection and addition is given by symmetric difference. If we choose D_1 for a metric on this space, we may define a norm (which we call the C_1 norm) on a graph G by $\|G\| = 1 - D_1(G, I) = C_1(G, I)$. With this definition, the following properties hold for all graphs A, B, and C over node set V:

$$\|A \cup B\| + \|A \cap B\| = \|A\| + \|B\|$$

$$1 - \|\bar{A}\| = \|A\|$$

$$\|A \oplus B\| \le \|A\| + \|B\|$$

$$\|A \cap B\| \le \|A\|$$

$$\|A \oplus B\| + \|A \oplus C\| \le \|B \oplus C\|$$

where $\bar{A}$ is the *complement* of A with respect to I. This last inequality allows us to define a metric $D(A, B) = \|A \oplus B\|$, which turns out surprisingly to be D_4.

5 Graph Similarity

Appendix B

An Extension of Neighborhood Similarity Measures

The weight of an edge (v,v') in a graph G will be denoted by $w_G(v,v')$ and its color by $x_G(v,v')$. The weight of a node v in G will be denoted $w_G(v)$ and its color by $x_G(v)$. Next we define a Kronecker delta function on the set of colors in a graph by $\delta(c,c') = 1$, if c is the same color as c' and is 0 otherwise. All of our measures then take the form

$$\frac{1}{v_1} \sum_{v \in V} \left(w_A(v) \, \theta \, w_B(v) \right) \left(\delta(x_A(v), x_B(v)) \right).$$

$$\frac{1}{v_2(v)} \sum_{v \neq v'} \left(w_A(v,v') \, \theta \, w_B(v,v') \right) \left(\delta(x_A(v,v'), x_B(v,v')) \right)$$

where v_1 and v_2 are normalizing functions. For example

$$\frac{1}{n} \sum_{v \in V} \delta(x_A(v), x_B(v)) \left[\frac{1}{|A_v \cup B_v|} \sum_{v \neq v'} \left(w_A(v,v') \, \theta \, w_B(v,v') \right) \right]$$

is a measure of similarity between two node-colored, edge-weighted graphs. Notice that this measure reduces to C_1 for graphs that are not colored or weighted. With careful normalization, a number of potentially useful measures are definable.

Chapter 6

A Graph-Theoretic Approach to Concept Clustering

Chris Esposito

Interobject proximities are an essential input to any of the Pathfinder algorithms that have been developed. However, the role of these proximities or weights after the network has been generated is less clear. There are a variety of viewpoints on this iss·e, all of which can be cast as different points on a continuum.

At one endpoint is the view that the interobject proximities are still the focus of attention, and that the edges in a PFNET are merely places to hang the most salient of these proximities. Since the general notion of distance is intimately related to the notion of a space, this view of PFNETs may often be adopted when aspects of a spatial model are appropriate and useful. The distance between two nodes is primarily determined by the weights attached to the edges that join them; structural issues, such as the number of edges in the connecting path, are secondary considerations.

At the other endpoint there is the view that the role of the edge weights is largely confined to determining network structure during generation. Such a view treats the PFNET simply as a graph. While it is clearly possible to define the distance between two nodes as the minimum number of edges needed to connect them, this measure is driven entirely by the structure of the network, and there is no guarantee (especially for directed networks) that distance measures so derived will satisfy the axioms for a given type of space, such as a metric space.

To a certain extent, the relative importance assigned to edge weights and network structure depends on the application, or the question to be answered. There are several applications which adopt a view that is in between the two endpoints described above, and use both edge weight and structural information. The work described in Knoebel, Dearholt, and Schvaneveldt (1988) or the database searching work described in Dearholt and Gonzales (1987) are examples of this middle view. This chapter examines how well network solutions fit the original data and compares measures of fit based on graph structure with measures based on edge weights. Also, a study is presented that examines the issue of edge weights versus structure from another perspective, the subjective ratings of clusters produced using edge-weight information (hierarchical clusters) are compared with ratings of clusters produced by some graph-theoretic clustering methods that ignore edge weights.

Parameter Choice and Network Fit

A general question that applies to most data-scaling methods (Pathfinder included) is this: Given a particular dataset and a scaling solution, such as a multidimensional scaling (MDS) layout or a PFNET, how good is the solution? There are several points to consider in answering this question. The first is the simplicity of the solution. If we start with an

$n \times n$ proximity matrix for n objects, then an MDS solution with some small (usually 2 or 3) number of dimensions is preferable to a solution with $n-1$ dimensions. A PFNET with a significant portion of the original $n^2/2$ edges remaining is, for most exploratory purposes, too dense and complex to be very useful. The experience of looking at large numbers of networks suggests that those where the number of edges is no more than about twice the number of nodes are the easiest to deal with.

However, simplicity is not the only consideration in assessing a solution. We would also like the solution to fit well with the original data and for the essential structure to be revealed with nonessentials carved away. There is a certain tension between these two criteria. The simplest solutions (usually trees in Pathfinder) can be too simple and not represent the latent structure in the original data very well. On the other hand, a trivial way to get a good fit is to essentially reproduce the original *complete network*, deleting few, if any edges. Of course, such a solution is often too complicated to be of much use. The desirable middle ground is a solution that is both simple and fits well with the original data.

Weights vs. Edges

In order to discuss how good a network solution is for a given dataset, we must first examine the factors that determine the simplicity and fit of the network. The most obvious determinants are the network generation parameters q, r, and z, since these completely determine the network for a given dataset. Factors that may be less obvious, but are just as important, are the choice of a correlation measure and the choice of what measurements to take on the network in order to compute the correlation measure. The q values used in the study reported in this chapter were 2 to $n-1$. The r values were 1 to 8. The z values[1] used were 0.0 to 3.0. The correlation measures used were the Pearson r_p (r_p will be used here instead of r to distinguish the statistic from the Pathfinder r parameter) and Spearman's ρ. The principal difference between these two is that r_p uses interval-scale properties while ρ only makes ordinal assumptions. As it turned out, they produced very similar correlations (as they often do), and so if we perform similar studies in the future, we would probably use only one of them.

Five domains were chosen, and for each domain, a set of representative terms was chosen. The domains were countries (9 terms); pieces of fruit (11 terms); items of clothing (11 terms); cities in New Mexico (20 terms); and a set of words extracted from a machine-readable copy of *Longman's Dictionary of Contemporary English* that were related to the word *bank* (20 terms). For each set of terms, pairwise relatedness ratings were obtained from human subjects to get a proximity matrix. From the individual matrices, an average proximity matrix and a standard deviation matrix were created for each domain. For each domain, three sets of networks were generated. In the first set, q was held to be $n-1$, z was set to 0.0, and r was varied from 1 to 8 (higher values of r usually produced networks identical to when $r = \infty$). In the second set, r was held at infinity, z was set to 0.0, and q was varied from 2 to $n-1$. In the third set, r was set to infinity, q was set to $n-1$, and z was varied from 0.0 to 3.0 in 0.1 increments.

For each network generated, two distance matrices were computed and each was correlated with the original data matrix using the two measures of fit noted above. The first distance matrix was computed by assuming that the distance between two nodes in the network is primarily a function of the weights attached to the edges on the path that joins them. The r value used to compute this distance matrix was always the same as the r value used to generate the network. While it is certainly true that the structure determines what

---

[1] For a discussion of how the z parameter is used in network generation, see Esposito (Chapter 3, this volume).

6 Graphs and Concept Clusters

paths exist between any pair of nodes, there is no necessary correlation between the distance from one node to another and the number of edges that separate them. Let this set of matrices be called *weight-based*. The second distance matrix was calculated by assuming that the structure of the network is essential. The distance between a pair of nodes is defined to be the minimum number of edges that separate them. Let these matrices be called *edge-based*.

Across the five domains, 353 networks and 706 network distance matrices were generated. The correlations between these network distance matrices and the five averaged proximity matrices provided some interesting results. In 291 of these networks (83%), the edge-based matrices correlated more highly with the raw data than the weight-based ones. As Figures 1 through 3 demonstrate, this was often by a substantial margin.

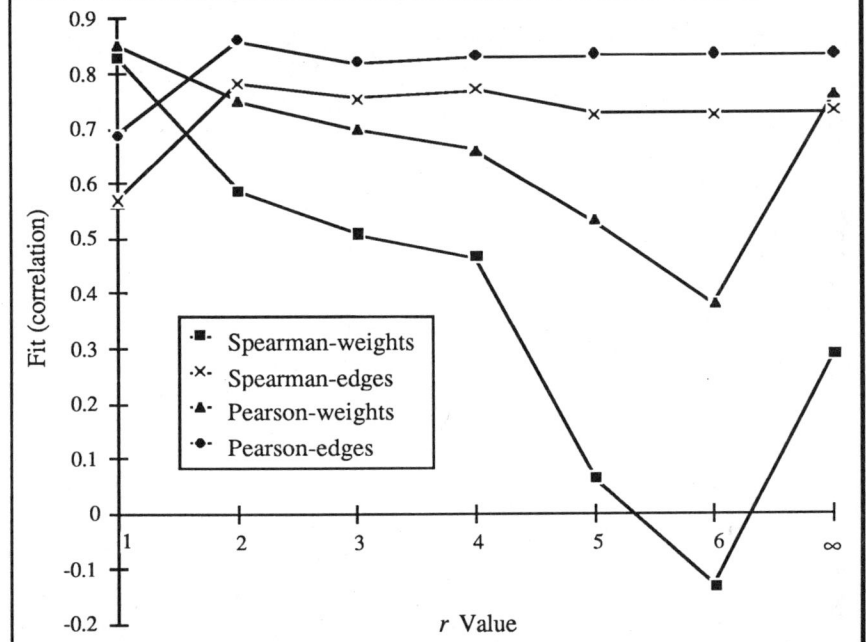

Figure 1. Fit as a function of r value for the Countries dataset, PFNETs($q = n\text{-}1$, $z = 0$).

In 62 networks (17%), the weight-based matrices correlated more highly, although the difference between the correlations was often much less. Part of this dominance by the structural view of the networks is explainable if we consider that in 328 of the 353 networks, r was held at infinity. This means that the length of a path is the length of the longest edge on it, so the largest weights occur far more often (and the smallest weights far less often) in the weight-based distance matrices than they did in the original proximity matrices. Since the edge-based approach ignored edge weights, no such skewing was present there. It is therefore not too surprising that the weight-based matrices correlated more poorly with the original data matrices than did the edge-based ones.

It is probably more revealing to look at those 35 networks where r was allowed to vary. In these cases, $q = n-1$ and $z = 0.0$. In 28 (80%) of these networks, the structural (i.e., edge-based) view still correlated more highly with the original data than the weight-based approach. In 7 networks (20%) it does not. Figure 1 shows some representative results. One conclusion that can be drawn from this analysis is that the principal role of the weights is in creating the right structure for the network. After generation, many uses of the network would be much better off concentrating on the graph-theoretic aspects of the network.

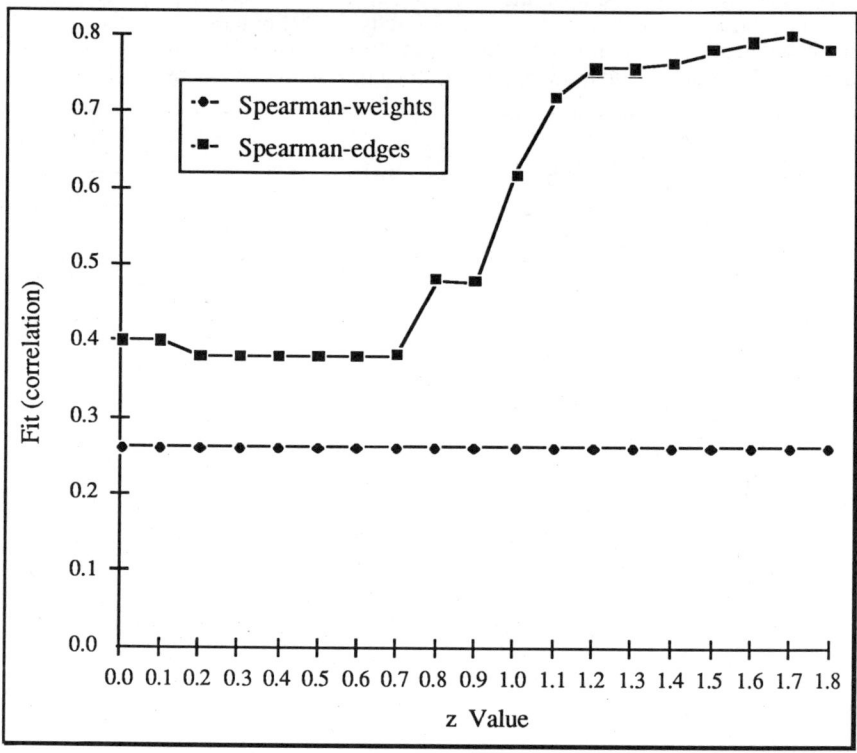

Figure 2. Fit as a function of z value for the Cities dataset, PFNETs($r = \infty$, $q = n-1$).

An Application of Network Structure

One of the results from the last section was that when measuring the fit between a Pathfinder network and the original data, measures of fit based on *graph-theoretic distance* proved generally superior to those measures based on edge weights.

In this section we will examine the issue of edge weights versus structure from a different perspective. A common approach when investigating the structure of a domain is to see how the concepts in that domain are organized into categories or clusters. We describe a study that compares the ratings of clusters produced by using edge-weight information

6 Graphs and Concept Clusters

(hierarchical clusters) with ratings of clusters produced by several graph-theoretic methods that ignore edge weights. A key part of this study is a different approach to clustering validity than is usually adopted, so it is worth discussing some of the validity issues that influenced the design of the study.

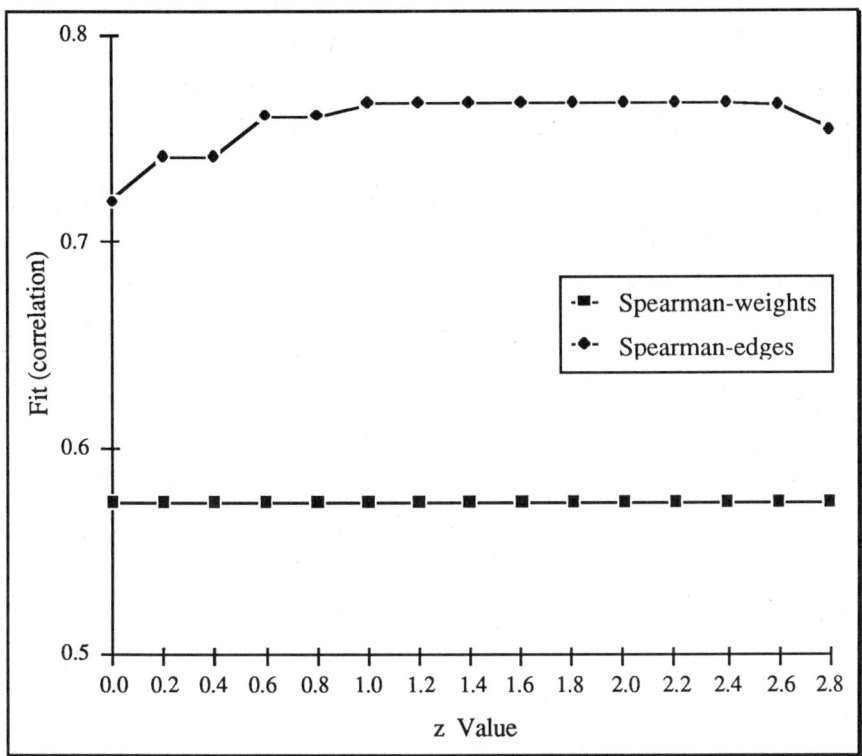

Figure 3. Fit as a function of z value for the Clothing dataset, PFNETs($r = \infty$, $q = n–1$).

General Validation Issues

One of the primary problems in cluster analysis is that the methods for evaluating the validity or usefulness of the generated clusters lag far behind the methods for producing them. The validation process generally proceeds down one of two roads that, unfortunately, rarely intersect. The first road is the researcher's purely subjective evaluation of the produced clusters. In the early stages of exploratory data analysis, it is reasonable for a researcher to use high subjective ratings of clusters in a familiar domain as an indication that the clusters are worth a closer and much more objective look. However, it is embarrassingly easy to unconsciously impose a structure on a two-dimensional plot of a dataset (for example) and then "find" evidence for the "natural" existence of the structure imposed. If these subjective ratings are the only factors considered when assessing clustering validity, there is an unfortunate tendency to treat them as if they are much more conclusive than they really are.

The second road is in many ways the more seductive of the two, in that clustering validity, for which there is a substantial body of literature, is cast in terms of optimizing a criterion function. The assumption implicit in this approach is that any clustering produced by such an optimization will also be optimal, or nearly so. This approach has several problems. First, the criterion is often chosen with little (if any) justification as to why optimizing it leads to better clusters. Second, as Dubes and Jain (1977) point out, many clustering techniques have no problem in producing clusters whether or not they are real. To use some criterion to form the clusters and then also use it to judge their validity effectively sidesteps the entire issue. Some concrete examples from the clustering literature should make the problems often encountered with this second approach a bit clearer.

One class of clustering algorithms (examples are Ball & Hall, 1970; Kennard & Stone, 1969; and MacQueen, 1967) begin with all of the items embedded in some p-dimensional space and then partition this space into some usually prespecified number of clusters. Some sort of global measure of fit (often a squared-error minimization) is used as the criterion function that guides placement of the items into the partitions.

The first problem encountered when using this sort of clustering method is picking the "right" number of clusters or partitions. The usual response is to plot the criterion function value versus the number of clusters, and hope (often in vain) that a sharp and clearly marked *elbow* will appear, thus marking what is presumably the right number of groups. However, as Thorndike (1973) points out, unless one has already demonstrated that optimizing the chosen criterion leads to *better* clusters (i.e., ones with higher subjective ratings), identifying the elbow often gives unsatisfactory results.

The second problem is that choosing a criterion function to optimize generally entails the adoption of some assumptions about the structure of the clusters to be produced. Many of the criteria that have been proposed (Friedman & Rubin, 1967; McRae, 1971) produce spherical clusters. A less restrictive but still common assumption is that all of the clusters have the same shape (Marriot, 1971). The basic problem here is one of imposed structure versus revealed structure: Clusters whose shapes are determined by these assumptions will be produced whether or not they naturally exist, and naturally existing clusters whose shapes are not in accordance with these assumptions will be largely ignored.

Method

The issue at hand is this: We are interested in assessing the value of several graph-theoretic structures as clustering methods by comparing their performance both with a control group of clusters and with a well-accepted hierarchical clustering method. A list of these structures will be given shortly. The domain for this pilot study is the UNIX[2] command set (McDonald & Schvaneveldt, 1988). The set of terms contains 152 UNIX commands. The relatedness data for the generated network were collected in the following way. Each command was put on a card and experienced UNIX users were asked to put related commands in the same pile. There was no constraint on the total number of piles, the size of each pile, or the number of piles a command could appear in (subjects were allowed to make duplicate cards). The proximity between a pair of commands was defined to be the conditional probability of one command in a pair being in a pile given that the other command was already in that pile. For the details on the computation of these conditional probabilities, see McDonald and Schvaneveldt (1988).

This proximity matrix was used in two ways. The first use was as input to a single-link hierarchical cluster analysis. This procedure produced 83 clusters that four subjects were asked to rate. In order to make the later comparisons more meaningful, only the

---

[2]UNIX is a trademark of AT&T Bell Laboratories.

6 Graphs and Concept Clusters

clusters in the same size range as the graph-theoretic structures (3 to 8 nodes) were chosen, for a total of 36 clusters. For each cluster, the subjects were asked to rate its quality on a scale of 0 to 5, with 0 as "don't know" (usually due to unrecognized terms in the cluster), 1 as "very bad," and 5 as "very good." The subjects were also asked to name the cluster if the rating was ≥ 2. The assumption was that if the rating was either "don't know" or "very bad" there was little point in asking for a name.

The other use for this proximity matrix was as input to Pathfinder. Since no estimates of rating variability could be obtained from these data, the z parameter was set to 0. The q parameter was set to 151 ($n-1$) and the r-metric to ∞. For each cluster generated from this network, we collected the same information (rating and name) as we did for the hierarchical clustering case, although we used a separate group of eight subjects for this rating process. A list of the cluster structures extracted from this network is given below.

Cliques. A complete graph is one in which every pair of nodes is *adjacent*. A subgraph S is *maximal* with respect to a property P if there is no node v such that $S + v$ also has property P. A clique is a maximal complete subgraph. For the purposes of this study, there were 10 subgraphs that would have been cliques (specifically K_4s) except that a single edge was missing. These were judged to be sufficiently clique-like so that they were included in this category.

Blocks. A *cutpoint* of a graph is a node whose removal increases the number of components, and a bridge is such an edge. A *nonseparable* graph is connected, nontrivial, and has no cutpoints. A *block* of a graph is a maximal nonseparable subgraph (Harary, 1969). Preliminary examination of some PFNETs reveals that many blocks correspond to conceptually coherent categories, such as the set of UNIX mail commands.

Stars. If entities in a basic-level category (Rosch, Mervis, Gray, Johnson, & Boyes-Braem, 1976) are presented to subjects, and relatedness estimates are used as the basis for a PFNET, the resulting network is organized as a *star* (Schvaneveldt, Durso, & Dearholt, 1985) with the category name in the center and exemplars at the tips. The high degree of the center node and the low degree of other nodes adjacent to it provide the basis of a metric for detecting starlike structures. For the purposes of this study, any node with a degree greater than 4 (along with its neighbors) was considered to be a star.

In addition to these structures, a group of random subgraphs was presented for subjects to rate. The presence of random clusters complements the use of subjective ratings as an evaluation measure in that the average rating for this group should provide an idea of what subjective numeric rating corresponds to a verdict of *no clustering*.

There are several ways of constructing random subgraphs. Perhaps the simplest way is to generate random sets of nodes, without concern for whether any pair of nodes or the set of nodes as a whole is connected. However, the size and diversity of the concept set, combined with the relative sparseness of the network, mean that the large majority of random clusters generated without regard to connectedness will be of such low quality that if comparing the average random cluster rating with the average rating for one of the nonrandom types is a test, then it is a test that is very difficult to fail. Passing such a test would then be of little value. A more restrictive criterion that leads to a stricter test is to require that each of the random clusters form a connected subgraph.

The first step in this assessment is to generate a network where reasonable numbers of these nonrandom structures exist for examination. This, in turn, requires that the network be neither too dense nor too sparse. The second step is to look at knowledgeable subjective ratings of these clusters. If these structures have any value as clustering methods then it is reasonable to expect the average subjective rating for each type to exceed the average rating for random connected subgraphs.

The generated network had 184 edges and contained 34 cliques, 23 starlike structures and 45 blocks, so there appeared to be enough of these structures to obtain reasonable averages for each type in this dataset. The number of nodes in these structures ranged from 3 to 8, distributed as follows: 3 nodes (41%), 4 nodes (23%), 5 nodes (17%), 6 nodes (14%), 7 nodes (3%), 8 nodes (1%). In order to be able to make some meaningful comparisons, we took the ratings of 30 randomly chosen connected subgraphs as a baseline for comparison. The size distribution for this set of subgraphs was the same as the above size distribution for the graph-theoretic structures.

Many of the criticisms of other clustering methods discussed earlier stem from two sources. First, these methods are often applied with little regard to the fit between the assumptions made by the scaling model and the properties of the data. Applying a method that makes interval assumptions to ordinal data is a classic example of this. Second, it is often very difficult to assess the validity of a solution. As previously noted, a common problem is to use the same criterion for both cluster generation and cluster evaluation. In this section we deal with both of these criticisms as they might apply to the present work.

With regard to the first criticism, the construction of the networks (with $r = \infty$) depends only on ordinal properties of the data. The graph-theoretic structures depend only on the presence or absence of edges in these networks and so also make only this assumption.

As for the second criticism, the present work clearly separates the generation and evaluation criteria. The generation criterion is that each cluster possess some specified structural properties. The evaluation criterion is that the average subjective rating for that structure type be greater than the average rating for random subgraphs by a statistically significant amount.

Results

As discussed in the previous section, there are five cluster types being examined in this study: random, single-link hierarchical, stars, blocks, and cliques. In addition, each structure had several different sizes. For each subject, we computed an average goodness rating for each type. For each type, we computed an average and standard deviation for each size across subjects and an average across all sizes and subjects. In addition, we combined the ratings for the stars, cliques, and blocks into a single group and computed an average and standard deviation for this group. The mean values are shown in Table 1.

One can frame the analysis in terms of three broad questions. First, which nonrandom cluster types outperform the random clusters? Second, how do the ratings of the graph-theoretic (star, block, and clique) clusters compare with the ratings of the hierarchical clusters? Third, are there substantive rating differences between the graph-theoretic clusters?

In answering these questions, several factors play a role in determining precisely what comparisons will be made and what statistics will be used to evaluate them. The small sample sizes require that the t statistic be used. We can answer the three broad questions given earlier by doing seven comparisons and t tests. Four of these tests are individual comparisons between random cluster ratings and the ratings of each of the nonrandom cluster types. The fifth comparison is between the hierarchical ratings and a combined average rating for the graph-theoretic structures. The sixth comparison is between cliques and blocks. The last comparison is between stars and blocks. Since one group of subjects rated the hierarchical clusters and a separate group rated the graph-theoretic and random clusters, comparisons involving hierarchical cluster ratings used t tests for between-group comparisons with independent sampling. All other comparisons used t tests for correlated samples. For each comparison, the size of the clusters was controlled. For example, cliques were compared only against random clusters of sizes 3 and 4 nodes, since there were no larger cliques.

6 Graphs and Concept Clusters

Table 1. Subjective rating means by cluster type and size.

Size (nodes)	Cluster Type				
	Random Connected	Star	Block	Clique	Single-Link Hierarchical
3	3.01	-	3.61	4.07	3.96
4	3.38	-	3.63	4.15	4.16
5	2.98	3.59	3.62	-	4.00
6	2.55	3.80	3.07	-	3.83
7	2.91	2.87	3.09	-	2.75
8	2.71	-	3.71	-	4.25
Average	2.92	3.38	3.46	4.12	3.83

For simplicity, let us first deal with the two between-group comparisons, hierarchical ratings versus random ratings, and hierarchical ratings versus average graph-theoretic ratings. Table 2 shows the t value and level of significance for each of the two tests.

The fact that hierarchical clustering significantly outperformed random clustering is not very surprising; the method, when properly used, has proven its value over the years. Given the good showing of the hierarchical clusters, it is also encouraging to note that there is no significant difference between the ratings of the hierarchical clusters and those of the graph-theoretic cluster types, taken altogether. For a more detailed look at each of these cluster types, we must examine the within-subject ratings.

Table 2. Between-group t-test results.

Cluster Type Comparison	t (10 df)	Probability
Hierarchical vs. Random	3.902	$p < 0.005$
Hierarchical vs. Average Graph-Theoretic	0.869	$p > 0.10$

Table 3 shows the results of the remaining five comparisons. Since there were eight subjects, there are seven degrees of freedom. One bit of notation worth explaining is that the numbers following a cluster type name indicate that only clusters of that size were used in that comparison; for example, "Clique vs. Random (3,4)" means that only random clusters of sizes 3 and 4 were used in that comparison.

Table 3. Within-subject *t*-test results.

Cluster Type Comparison (Number of Nodes)	t (7 df)	Probability
Clique vs. Random (3,4)	16.12	$p < 0.005$
Block vs. Random	8.68	$p < 0.005$
Star vs. Random (5,6,7)	6.06	$p < 0.005$
Clique vs. Block (3,4)	7.97	$p < 0.005$
Star vs. Block (5,6,7)	2.41	$p < 0.025$

The results of the top three comparisons in the table confirm a conclusion reached from the between-group tests, that the graph-theoretic clusters significantly outperform random clusters. The results of the last two tests suggest that there are significant differences in how well the graph-theoretic clusters perform as clustering methods. The average clique rating is substantially higher than either the block average or the star average, so among this group it emerges as a fairly clear winner. Stars are in second place, outperforming blocks. The dominance of cliques as a clustering method is not unique to this work; Shepard and Arabie (1979) report similar results and characterize a clique as a collection of entities that all share one or more common properties. The fact that some of the structures used in this study were actually *near-cliques* (since one edge was missing) suggests that this criterion is somewhat robust in the face of small violations of the structural requirements.

The cluster types with the two highest averages are cliques and hierarchical clusters. If, as the results just obtained suggest, we accept the proposition that these are generally good clustering methods, it is reasonable to ask what sort of risks are involved in using these methods. More specifically, what is the likelihood that the rating for some particular clique, for example, will be low? One possible way to answer this question is outlined below. For each of the two cluster types, compute an average rating for each cluster across subjects, split the rating scale into equal-sized intervals, and plot the distributions of frequency versus average rating for the hierarchical clusters and the cliques. Figure 4 shows these distribution plots.

As shown in Table 1, the average rating for the random or *no clustering* case was 2.92; hence it is reasonable to assert that any cluster whose rating does not exceed this value is a *bad* cluster. The average ratings for four (11%) out of the 36 hierarchical clusters did not meet this minimal rating; for cliques, only three (8.6%) did not meet or exceed this minimal rating, so the efficiency, or the ratio of good clusters to total clusters, is fairly high for both of these methods.

Conclusions

Earlier we examined an issue relating to network fit: The role of the edge weights after the network has been generated. For any network, it is possible to compute distance between any pair of nodes in two different ways. The first way is the graph-theoretic distance. The second method relies solely on edge weights; this is the distance that is computed by algorithms like Djikstra's Single Source Shortest Path or Floyd's All Pairs Shortest Path (see Aho, Hopcroft, & Ullman, 1974). Across all five domains and a wide range

6 Graphs and Concept Clusters

of q, r, and z values, the distance matrices derived from network structure correlated much more highly with the original data than did distance matrices derived from edge weights. This suggests that the primary role of the edge weights is in creating the right structure for the network.

In this chapter, we also compared several different clustering methods: hierarchical, cliques, stars, and blocks. In addition, all of these were compared against a group of random connected subgraphs. One result is that all four nonrandom methods were significantly better than the random clusters and so have some value as clustering methods, although for the hierarchical clustering method this is not a surprise. Additionally, cliques appear to be the best graph-theoretic clustering method, followed by stars and then blocks.

The work described here represents only a beginning effort. While the results are encouraging, more and larger studies should be done using different domains, additional clustering methods, and more subjects.

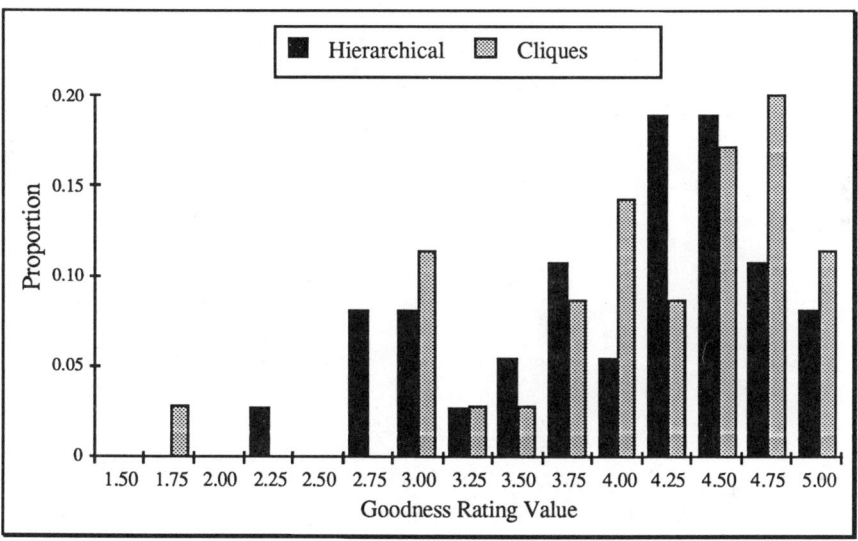

Figure 4. Frequency versus average rating for hierarchical clusters and cliques.

Chapter 7

Empirically Defined Semantic Relatedness and Category Judgment Time[*]

Nancy Jaworski Cooke

Cognitive psychologists generally believe that information is represented in memory in an organized fashion (but see Landauer, 1975) and that the basis for this organization is association. If this is the case, tasks that require memory retrieval should be affected by associations among the items. In fact, much empirical evidence has suggested that the organization of stimulus materials affects performance on memory-related tasks. It has been shown that list recall is facilitated when the list is organized according to common associations (Jenkins, Mink, & Russell, 1958; Jenkins & Russell, 1952). In addition, for cases in which common associations are not obvious, subjects will impose their own organization on the material in order to remember it (Tulving, 1962). Others have found significant correlations between recall order and judgments of pairwise relatedness, again strengthening this link between memory retrieval and association (Caramazza, Hersh, & Torgerson, 1976; Cooke, Durso, & Schvaneveldt, 1986; Schwartz & Humphreys, 1973). Finally, Meyer and Schvaneveldt (1971) found that subjects judged the lexicality of pairs of related words faster than unrelated pairs, suggesting that the related context provided by the first word primed or activated associated concepts in memory, thus facilitating retrieval of the second word.

In recent years various models of memory organization have been proposed to describe how concepts and relations could be represented in order to account for relatedness effects in memory tasks. In network models of memory, concepts are represented as nodes in a graph structure, and relations between concepts as links between nodes (Anderson, 1983; Anderson & Bower, 1973; Collins & Loftus, 1975; Collins & Quillian, 1969; Glass & Holyoak, 1975; Quillian, 1969; Sowa, 1984). The degree of relatedness between two concepts can be represented by either the strength of a link, or the number of shared links. Links are also labeled according to the type of semantic relation (e.g., property relations are represented by *has-a* links and superordinate relations by *is-a* links). Although specific instantiations of these network models differ, the general network structure has received empirical support (e.g., Collins & Loftus, 1975; Collins & Quillian, 1969).

Alternative models of memory have been proposed by other investigators (McCloskey & Glucksberg, 1979; Smith, Shoben, & Rips, 1974) that involve a comparison process operating over feature sets. In their model, Smith et al. (1974) represent a concept by a set of features, some of which are defining features possessed by all category members (e.g., the presence of *wings* is a defining feature of *birds*). Characteristic features, on the other hand, are shared by many, but not all, category members (e.g., *flying* is a characteristic feature of *birds*). In this model specific relations, such as *is-a* and *has-a*, are not

---

[*]I would like to thank Russell Branaghan and Roger Schvaneveldt for their assistance in designing and carrying out this study.

represented explicitly as they are in network models, but relations are derived by assessing the overlap among feature sets. Like network models, feature comparison models can also account for a variety of empirical findings, such as typicality judgments (Rosch, 1973).

In summary, empirical evidence exists that supports both network and feature comparison models. Furthermore, in cases in which the models do not adequately explain the data, they can be altered slightly by changing the structure (i.e., adding features or links) or the processing assumptions. Which model presents the most accurate description of how memory is actually organized? Unfortunately, if these models are intuitively or logically based, it is difficult to evaluate them in terms of psychological meaningfulness. That is, if a particular model fails to account for certain experimental results it could be due to misguided intuitions on the part of the creator of the specific model, a discrepancy between relations used by humans and logical relations, or inadequacies of the general model.

One solution to the problem of evaluating various arbitrary models of memory organization is to generate the models empirically. Psychological scaling techniques, such as multidimensional scaling (MDS) and Pathfinder network scaling, generate various structures from subjects' judgments of relatedness and it is assumed that these judgments, and consequently the resulting structure, reflect information about memory organization. To the extent that the empirically derived structure adheres to the definitions and assumptions of the general memory model, it can be used to evaluate the psychological meaningfulness of that model.

Empirically Derived Structures

MDS algorithms take pairwise proximity estimates for a set of concepts and generate d-dimensional spatial layouts of those concepts, where the value of d is decided by the experimenter (Kruskal, 1977; Kruskal & Wish, 1978; Shepard, 1962). Dimensions are assumed to reflect features along which the concepts vary, and psychological proximity is represented by distance between concepts in the spatial representation. MDS solutions share many of the characteristics of feature comparison models and in fact have been used by theorists who adhere to this type of model. For instance, several investigators have used MDS procedures as means of empirically identifying semantic features (Rips, Shoben, & Smith, 1973; Rumelhart & Abrahamson, 1973; Shoben, 1976). Additionally, results from various studies have indicated that the distances derived from MDS solutions are psychologically meaningful in that they correspond to free recall order (Caramazza et al., 1976), analogy completion (Rips et al., 1973; Rumelhart & Abrahamson, 1973), similarity judgment time (Hutchinson & Lockhead, 1977), categorical judgment time (Caramazza et al., 1976; Rips et al., 1973; Shoben, 1976), and judgments in an inductive reasoning task (Rips, 1975).

Recent development of techniques like Pathfinder for empirically deriving network structures (Chi & Koeske, 1983; Fillenbaum & Rapoport, 1971, Friendly, 1977; Hutchinson, 1989; Schvaneveldt & Durso, 1981; Schvaneveldt, Durso, & Dearholt, 1985) has enabled researchers to investigate the psychological meaningfulness of network representations of memory. It is important to emphasize the fact that Pathfinder can use the same type of input as MDS. As a result, the same set of proximity estimates can be used to derive both spatial and network representations which can then be compared in terms of psychological meaningfulness. Because the models are algorithmically based, the definitions and assumptions of each model are well-specified and thus, testable.

There have been several studies conducted (many of which are discussed in this book) that have compared MDS structures to Pathfinder network structures by evaluating the effectiveness of the respective structures in terms of a particular application, such as training, selection, or interface design (e.g., Roske-Hofstrand & Paap, 1986b; Schvaneveldt,

Durso, Goldsmith, Breen, Cooke, Tucker, & DeMaio, 1985). For the most part, results have indicated that each representation has particular strengths and weaknesses depending on the application. In this chapter, a theoretically driven comparison of MDS and Pathfinder representations is presented. By evaluating the two types of structures in terms of their correspondence to basic memory-related tasks, one may gain an understanding of the differences between the two scaling techniques that impacts on their usefulness in certain applications. In addition, such an understanding should shed light on the basic question of memory organization as defined by network and feature comparison theories.

Empirically Derived Structures and Recall

In a recent study, Cooke et al. (1986) compared MDS and Pathfinder structures in terms of their correspondence to performance on serial and free recall tasks. For the serial recall task, lists of 13 words were constructed so that they corresponded to either a Pathfinder network or MDS representation of those words. That is, successive list items were either linked in the network and distant in MDS, or close in MDS and not linked in the network. Unorganized control lists were also constructed in which successive items were both unlinked and distant in MDS. Results indicated that subjects learned the network list in fewer trials than its control, but there was no difference between the MDS list and its control. Furthermore, subjects learned the network list in fewer trials than the MDS list. These results suggested that the network organization captured information about memory organization that was particularly useful for serial recall.

In a second study, Cooke et al. (1986) extended these findings to a free-recall paradigm in which subjects recalled a list of 13 items in any order. Proximities derived from recall order and averaged across all subjects correlated significantly with similarity ratings, proximities derived from the MDS solution, and proximities derived from the Pathfinder network. However, partial correlations revealed that unlike the MDS proximities, the Pathfinder proximities correlated significantly with recall order even with the effect of the original similarity ratings partialled out. These results suggest that although both Pathfinder and MDS structures correspond to performance in recall tasks, Pathfinder extracted psychologically valid information about the structure of memory that was not explicit in the original ratings.

In speculating about the extra information that Pathfinder captures, Cooke et al. (1986) pointed out that the Pathfinder algorithm weights judgments about related pairs more heavily than unrelated pairs. MDS on the other hand, weights all judgments equally. Secondly, it is possible that subjects provide more accurate estimates of relatedness for related pairs than unrelated pairs (cf. Roske-Hofstrand & Paap, Chapter 4, this volume). Thus, the Pathfinder advantage could be due to an emphasis on relatedness judgments that are the most accurate. In general, Pathfinder tends to emphasize local relations between pairs of concepts, and MDS stresses global relations among a set of concepts in the form of dimensions. Whereas the local information represented by Pathfinder corresponds well to performance in recall tasks, it is possible that performance in other tasks (e.g., analogy completions, categorical judgments) is better captured by the global information emphasized by MDS.

Empirically Derived Structures and Category Judgment

Indeed, numerous studies involving category judgment have indicated that semantic relatedness affects the time it takes subjects to decide whether two items belong to the same category or different categories (Caramazza et al., 1976; Herrmann, Shoben, Klun, & Smith, 1975; Rips et al., 1973; Schaeffer & Wallace, 1969; Schvaneveldt, Durso, & Mukherji, 1982). Relatedness has been defined in a variety of ways. For example, Schaeffer and Wallace (1969) had subjects judge items to be the same if they were both living things or both nonliving things. Stimulus items consisted of types of mammals, flowers, metals, or fabrics. Subjects were faster to make a "same" judgment if both items were members of the same category (e.g., both flowers or both metals). Thus, in this case, relatedness was defined in terms of logical relations.

Rips et al. (1973) defined semantic relatedness according to either subjects' ratings of similarity or derived distance in an MDS solution. Items were either birds or mammals and subjects were asked to make a "same" judgment whenever the two items presented referred to one of the categories (half of the subjects responded "same" to two birds and the other half responded "same" to two mammals). They found that both the original ratings and the MDS distances predicted reaction time for this task, although the MDS distances predicted slightly better than ratings for mammals and slightly worse than ratings for birds. Interestingly, neither interitem ratings nor interitem MDS distances accounted for reaction time, however, ratings and distances between the instance and the category did. Similar results were obtained by Herrmann et al. (1975) who found that similarity as represented in a hierarchical cluster analysis was predictive of same/different reaction time.

In summary, several studies have supported the relationship between time to judge that two items are from the same category and degree of similarity or semantic relatedness. This relationship has been demonstrated using various measures of semantic relatedness, including logical relations, similarity ratings, and derived distance from a multidimensional scaling solution or cluster analysis. As mentioned previously, whereas Pathfinder networks are better than MDS at representing relations useful for recall, MDS solutions may be better than Pathfinder networks at representing relations that are useful for tasks involving categorical judgments. Therefore, in the following study MDS and Pathfinder representations were compared in terms of their relation to judgment time in a categorization task. The first step involved identifying a concept set and generating the Pathfinder and MDS structures.

Construction of the Structures

The 30 stimulus items that were selected for this experiment consisted of the superordinate category *animal*, six categories subordinate to animal (e.g., *pet, bird, fish*), and 23 instances subordinate to those categories (e.g., *dog, sparrow, bass*). These items are presented in Table 1. Care was taken to ensure that the total set of 30 items comprised a fairly broad category (i.e., *animals*) in order to increase variance among the pairwise proximities. Rips et al. (1973) noted that their failure to obtain a significant effect of interitem distance may have been due to the fact that the categories were fairly narrow (i.e., *mammals, birds*) and thus, within-category variance was small.

7 Category Judgment Time

Table 1. Items used to construct the Pathfinder and MDS structures.

animal	dog	horse	trout
farm animal	cow	lion	wolf
reptile	turtle	dolphin	robin
pet	donkey	bat	mouse
bird	lamb	rat	bull
fish	bass	cat	sheep
wild animal	sparrow	tiger	whale
	eagle	lizard	

Fifteen introductory psychology students at New Mexico State University voluntarily participated in the rating part of this study in order to fulfill partial course credit. They were seated in front of an IBM PC and presented with instructions about the rating task, followed by a randomized list of the 30 items so that they would get an idea of the scope of the items that they would be rating. They were then presented with all 435 pairs of concepts, one pair at a time. The pairs were randomized for each subject and the order of the items in each pair was counterbalanced across subjects. A rating scale appeared on the screen along with the pair. The values on the scale ranged from 1 (slightly related) to 5 (highly related) and were selected by moving a bar marker over the value and entering the response by pressing the SPACE BAR. Subjects also had the option of entering a "U" (unrelated) for a pair that they felt was completely unrelated. This "U" option was included to reduce variance in the ratings at the lower end of the scale.

The mean intersubject rating correlation was .571. Relatedness ratings were inverted (ratings of U translated to a 6) and averaged across the 15 subjects. These average dissimilarities were then submitted to Kruskal MDS (Kruskal, 1977; Kruskal & Wish, 1978) and Pathfinder network algorithms. The resulting MDS and network representations are presented in Figures 1 and 2, respectively. Three dimensions were chosen for the MDS solution based on the Isaac and Poor (1974) procedure. Also, the stress tended to level out at .149 for three dimensions. Spatial proximities used to predict judgment time in the next part of the study were derived by taking the euclidean distance between each pair of items in this three-dimensional space. In generating the Pathfinder network the options $r = \infty$ and $q = 2$ were selected because these parameter values required only ordinal assumptions about the data, yet they yielded a network of greater density (38 links) than the tree (29 links) that was generated when q was set to $n-1$. Pathfinder proximities were equal to the length of the shortest path between each pair of nodes in the network where path length was computed by assigning ranks to each link weight and summing the ranks of the links in the path. Summing ranks required only ordinal assumptions about the link weights. In this way, three sets of 435 proximity estimates (one estimate per pair of 30 items) were generated in this part of the study. One set was obtained by averaging the original relatedness across subjects and the other two sets were obtained from the MDS and network structures derived from these average ratings.

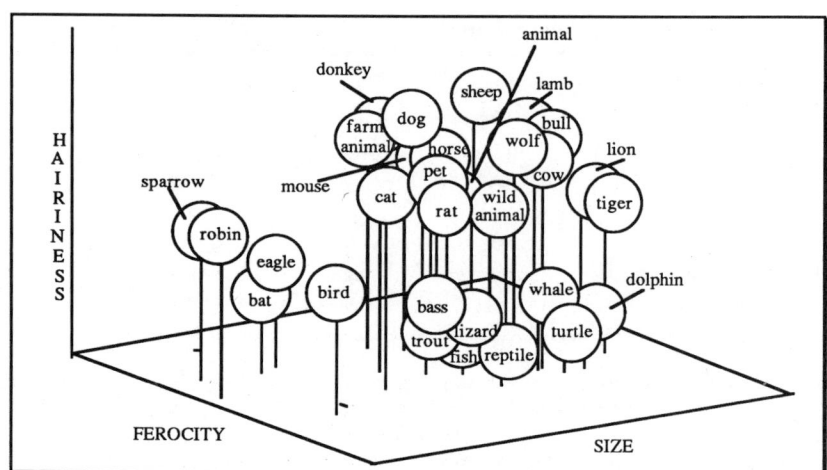

Figure 1. The Kruskal three-dimensional MDS representation. The dimension labels correspond to the author's interpretation of the space.

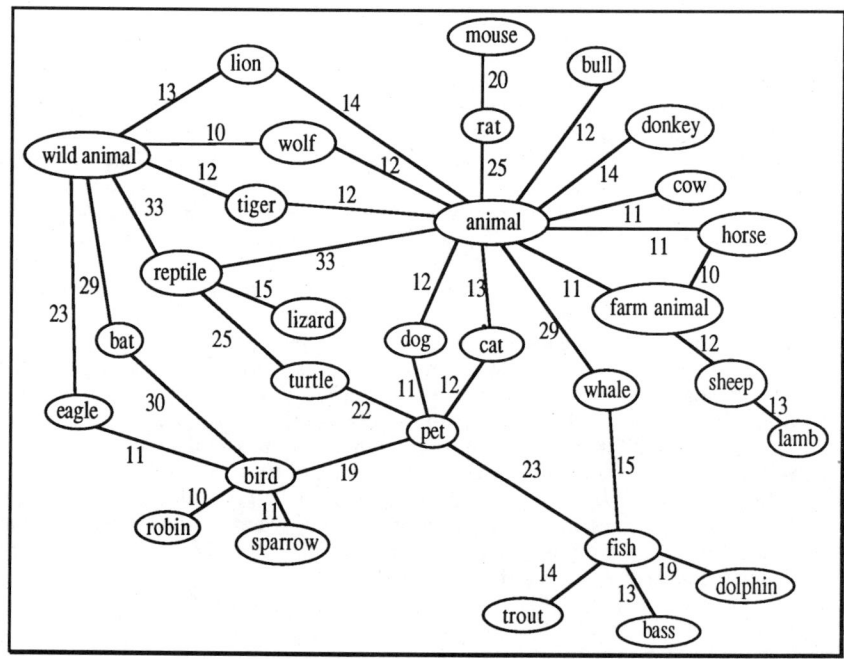

Figure 2. The Pathfinder network representation: PFNET (∞, 2).

Categorical Judgment Study

Method

Subjects. Fifty-four introductory psychology students from New Mexico State University volunteered in partial fulfillment of a research familiarization requirement.

Materials. The materials used in this task consisted of the 23 instances from the total set of 30 previously scaled. Because of the nature of this task the superordinate concepts were not used in this study. An additional set of distractor items was chosen that contained nonanimals (e.g., *carrot, carpet, road*), as well as some additional animals not included in the set of 30.

Procedure. Subjects were seated individually in front of a Terak microcomputer on which instructions about the task were presented. They were told that pairs of words would be displayed and that they were to respond by pressing the key marked "YES" if both words referred to animals. Otherwise, they were to respond by pressing the key marked "NO." Subjects were then presented with 25 practice trials. Animal concepts that were used in the practice trials were taken from the distractor item set. After the practice session subjects were able to ask the experimenter questions.

Out of the total 392 trials, 253 pairs (63%) required a "YES" response. These pairs were constructed from all possible pairs of the 23 animal instances. The remaining 149 (37%) of the trials required a "NO" response. Seventy of these pairs consisted of 2 nonanimals and 69 consisted of an animal from the set of 23 paired with a nonanimal. Therefore, subjects could not simply base responses on the presence of an item from the set of 23. Trials were divided into 8 blocks of 49 with a filler trial for warm-up inserted at the beginning of every block. After each block subjects were presented with feedback concerning their error rate and average judgment time for that block. In addition, they could take a break in between blocks if they wished.

Results

Response times to the 253 "YES" pairs were included in the analysis along with the 253 corresponding proximities from each of the three sets. A $\log_{10}$ transform was conducted on the correct response times. Average error rate for these trials was quite low (1.4%), so only correct trials were included in response time averages.

In order to determine whether the interitem proximity estimates could account for differences between long and short judgment times, a median split was performed on each of the three sets of proximities (ratings, MDS, and network). Then, for each proximity set, the average judgment times for each subject associated with items having either high or low proximities were calculated. It was expected that times corresponding to high interitem proximities would be longer than times corresponding to low proximities in accord with previous results investigating similarity and same category judgments. Differences between "high" times and "low" times are significant for all three comparisons at the .01 level: ratings, $t(53) = 4.8$, $SE = .002$; MDS, $t(53) = 4.17$, $SE = .002$; and network, $t(53) = 6.16$, $SE = .002$. Mean response times (converted from log to original scale) are presented in Table 2. These results do not discriminate between the spatial and network structures, but it is encouraging that each representation captures information that corresponds to either long or short judgment times.

Table 2. Mean response times (msec) associated with high and low proximities.

Proximity Measure	Less Related	More Related
Ratings	652	638
MDS	650	639
Network	653	637

A fine grain analysis was also conducted by investigating correlations and partial correlations between proximities and response time. Correlations for the three sets of proximities and average log response times are all significant at the .01 level with 251 degrees of freedom. The correlation matrix is presented in Table 3. When split-half reliability of response times ($r = .502$) is taken into account, the resulting corrected correlations between proximities and response times are .320, .336, and .413 for ratings, MDS, and network, respectively. The magnitude of these correlations is slightly lower than those found in similar studies. For instance, Rips et al. (1973) obtained multiple correlations between MDS distances and same response time of .55 and .72; however, these distances were based on an MDS solution that was rescaled using only the subset of rating estimates relevant to the categorization task. Because rescaling makes use of all information in the solution, it may result in distortions due to contextual changes.

Results of partial correlations replicated those results found by Cooke et al. (1986). The correlation between network interitem proximities and response time after removing the effect of the ratings is significant, $r(250) = .208$, $p < .01$, but the correlation between MDS proximities and response times without the rating effect is not, $r(250) = .087$. In other words, the network-derived proximities tend to capture information relevant to response times that is linearly independent of the original relatedness ratings used to generate the network structure. The MDS proximities are not independently predictive of response times as would be expected given the high (.853) correlation between MDS proximities and ratings. Again, it seems that the network representation extracts some information that is not directly represented in the original ratings.

Table 3. Correlation matrix for average ratings, MDS interitem proximities, network interitem proximities, and average response time.

	Ratings	MDS	Network	Response Time
Ratings	1.000	.853	.544	.227
MDS	.853	1.000	.500	.238
Network	.544	.500	1.000	.293
Response Time	.227	.238	.293	1.000

Unlike the results found previously using free recall (Cooke et al., 1986), the partial correlation between ratings and response time after removing the network effect ($r(250) = .084$) or the MDS effect ($r(250) = .047$) is not significant. Also, beyond the shared ratings there is virtually no overlap between the MDS and network solution

(r network, MDS/ratings (250) = .082). Thus, although the MDS solution represented the critical rating information, the network captured all of this information plus additional information critical to the categorical judgment task that is not directly represented in the ratings. The possible content of this additional information is discussed below.

Whereas the above analysis was performed on interitem distances, an analysis based on the proximity between the first item in a pair and the superordinate *animal*, or between the second item presented and *animal*, did not produce radically different results. Correlations between response times and either first item-*animal* proximity or second item-*animal* proximity were of a lesser magnitude than interitem distance correlations and ranged from $r = -.091$ to .189. Consequently, multiple correlations based on all three types of proximity (interitem, first item-*animal*, and second item-*animal*) were of magnitude comparable to interitem correlations alone.

Conclusions

The results of this study support the psychological meaningfulness of Pathfinder structures. Although long or short response times corresponded to gross distinctions (high vs. low proximity) in the ratings, MDS solution, or Pathfinder network, a more detailed analysis revealed that the Pathfinder network accounted for approximately 17% of the variance in the response times, whereas the ratings and MDS accounted for only 10% and 11% of the variance, respectively. Furthermore, partial correlations reveal a pattern of results similar to those found using a free recall task (Cooke et al., 1986). That is, unlike MDS, the Pathfinder network is predictive of category judgment time even when the rating effect is partialled out. In particular these results are interesting, given that a categorization task seems compatible with a spatial representation. In general, these results lend some support to the generalizability of the Cooke et al. (1986) results. That is, the advantage of the network representation over the spatial one appears to be more than an artifact of the recall task.

It seems that the Pathfinder algorithm captures and represents information in the original ratings that is not represented explicitly in the ratings or in the multidimensional scaling solution. It was previously mentioned that this "extra" information may arise from Pathfinder's emphasis on local, instead of global, relationships. This local/global distinction between the two scaling procedures parallels a distinction made by Lorch (1981) between retrieval (network) models and feature comparison models of memory. He points out that most comparison models describe relatedness effects in terms of semantic overlap among features and in doing so emphasize the comparison process that is required to compute relations. On the other hand, retrieval models emphasize the direct storage of relations and explain relatedness in terms of either semantic overlap (number of shared connections) or relation strength, "the strength of the most accessible subject-predicate connection which is sufficient to determine a response" (Lorch, 1981, p. 595). However, in retrieval models, as in the Pathfinder procedure, strength is emphasized over semantic overlap.

Lorch (1981) suggested that these two accounts of relatedness, semantic overlap and relation strength, could explain different patterns of results for false items in a category verification task. In one of his experiments subjects were required to respond "true" if the first item of a pair was a member of the *category* indicated by the second item (e.g., *bee-insect*). False pairs consisted of items that had strong (e.g., *bee-wings*) or weak (e.g., *bee-stinger*) *property* relations. Subjects were faster to make false responses to strongly related pairs than weakly related pairs. However, this finding was reversed in an experiment in which unrelated false items (e.g., *bee-hair*) were included. Lorch (1981) concluded that subjects based their judgments on different information in each case. In the first case, judgments were based on relation strength and thus, strong property relations resulted in rapid judgments. In the second case, subjects based their decisions on overall semantic overlap, and accordingly, low overlap resulted in a quick one-stage response, whereas high overlap often resulted in a second stage of additional processing.

Assuming that relatedness does consist of semantic overlap and relation strength, then it is possible that network and spatial representations differ to the extent that they capture information about each of them. MDS representations tend to define relatedness in terms of semantic overlap and not specific relation strength. The Pathfinder proximities, however, were based on relation strength (i.e., link weights) as opposed to the number of shared connections. To the extent that relation strength is relevant to the task, Pathfinder's emphasis on this component of relatedness could provide it with a predictive advantage over MDS.

It would be interesting to investigate the correspondence between response time and a measure of semantic overlap based on the Pathfinder network. Such a measure of semantic overlap might be calculated by counting the number of paths consisting of q or fewer links that connect two nodes, where q is the maximum number of links allowable in any path. In other words, instead of calculating the shortest path between two items (i.e., relation strength), the number of short paths would be counted. Based on the network in Figure 2, the semantic overlap ($q = 2$) between *wild animal* and *animal* equals 4 and is greater than the overlap of 2 between *farm animal* and *animal*. However, the relation strength based on the sum of the ranked link weights equals 15 and 6.5 for *wild animal-animal* and *farm animal-animal* respectively. Keeping in mind that smaller link weights correspond to a higher degree of relatedness, then these two measures of relatedness would make opposite predictions about response times to these pairs.

It would also be useful to compare MDS and Pathfinder representations in terms of some other tasks, particularly some that are likely to require interpreting relatedness in terms of semantic overlap (e.g., analogy completion). It is also possible that the method of obtaining proximities from multidimensional space does not adequately represent the concept of semantic overlap captured by MDS. An alternative metric may be better suited in describing the strengths of MDS. These are all issues requiring future research.

Chapter 8

Pathfinder Networks and Multidimensional Spaces: Relative Strengths in Representing Strong Associates

Russell J. Branaghan

Multidimensional Scaling (MDS) (Kruskal, 1964) and the Pathfinder algorithm (Schvaneveldt, Durso, & Dearholt, 1985; Dearholt, Schvaneveldt, & Durso, 1985) share the goal of reducing large amounts of proximity data to an interpretable form. In psychological research, the resulting representations can reveal interesting relationships among concepts in memory. The two techniques, however, achieve their goals by way of different mechanisms. As a result, the two methods often produce very different representations of memory.

One goal of MDS is to represent the semantic dimensions underlying a domain of concepts. The dimensions revealed can provide information about how concepts are organized in memory. MDS positions points corresponding to the domain items into a space with k-dimensions. The euclidean distance between the points represents the psychological distance between the concepts.

Pathfinder uses a graph-theoretic technique which judges the importance of the relationships between items in each pair of concepts. Pathfinder produces a network representation of the concepts in a domain. It includes a link between two concepts in a network if and only if the link is a minimum length path between the two concepts. It defines a network which includes important links as indicated by the proximity data.

Both MDS and Pathfinder require estimates of psychological dissimilarity as input. These estimates are often obtained by having subjects rate the pair-wise relatedness of items in a domain. For example, subjects may rate the relatedness of pairs of animals on a scale of 1 to 5 (with 1 being extremely related and 5 being extremely unrelated). In doing so, subjects frequently find it easier to assign number ratings to related pairs than to unrelated pairs. For instance, subjects have no trouble determining that the pair *lion-tiger* should receive a rating of 1. They are very strong associates. Further subjects know that they are more related than say *lion-monkey* which may receive a rating of 2 or 3.

In contrast, subjects are often uncomfortable assigning numbers to unrelated items. These items are simply unrelated. It is difficult to give a number to this unrelatedness. Should the pair *whale-lion* receive a 4 or a 5? Is the pair more or less related than *elephant-penguin*? Assigning a number to the degree of unrelatedness often does not make sense. As a result, ratings for related pairs may contain more meaningful information than ratings for unrelated pairs. Further, subjects rate strong associates quickly, easily, and with a high degree of intersubject agreement. This indicates that these ratings may be particularly meaningful, as well as informative, about the structure of memory.

Although the ratings for strong associates may be particularly meaningful, MDS does not weight these ratings differently in determining the representation. This may cause some strong associates to be greatly distorted in MDS representations. MDS uses a least-squares technique in determining the arrangement of all concepts in some k-dimensional space. Each rating datum, whether it represents the relationship between *lion* and *tiger* or between *lion* and *trout*, exerts the same level of influence or constraint on the spatial solution. MDS tries to mutually satisfy these constraints by minimizing a least-squares measure called stress. Essentially, MDS may distort the representation of very psychologically meaningful strong associations to fit all ratings data, meaningful and otherwise, to some multidimensional space.

A paired-associates learning task was used in this study because it stresses relationships between individual pairs of concepts. Subjects are asked to recall the second word in a pair after being presented with the first word. The subject must form an association between the two words in order to be successful. Subjects learned lists of paired-associates which were linked in a Pathfinder solution or had very low interitem MDS distances. Lists were constructed using two sets of materials. One set was a relatively homogeneous domain which consisted of many strong associates. The other was a heterogeneous domain which had fewer strong associates. It was expected that many individual pairs would be distorted in the MDS space to achieve optimal fit of all concepts. This would cause some strongly associated items to be placed far away from each other in the MDS space. As a result, these strong associates would be excluded from the MDS list because of their large interitem distances. On the other hand, Pathfinder determines whether to link items on a pairwise basis. It has no global goodness-of-fit measure to optimize. As a result, these distortions should not take place in the Pathfinder solution. Linked items should be the strongest associates present in the materials.

The associates paired according to Pathfinder should be easier to learn than ones paired according to MDS solutions. Further, associates which are linked in Pathfinder and have small (highly related) link weights should be learned more easily than linked items with large link weights. This is simply because strong associates will have smaller weights due to smaller ratings values.

It was hypothesized that a paired-associates list organized according to the Pathfinder solution would be learned more quickly than one organized according to an MDS solution. Further, a Pathfinder list containing linked pairs with small link weights should be learned more easily than a similar list with larger link weights. However, all three of these lists should be learned more easily than a list of randomly selected word pairs.

Method

One group of subjects learned a list in which the word pairs had the closest interitem distances in the MDS solution (MDS group). Two other groups learned lists with items which were linked by the Pathfinder algorithm. One of the Pathfinder lists consisted of items with the smallest (i.e., most related) link weights in the Pathfinder output (Short-linked). The other had items which were linked in Pathfinder but had the largest link weights (Long-linked). The remaining condition served as a control, containing lists with items randomly selected (Random group) from the domains shown in Tables 1 and 2 at the beginning of each trial. Planned comparisons were performed. Performance on the Short-linked list was compared with that on the MDS list. Additionally, Short-linked was compared with Long-linked to determine if rating values (i.e., link weights) add information important for organization. Long-linked was compared to MDS to determine how the theoretically worst Pathfinder list compares to the best MDS list. Finally, the average of all structured lists was compared to the Random list to determine if semantic structure, in general, facilitates learning of paired-associates.

8 Representing Strong Associates

Additionally, many correlations among structural measures, as well as between structural and performance measures, were performed using the data collected from the Random group. Since random pairs were selected independently for each subject in the Random condition, data are available for many different word pairs. Any given word pair can be described by rating values, MDS interitem distances, Pathfinder links, and so on. Various correlations were performed to determine how predictive structural and ratings information is of paired-associate learning.

Construction of the Scaling Representations

Two sets of materials were used. Set I is fairly homogeneous, and includes a large number of strong associates. All of the items in this domain were animals of some type. The set is shown in Table 1. Set II consists of the materials used by Cooke, Durso, and Schvaneveldt (1986). This set, shown in Table 2, is more heterogeneous. It includes animals, plants, properties, parts of animals, parts of plants, and subordinate-superordinate relations. Further, it contains fewer strong associates than the materials in Set I.

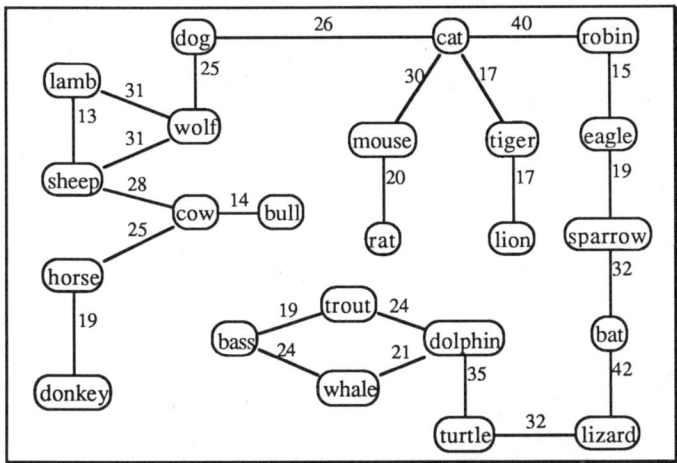

Figure 1. PFNET($r = \infty$, $q = n-1$) for the Set I materials.

Set I

Collection of Ratings. Fifteen students, enrolled in an introductory psychology course, rated the relatedness of all pair-wise combinations of animal concepts. Ratings were made on a scale from 0 to 5 with 5 being extremely related and 0 being unrelated. The concepts that were scaled are shown in Figures 1 and 2.

The scaling solution derived from these ratings was originally used for a different task in which subjects were asked to judge whether an X is an animal (see Cooke, Chapter 7, this volume). To make the set of items more homogeneous, ratings data of superordinates such as *fish*, *bird*, and *animal* were removed. There were 7 of these superordinates. Pathfinder and MDS representations were constructed using the remaining 23 basic-level concepts.

The interitem MDS distances and the structure of the Pathfinder solution may have been somewhat different if the superordinates were included. However, both scaling solutions were constructed from the same remaining data. There is no reason to believe that removing the superordinates had differential effects on Pathfinder and MDS solutions.

Scaling Solutions. The Pathfinder network for the Set I concepts is shown in Figure 1. This is the solution when $r = \infty$ and $q = n-1$, the sparsest network which can be derived. The assumption is that this network includes the most important links. The MDS solution for the concepts is shown in Figure 2. Three dimensions were chosen as the optimal dimensionality for the solution because stress and r^2 seemed to elbow at three dimensions. The Kruskal stress of the solution was .12.

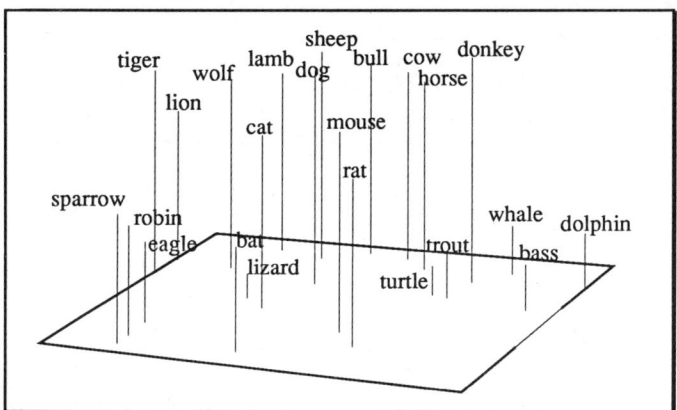

Figure 2. Three-dimensional MDS solution for the Set I materials.

Set II

Lists were also constructed from the materials and scaling solutions in Cooke et al. (1986). These materials are referred to as Set II and are shown in Figures 3 and 4.

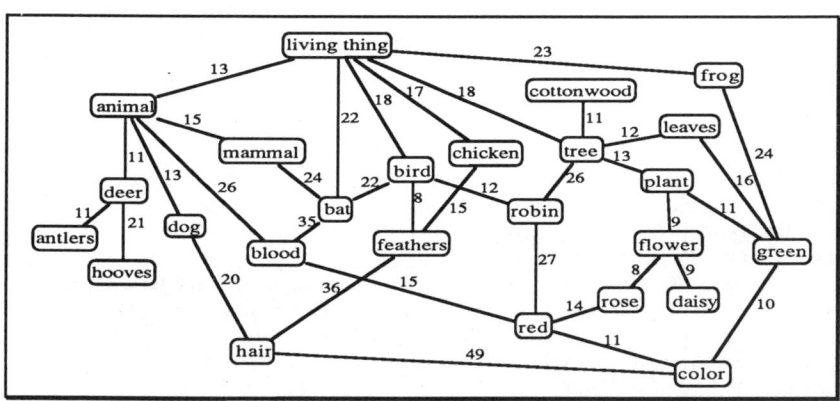

Figure 3. PFNET (parallel method) for the Set II materials.

8 Representing Strong Associates

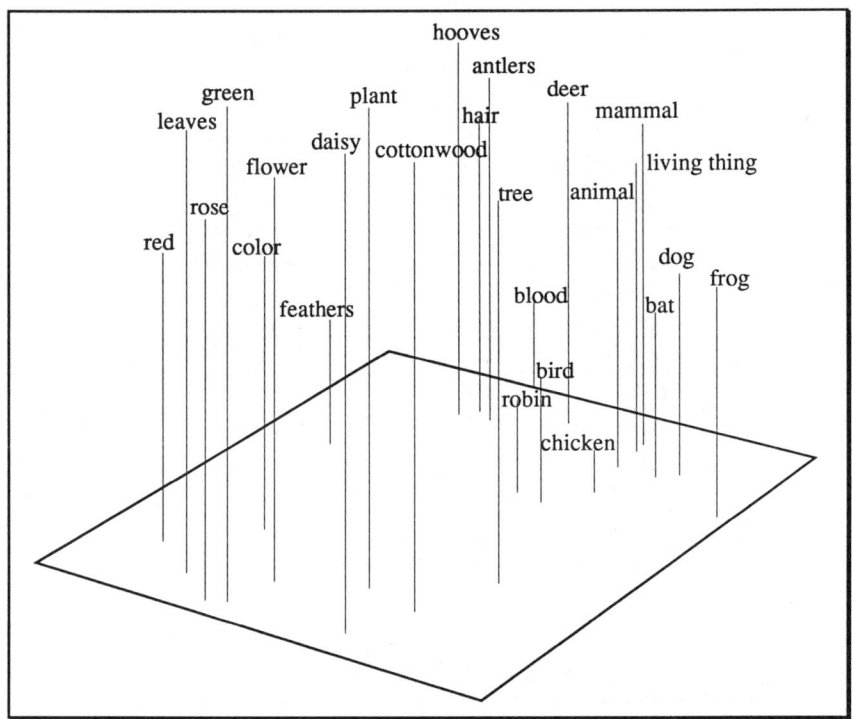

Figure 4. Three-dimensional MDS solution for the Set II materials. (From "Recall and Measures of Memory Organization," by Cooke, Durso, & Schvaneveldt, 1986, *Journal of Experimental Psychology: Learning, Memory, and Cognition, 12*(4), p. 542. Copyright 1986 by APA. Adapted by permission.)

Scaling Solutions. Cooke et al. (1986) used the parallel option of Pathfinder (Schvaneveldt et al., 1985) to construct their Pathfinder solution. The Pathfinder solution is shown in Figure 3. The ALSCAL-S (Young, Takane, & Lewyckyj, 1978) MDS solution (used in Cooke et al., 1986) for the materials is shown in Figure 4. Three dimensions were chosen as the optimal dimensionality for this because stress and r^2 seemed to elbow at three dimensions. The Kruskal stress for this solution was .21.

Construction of the Paired-Associate Lists

Selection of the items in each list was restricted in the following way. Any word appeared only once in a paired-associate list. The Short-linked list had those items which were linked by Pathfinder and had the smallest (i.e., most related) link weights. The Long-linked list had the linked items with the largest link weights. The MDS list contained those items with the smallest interitem distances. Random lists consisted of randomly selected word pairs chosen independently for each subject from the domains in Tables 1 and 2. The organized paired-associate lists, along with their mean interitem MDS distances, Pathfinder weights, and mean ratings, are shown in Tables 1 and 2.

Table 1. The paired-associates lists for the Set I materials.[a]

Linked Short	Linked Long	MDS
lamb-sheep	lizard-bat	cow-horse*
cow-bull	robin-cat	robin-eagle*
robin-eagle	dolphin-turtle	bull-sheep
tiger-lion	wolf-sheep	wolf-lion
donkey-horse	cow-horse	rat-mouse*
bass-trout	whale-bass	bass-trout*
rat-mouse	rat-mouse	whale-dolphin*
whale-dolphin	sparrow-eagle	sparrow-bat*
wolf-dog	tiger-lion	donkey-dog
MDS = 41.9	MDS = 61.6	MDS = 38.8
PF = 18.1	PF = 28.1	PF = N/A
Ratings = 18.1	Ratings = 28.1	Ratings = 26.2

[a]The mean PFNET weights, ratings, and interitem MDS distances are shown below the lists. A * next to an MDS pair indicates that it was also linked in the Pathfinder network.

Table 2. The paired-associates lists for the Set II materials.[a]

Linked Short	Linked Long	MDS
flower-rose	hair-color	bird-robin*
bird-feathers	blood-bat	mammal-deer
color-green	robin-red	tree-cottonwood*
tree-cottonwood	frog-green	antlers-hooves
animal-deer	living thing-flower	dog-bat
blood-red	deer-hooves	plant-daisy
living thing-chicken	feathers-chicken	color-red*
hair-dog	animal-mammal	flower-rose*
mammal-bat	plant-tree	leaves-green
MDS = 13.6	MDS = 20.3	MDS = 4.3
PF = 12.5	PF = 26.3	PF = N/A
Ratings = 12.5	Ratings = 26.3	Ratings = 19.3

[a]The mean PFNET weights, ratings, and interitem MDS distances are shown below the lists. A * next to an MDS pair indicates that it was also linked in the Pathfinder network.

8 Representing Strong Associates

An additional set of lists, which reversed the order of the word pairs in Set II, was constructed. As can be seen in Table 2, all of the pairs which contain superordinate-subordinate relationships list the superordinate first. However, it should not be assumed that superordinate-subordinate relationships are bidirectional. Reversing the order of these pairs provides information about whether this directionality is important.

Procedure

One hundred twenty-eight (80 in Set I, 48 in Set II) New Mexico State University undergraduate students participated in partial fulfillment of an experimental familiarity requirement. An additional 36 subjects learned the reversed Set II lists. Subjects were randomly assigned to one of eight combinations of materials and list type. Subjects were seated in front of a TERAK 8510 computer terminal and instructions which explained the procedure of the experiment were displayed on the screen. The subjects then received a familiarization trial, in which the items to be learned were shown on the screen one pair at a time for five seconds each. During the experimental trials, subjects were shown pairs of words in randomized order. The first item in the pair was presented on the screen and the subject's task was to type the appropriate accompanying word within 15 seconds. The correct answer was shown as soon as the subject typed the return key. The answer was displayed whether subjects made a correct or incorrect response or no response. There was no dropout method used. Subjects were encouraged to respond as quickly and accurately as possible. The task concluded when the entire list of word pairs was learned correctly or when the list had been presented 20 times.

Results

Planned comparisons (Keppel, 1982) on data collected using the Set I materials will be followed by those for Set II. These will be discussed in terms of three dependent variables: number of trials to achieve a 100% correct criterion, number of errors committed in achieving criterion, and reaction time to the first keypress on the final trial for each item. Also, correlations will be briefly discussed.

Set I

Means for the four conditions are shown in Table 3. As predicted, subjects in the Short-linked group learned the word associations in significantly fewer trials, $F(1,38) = 4.3$, $p < .05$, and with fewer errors, $F(1,38) = 8.4$, $p < .01$, than the subjects in the MDS group. Further, subjects in the Short-linked group performed better than those in the Long-linked group. Again, this is evidenced by fewer trials needed to reach criterion, $F(1,38) = 4.7$, $p < .05$, as well as fewer errors, $F(1,38) = 9.8$, $p < .01$. As expected, lists which were structured according to a scaling solution were easier to learn than the random lists. Subjects in the Random condition required more trials, $F(1,76) = 54.8$, $p < .001$, and made more errors, $F(1,76) = 52.02$, $p < .001$, than subjects who received structured lists. There were no significant differences on any measure between the Long-linked and MDS groups. Additionally, analysis of the RT data revealed no significant differences between any of the conditions.

Table 3. Means for the four conditions with the Set I materials.

List	Trials	Errors	RT
Short	1.7	.85	1726
Long	4.2	8.60	2019
MDS	4.0	7.80	1928
Random	7.7	25.30	2036

Set II

The means for subjects' performance are shown in Table 4. Again, analysis of planned comparisons indicates that subjects who learned a structured list did so in fewer trials, $F(1,44) = 12.6$, $p < .001$, with fewer errors, $F(1,44) = 24.5$, $p < .001$, and with faster reaction times, $F(1,44) = 7.8$, $p < .01$, than subjects who learned randomly arranged lists. All other comparisons were not significant. Further, there were no significant differences among groups with the reversed lists, or when data from the two lists were combined.

Table 4. Means for the four conditions with the Set II materials.

List	Trials	Errors	RT
Short	3.1	4.9	1817
Long	2.6	3.8	1704
MDS	3.3	6.3	1837
Random	4.9	5.3	2387

Correlations

The Random lists used in the experiment provide data about many individual pairs of items. These include performance data, such as number of errors made on a particular pair during the experiment, as well as structural data, such as the pair's interitem MDS distance and whether the pair was linked in Pathfinder. Various correlations between structural and performance measures, as well as among structural measures were performed on this data.

Several variations of Pathfinder networks were generated for exploratory purposes to determine if they are predictive of paired-associate learning. A more dense (i.e., contains more links) Pathfinder network, with $q = 2$ (PF2) was generated. Also, two matrices of graph-theoretic distance, one with $q = n-1$ (Graph) and one with $q = 2$ (Graph2) were generated. In these matrices each entry equals the number of links separating the two corresponding concepts in the Pathfinder network.

Most of the correlations between structural and performance measures were fairly low. It is probable that this is because each correlation tries to predict only one particular performance score given one piece of structural information. The correlations would probably be higher if we were trying to predict a mean of scores rather than one individual score. It should be mentioned that given the number of correlations performed, the family-wise error rate may be fairly high. Therefore, one should not put too much faith in any one correlation, since some correlations may be significant by chance alone.

8 Representing Strong Associates

There was little difference between the ability of Pathfinder and MDS to predict performance. Further, when the effects of the original ratings data were removed from the correlations, most of the resultant partial correlations were not significant. This indicates that correlations between structural measures and dependent measures could be accounted for by the ratings alone. The only exception to this was the Graph2 structural measure with the Set II materials.

Set I

Ratings correlated with percent correct, $r(178) = -.24$, $p < .05$, errors $r(178) = .2$, $p < .05$, reaction time, $r(178) = .17$, $p < .05$, and number of trials to the first correct trial, $r(178) = .19$, $p < .05$. MDS correlated with errors, $r(178) = .17$, $p < .05$, percent correct, $r(178) = -.24$, $p < .05$, and reaction time, $r(178) = .17$, $p < .05$. Pathfinder correlated with errors, $r(178) = -.19$, $p < .05$, percent correct, $r(178) = .24$, $p < .01$, and number of trials to first correct, $r(178) = -.19$, $p < .05$. Pathfinder with $q = 2$ (PF2) correlated with errors, $r(178) = -.18$, $p < .05$, percent correct, $r(178) = 2$, $p < .05$, and number of trials to first correct trial, $r(178) = -.16$, $p < .05$. Graph-theoretic distance with $q = n-1$ (Graph) was correlated with reaction time, $r(178) = .2$, $p < .05$. Finally, graph-theoretic distance with $q = 2$ (Graph2) correlated with percent correct, $r(178) = -.17$, $p < .05$, and reaction time, $r(178) = .17$, $p < .05$.

Set II

Ratings correlated with errors, $r(106) = .23$, $p < .05$, percent correct $r(106) = -.30$, $p < .01$, and number of trials to first correct trial, $r(106) = .27$, $p < .01$. MDS correlated with errors $r(106) = -.32$, $p < .01$. Finally, Graph2 correlated with errors $r(106) = .33$, $p < .01$, percent correct $r(106) = -.40$, $p < .01$, and number of trials to first correct trial, $r(106) = .38$, $p < .01$.

When the effects of ratings were removed from these correlations, almost all of the partial correlations between a structural measure and a performance measure were no longer significant. The only significant partial correlations involved Graph2 ($q = n-1$). Graph2 correlated with errors, $r(106) = .24$, $p < .01$, percent correct, $r(106) = -.27$, $p < .01$, and number of trials to first correct trial, $r(106) = .28$, $p < .01$.

Discussion

As expected, semantically structured lists were learned in fewer trials and with fewer errors than lists of random word pairs. However, the hypothesis that the Short-linked Pathfinder list would be learned more easily than the MDS list was confirmed only with materials from Set I. These were the animal concepts which included a large number of strong associates. With Set II, the domain with fewer strong associates, subjects performed equally well on the Pathfinder and MDS lists.

Whereas there were no differences among scaled lists in Set II, in Set I subjects required two and one-half times more trials to learn the MDS list than to learn the Short-linked Pathfinder list. Moreover, they made seven times as many errors while doing so. On average, subjects learned the Short-linked Pathfinder list in less than two trials, and they averaged less than one error. This means that many of the subjects recalled the list immediately after seeing the familiarization phase.

The ease with which this Short-linked list was learned indicates that the structure of the pairings is strongly related to the structure of the subjects' memory. Inspection of Table 1 shows how simple the Short-linked Pathfinder list is. Each first item is accompanied by a second item which is a strong associate. This is particularly the case with *tiger-lion, rat-mouse, wolf-dog,* and *lamb-sheep.* The least associated pair in the Short-linked group according to the ratings is *wolf-dog.* The MDS list is also easy, but it has fewer strongly associated pairs and more weakly associated pairs. Weak pairs include *donkey-dog, wolf-lion,* and *sparrow-bat.*

In Set I, Pathfinder seems to be linking primary (or at least strong) associates. Set I had more strong associates than Set II. Pathfinder links those items with the smallest ratings. If strongly associated items exist in the domain, they will be linked. Conversely, in a domain with few strong associates, links may not be between strongly associated concepts. They are simply between the most associated items in the domain. When this is the case, Pathfinder's ability to isolate strong associates may not be realized. On the other hand, because of the distortion of local relationships, MDS may be less effective at identifying strong associates.

It was predicted that the Short-linked Pathfinder would be learned more easily than the Long-linked Pathfinder. Again, this was true for the Set I materials but not the Set II materials. In Set I, word pairs with smaller link weights were easier to recall than word pairs with larger link weights. Again, this may be due to the reasons explained above. Effects due to differences in ratings may be most sensitive to strong associates.

Data from Humphreys and Greeno (1970) indicate that paired-associates learning consists of two main subprocesses: storage of the pairs in working memory and learning to retrieve these pairs. These subprocesses, they speculate, are analogous to recall and recognition memory. When subjects are shown the word pairs, they store them in memory using imagery, mnemonics, rehearsal, or some mediating word, and it is believed that the strategy for storing the pair will have a large effect on the nature of its representation in memory. Once stored in memory, its representation may not be optimal for retrieving the pair. Learning to retrieve the word pair is tantamount to fine tuning the memory structure so that it is easily accessed.

The effect of strong associates on the first stage of paired-associate learning is not clear. Perhaps strongly associated word pairs exist as pairs in memory prior to the task, whereas weakly associated pairs do not. It is easier to imagine how strong associates might affect the second stage. If the second stage is one of fine tuning the memorial representation of the items, this process is not needed (or is greatly abbreviated) for strong associates. That strong associates are spontaneously remembered indicates that the memorial representations of them are already quite suitable for access. So, the superiority of strong associates may be the result of eliminating the second stage of learning.

The present study indicates that both Pathfinder and MDS are good organizers of paired-associate learning material. This is evident in the superiority of these methods over random selection of items. However, when paired with previous results, this study suggests that Pathfinder is better than MDS at organizing materials for tasks which emphasize pair-wise relationships. This is qualified by the results which suggest that the presence or absence of strong associates may impact strongly on the relative superiority of Pathfinder. Nevertheless, Pathfinder is a better organizer of materials for a serial recall task, and a better predictor of performance on a free recall task (Cooke et al., 1986). Further, when strong associates are present, Pathfinder is a better organizer of materials for a paired-associates task.

Chapter 9

Directed Graphs as Memory Representations: The Case of Rhyme*

David C. Rubin

This chapter describes an attempt to uncover the structure of rhyme categories, an attempt that provides evidence for the value of directed graphs as memory representations.

Psychologists know a great deal about the structure of semantic categories (Deese, 1965; Fillenbaum & Rapoport, 1971; Friendly, 1977, 1979; Gruenewald & Lockhead, 1980; Meyer & Schvaneveldt, 1976; Nelson, 1981; Rosch, 1975; Rubin & Olson, 1980). They know much less about the structure of rhyme categories even though rhyme is central to the understanding of retrieval in many domains (e.g., Hyman & Rubin, 1988; Wallace & Rubin, 1988a, 1988b). Only one laboratory has extensively studied the role of rhyme in memory (Nelson, 1981; Nelson, McEvoy, & Friedrich, 1982), and their view, based on cuing effects, is that words in rhyme categories are unstructured except that each word has a link to the rhyme sound that defines its category. Figure 1 is a hypothetical network of the *air* rhyme category based on Nelson's (1981) representation of a rhyme category; the individual words do not link to each other, and the strength of a word's membership is given by the length of its links to the central rhyme node. In contrast, semantic categories, such as *animals* or *parts of the body*, would show many links among the items as well as a link to the central concept (Schvaneveldt, Durso, & Dearholt, 1989).

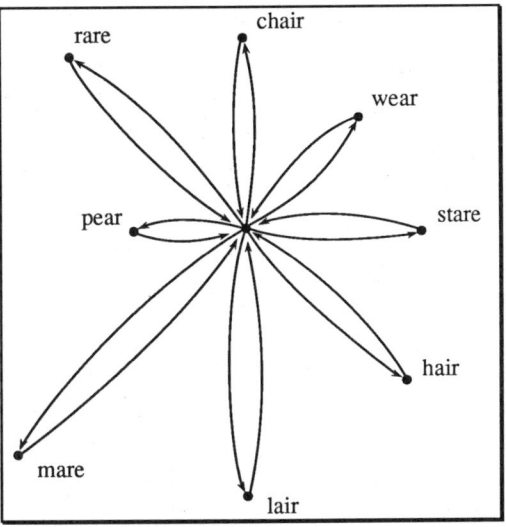

Figure 1. A hypothetical network for the rhyme category *air* based on Nelson (1981).

---

*I wish to thank Orest Zborowski for writing the programs that converted the recall protocols into similarity matrices, and Russ Branaghan, Doug Nelson, and Roger Schvaneveldt for their comments on the chapter. Support was provided by NSF grant BNS-8410124.

Multidimensional Scaling of Rhyme Categories

In order to investigate directly the way in which rhyming words are organized in memory, retrieval from long-term memory was studied using standard multidimensional scaling (MDS) techniques available before the invention of Pathfinder. To provide a comparison, semantic categories were also included in the study. People were asked to list all the words they could that referred to instances of a rhyme or a semantic category (Bousfield & Sedgewick, 1944), a task not very different from that discussed by the British Empiricists, and a task that can be viewed as tracing a path through memory. Under this view, words recalled next to each other are assumed to be related, and MDS techniques can be used to provide a picture of the associative memory structures (Rubin & Olson, 1980).

In the particular experiment reported here, 100 Duke University undergraduates were asked to list all the words they could in 60 seconds that had the rhyme sounds of *air, ear, ed*, and *ee*, as well as all the words they could that belonged to the semantic categories of *animals, beverages, furniture,* and *parts of the body*. These eight categories were chosen as a sample because they all had a large number of instances. A tape recording, read by an undergraduate native of North Carolina, was used to present the stimuli. Two random orders were presented with rhyme and semantic categories alternating. Subjects were asked to "please turn to page X and write down all the words that you can that ..."

All responses were compiled, with different responses being combined under the same word only if they were different spellings of the same word. Singular and plural words were not combined because they have different rhymes. Thus, *eye* and *eyes* were scored as separate responses. The most frequent 20 words in each category were then selected for further analysis. This ensured that at least 10 subjects recalled each of the 20 words used. For each category the number of times any of these 20 words were emitted in succession was counted as a measure of similarity (Rubin & Olson, 1980). That is, each cell in the lower triangular similarity matrices was indexed by two words, with the value of the cell equaling the number of subjects who recalled the two words immediately next to each other. Thus the cell for *dare-care* in the rhyme category *air* contained 39 because 39 subjects out of 100 recalled these words next to each other.

The resulting symmetrical similarity matrices were submitted to a smallest space analysis MDS solution (Lingoes, 1973). Two-dimensional solutions are included as figures because the greatest decrease in stress and increase in fit occurred in going from one to two dimensions and because the three-dimensional solutions failed to provide any additional information. The average coefficient of alienation values for the 1-, 2-, 3- and 4-dimensional solutions were .39, .23, .15, and .10, respectively.

The standard way to interpret MDS solutions is just to look at them. The solutions for the semantic categories are easily interpreted. For instance, the *animal* category is consistent with earlier work performed on this domain (see Rubin & Olson, 1980, for a review). The *parts of the body* category, for which there was no previous work, is provided in Figure 2 as an example. This domain can be divided into a head cluster on the right, a limb cluster on the left, and a torso cluster at the center bottom. *Eyes* and *ears* do not appear immediately adjacent to *eye* and *ear* because no subject ever said "*eye, eyes,*" or "*ear, ears.*" They are, however, close to each other because these words were often said next to *nose* and *mouth*. The axes could be labeled as dimensions, but this is a stronger claim than is warranted by the data (Rubin & Olson, 1980).

The rhyme domains are harder to describe in terms of obvious organized clusters of words, though some structure is apparent. For instance, where homonyms occur, they tend to be near each other. There are 15 distinct homonym pairs in the rhyme category

9 The Case of Rhyme

MDS solutions. In 70% of the 320 cases in which a homonym pair was present in a subject's recall, the homonyms appeared next to each other. There is also a hint of some semantic structure in the rhyme categories, for instance, in the *ee* category, *we, he, me*, and *she* cluster. The *air* category is shown as an example in Figure 3.

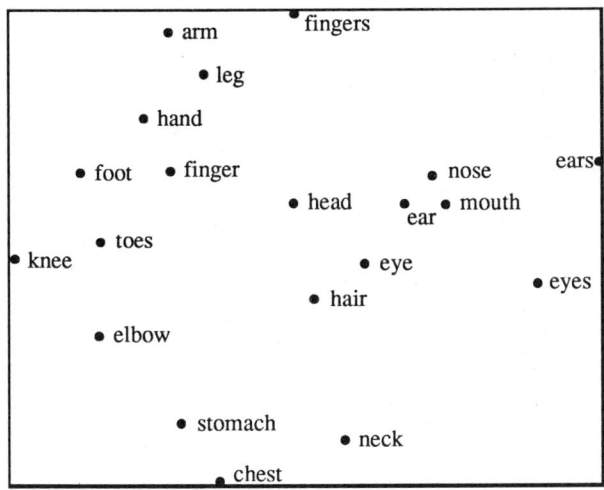

Figure 2. An MDS solution for the semantic category, *parts of the body*.

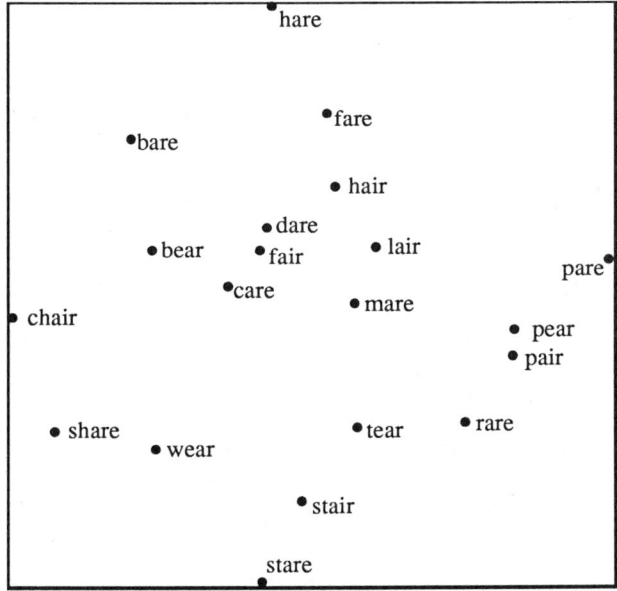

Figure 3. An MDS solution for the rhyme category, *air*.

Interpreting the Rhyme Spaces Quantitatively

The rhyme categories offer no simple visible interpretation, but perhaps it is possible that the structure of rhyme domains could be understood in terms of phonetic similarity. In order to investigate this possibility, the words listed in the rhyme categories were transcribed phonetically. To avoid making assumptions about the similarity of clusters of phonemes, only words consisting of a single phoneme plus the rhyme ending were included. This reduced the number of similarity values among the word pairs from 760 (i.e., the 20 × 19/2 comparisons among the 20 words of each of the four rhyme categories) to 516 similarity values. The pairings of homonyms were also removed to avoid confounding the special effects of their similarity with phonetic similarity. This further reduced the number of similarity values among word pairs to 502.

Two measures of phonetic similarity were formed. The first was the number of distinctive features distinguishing the two phonemes. This measure was taken from Fromkin and Rodman (1983). The second measure was the number of times two phonemes were confused in listening experiments. This was taken from the combined data of Tables 2 through 6 from Miller and Nicely (1955). Because all phonemes did not appear in these tables, the phonetic confusion measure could be obtained for only 204 of the word pairs.

It is also possible that the orthography of the words could have an effect. Therefore, in addition to the two phonetic measures, two parallel measures of visual properties were included for comparison. The number of distinctive features separating the initial letters was taken from Gibson (1969), and a visual confusion matrix was made by combining Tables 1 and 2 of Townsend (1971). Finally, the spelling of the rhyme sound was used to measure visual similarity. Word pairs with the same orthographic ending were assigned a value of 1.0, and word pairs with different endings were assigned a value of 0.0.

Two measures of similarity in memory were used. The first was the same number-of-times-next-to measure used for the figures. The second was this next-to measure divided by the number of subjects who recalled both words in the pair. The latter normalized measure was included as an additional check (Rubin & Olson, 1980).

None of the phonetic or spelling measures based on the initial phoneme or letter correlate with either similarity measure as highly as .07. The dichotomous measure of rhyme spelling has only small correlations with the next-to and normalized measures (.09 and .13, respectively). These correlations account for, at most, a negligible amount of structure. Thus, once the phonetic constraint of searching within a rhyme category is met, the phonetic and visual similarity of the words plays little further role.

Standard MDS solutions for the rhyme categories revealed no clear, interpretable structure as was found in the semantic domains, nor did phonetic or orthographic hypotheses, as tested by correlations. Some structure could be interpreted in terms of homonyms and semantics (e.g., *we, he, she, me*), but most of what appears in the MDS solutions cannot. Perhaps Nelson's claim is correct and the rhyme categories are not really structured. In order to test this possibility, a reliability check was performed to see if different groups of subjects recall the same items next to each other. The 100 subjects were divided into two groups of 50, and a correlation was calculated between the cells of the next-to similarity matrices that resulted from each group. As with the correlations performed on the rhyme categories, each matrix was treated as an ordered list of 190 (20 × 19/2) cells. The correlations for the four rhyme and the four semantic categories were .798, .730, .771, .794, and .935, .891, .881, .893, respectively, indicating that all eight of the domains were quite structured, though the semantic categories were more structured than the rhyme categories.

9 The Case of Rhyme

Another approach to measuring the structure of the rhyme similarity spaces is to examine the distribution of values in the cells of the matrix. If a next-to similarity matrix, such as those constructed here, was from a highly structured domain, it should contain many cells with high numbers and zeros, indicating pairs of words that often or never appeared next to each other. A similarity matrix from an unstructured domain, given the nature of random occurrences, should contain many moderate values. In order to provide values from unstructured categories that otherwise resemble the categories tested here, the recall protocols of each subject were randomized and similarity matrices formed from the random orders. Three random orders were sufficient to provide a smooth distribution of values for the unstructured matrices. The four rhyme and the four semantic matrices were combined and compared with their respective random order matrices as shown in Figures 4 and 5.

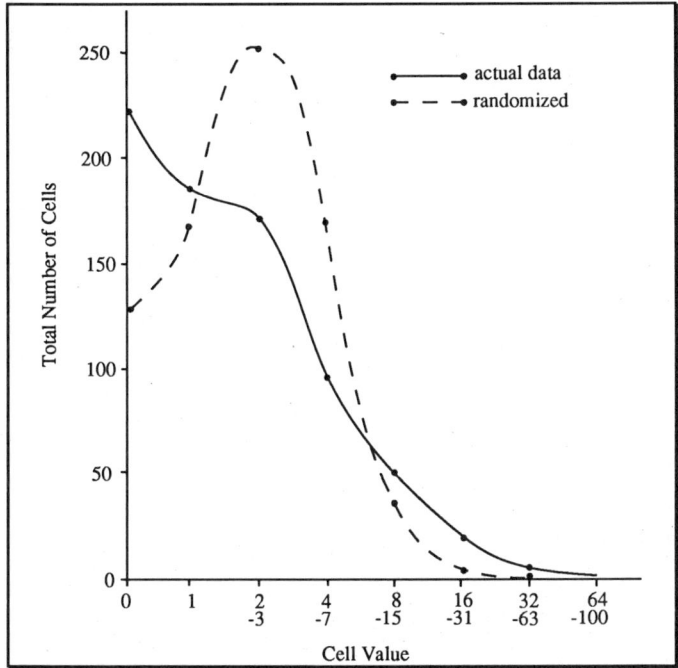

Figure 4. The distribution of actual and randomized next-to values in the cells of the semantic category similarity matrices.

The rhyme category appears to be highly structured. Compared to the random order matrices, the actual matrices had more cells with 0 and 1 entries, less cells with entries at each value between 2 and 10, and the same or more cells with entries at each value above 10. The semantic categories showed the same pattern with more cells with 0 and 1 entries, less cells with entries at each value between 2 and 9, and the same or more cells with entries at each value above 9. The highest value in any cell of the three random orders of the eight matrices was 22. The rhyme categories had 9 cells with values over 22, the semantic category had 11.

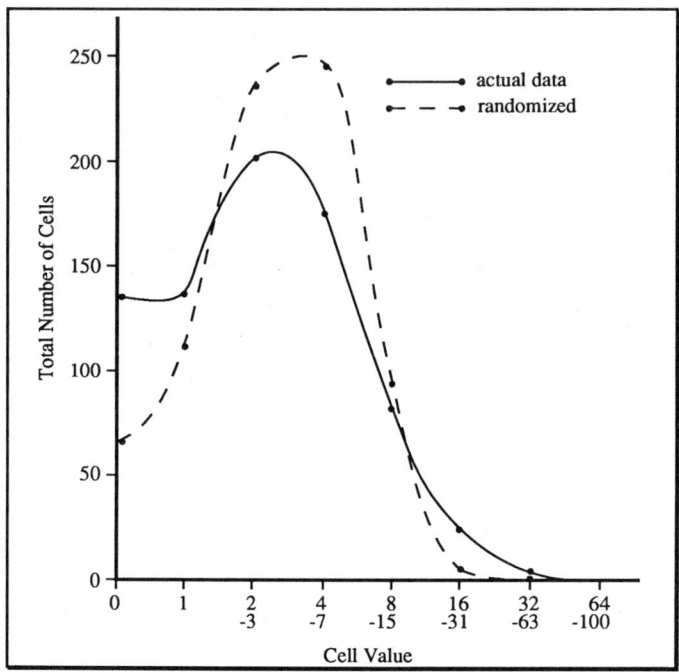

Figure 5. The distribution of actual and randomized next-to values in the cells of the rhyme category similarity matrices.

A simple summary statistic of the difference between the observed and random matrices is the sum of the absolute values of the differences between the observed and random frequencies added over all possible values. This sum is 266 for the rhyme categories and 324 for the semantic categories. A more standard measure of goodness of fit is chi-square. In order to keep the random (or expected) frequencies greater than 2 in each cell, all values greater than 12 were combined into one category. The chi-square values for the resulting 14 groupings for the combined rhyme and semantic categories are 195 and 309, respectively ($df = 13$; chi-square values greater than 12 would have $p < .001$ under the liberal assumption that the matrix cell entries are independent observations).

One difference between the rhyme and semantic categories is that rhyme categories have no cells for which more than 40 subjects out of 100 recalled words next to each other, whereas the semantic categories have two cells over 50. *Cat* and *dog* were recalled next to each other 89 times, and *lion* and *tiger* were recalled next to each other 53 times. It should be noted that neither the absolute sum nor the chi-square statistic is sensitive to the magnitude of these two high-cell values in that neither of these summary statistics would change if the two high-cell values were changed from their current values of 89 and 53 to values of 13. This is because the sum-of-the-absolute-differences statistic is not affected by which cell leads to the difference, and the chi-square statistic is calculated after all values greater than 12 are combined into one category.

9 The Case of Rhyme

In summary, it appears that both semantic and rhyme categories tested are structured, though there is more structure in the semantic categories. The structure of the semantic categories can be interpreted in terms of semantic similarity among the items. The structure of the rhyme category, with the exception of homonyms and occasional semantic structure, cannot be easily interpreted.

A Pathfinder Solution

The resources of standard MDS have been exhausted, and no answer has been found to the question of how rhyme categories are organized. Nelson (1981) hypothesized that there is no structure, but this is not the case here. On the other hand, there is no clear way to describe the structure that does exist. This research project remained a paradox for some time, until two occurrences combined to provide a resolution. First, Bruce Ammons, who was a graduate student at Duke at the time, observed that he had no strong organization for rhyme categories and so would do the task given to the subjects by using the alphabet to prompt himself to find rhyme words. Second, Pathfinder became available.

The task given to the subjects required the output of rhyming words. If no clear organization was available in long-term memory, perhaps some temporary search strategy, like using the alphabet to cue words, would have been used. This is a common strategy when trying to think of people's names, so Ammons's suggestion made intuitive sense. Moreover, the strategy is especially effective for rhyme categories. All the subjects would have had to do is combine each letter of the alphabet in turn with the rhyme ending and decide whether or not a word resulted.

An examination of the unanalyzed recalls supported the hypothesis for at least some subjects. The following two subjects' recalls from the *air* category are clear cases. The recalls begin with words formed from alphabetically ordered initial letters paired with the "air" rhyme. When the end of the alphabet is reached there is a search for more complex sounding words. The first subject recalled "*bear, care, dare, fair, hair, mare, pear, pair, rare, tare, tear, wear, share, Blair, flair, snare.*" The second subject recalled "*bear, care, dare, fair, hair, hare, lair, mare, pare, pair, rare, tear, where, impair, affair, there, stair.*" The problem remained of how to explore this possibility in a clear quantitative fashion.

If the alphabetical strategy was being used, the symmetrical next-to similarity measure would not be ideal; rather a measure that noted directionality is needed. For this reason a directional next-to, or follows, measure was formed from the data from the four semantic and four rhyme categories. The entry in each cell was the number of subjects who recalled the word defining the row followed by the word defining the column. Thus the cell *dare-care* contained seven because seven subjects recalled *dare* followed by *care*, whereas the cell *care-dare* contained 32 because 32 subjects recalled *care* followed by *dare*. This resulted in square rather than lower triangular similarity matrices, in which the main diagonal was undefined and set to zero for purposes of calculation. If these new matrices were collapsed along the main diagonal to form symmetrical lower triangular matrices by summing the values from cell (i,j) with cell (j,i), the previously analyzed eight matrices would result.

The square matrices were submitted to Pathfinder. Because Pathfinder requires dissimilarities, each cell was transformed by subtracting its value from 99. In addition, the resulting remainders that were greater or equal to 98 were considered as infinite, so that a word had to follow another word more than once in the set of 100 subjects' recalls in order to be counted in the solution. Minimally connected networks were obtained by setting the r value of the Minkowski r-metric to infinity. No limit was set on the possible number of links allowed in a path. Figures 6 and 7 present solutions to the *parts of the body* and the *air* domain. The placement of the nodes is identical to that in Figures 2 and 3. The links come from the Pathfinder solution.

Figure 6 reveals some interesting structure. The network divides into three major areas: the head, the extremities, and the body. The node *nose* provides a high-degree node for the head, whereas the node *head* provides a high-degree node for the network as a whole. The nodes *hand* and *leg* provide high-degree nodes for the extremities. The recall does not traverse the body in an orderly fashion based on location; rather, analogy seems to be the key. The pairs *fingers* and *toes, elbow* and *knee, arm* and *leg*, and *hand* and *foot* are all connected. The MDS solution on which the Pathfinder solution is superimposed does not provide this information as clearly.

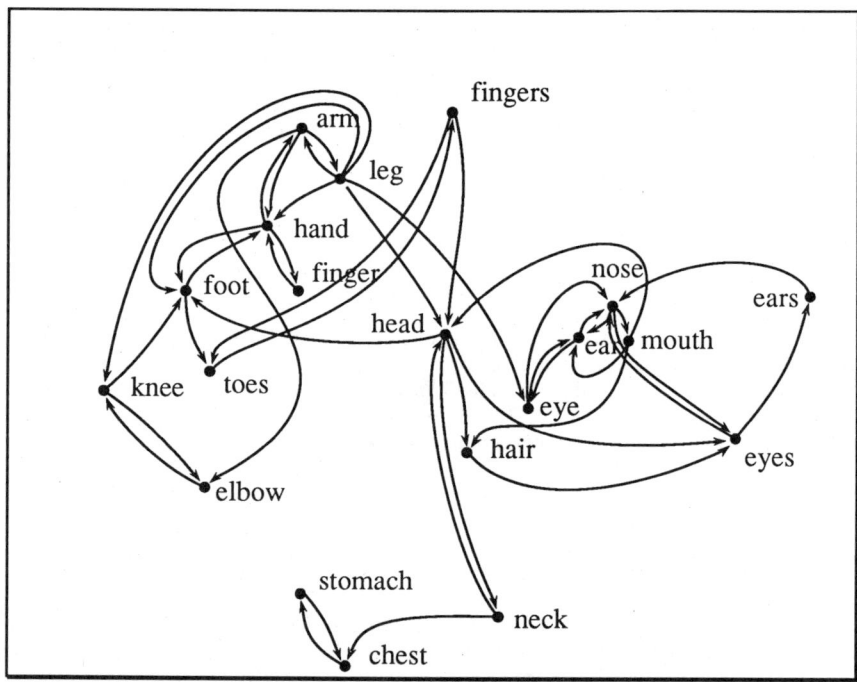

Figure 6. A minimally connected Pathfinder solution for the semantic category *parts of the body* superimposed on the MDS solution of Figure 2.

9 The Case of Rhyme

Figure 7 provides the solution to the question of the structure of the rhyme categories, something more traditional scaling techniques failed to do. Starting at the node *bare*, the network follows an alphabetical path. This can be seen more clearly in Figure 8, which is a replotting of Figure 7. This time, however, the solution and the figure include only those nodes that consist of a single letter followed by the *air* sound. A "U" was chosen instead of a straight line, partly because upon reaching the end of the alphabet the subjects at times return to the beginning for a second try and partly to make the figure clearer. Most of the cycles and jumps over nodes are caused by homonyms.

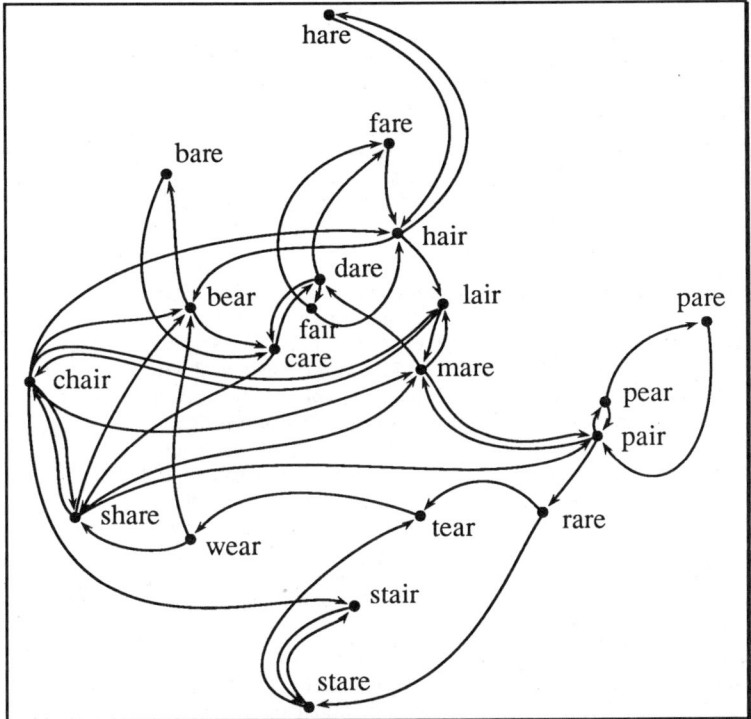

Figure 7. A minimally connected Pathfinder solution for the rhyme category *air* superimposed on the MDS solution of Figure 3.

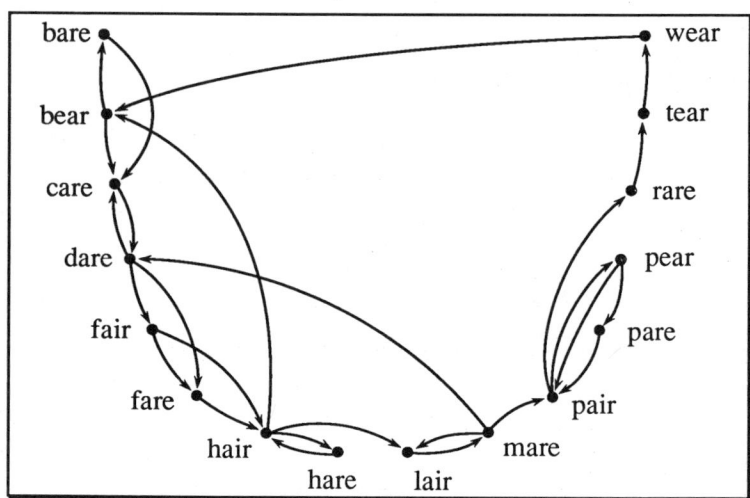

Figure 8. The Pathfinder solution of Figure 7 plotted to show the alphabetical organization of the words formed by a single letter followed by the *air* rhyme ending.

Figures 9, 10, and 11 present the nodes that consist of a single letter followed by the rhyme sound for the remaining three rhyme categories. First, in the figures for all four rhyme categories there is a clear alphabetical organization. Second, there are often words near the end of the alphabet linking back to the first alphabetical word of the category (e.g., *wear* to *bear*, *rear* to *bear*, *Ted* to *bed*, and *see* to *bee*). This probably results from recalls in which the alphabetical search started from the last of a series of easy-to-access words rather than from the beginning of the alphabet. Third, there are several frequently recalled words that serve as high-degree nodes in addition to their role in alphabetical search. The observation that some words are easier to recall than their alphabetical neighbors argues against a strict alphabetical search; words recalled more frequently than their alphabetical neighbors must have links that are not alphabetical. In particular, the two words in each of the four rhyme categories that were most frequently recalled each have five or more links with at least one of those links to a word not near it alphabetically. For Figures 8, 9, 10, and 11 these high-degree node words are *hair* and *dare*, *fear* and *near*, *bed* and *said*, and *see* and *me*. Fourth, there are other assorted associations that are consistent among subjects but that do not result from alphabetical search. The third and fourth types of links indicate that forms of search or association other than alphabetical search are also functioning.

Thus, Nelson's (1981) hypothesis based on cuing data is supported. Undergraduates seem to have no interpretable fixed structure for rhyme domains, but rather produce rhyme categories when needed using an algorithm based on the alphabet with additional rules to look for homonyms and to take advantage of any semantic structure that exists. Such constructions of semantic categories have been noted by Barsalou (1983, 1987) for ad hoc categories, such as *things to take on a camping trip*. Figures 8, 9, 10, and 11 reveal that such constructive techniques are used even for the major rhyme categories.

9 The Case of Rhyme 131

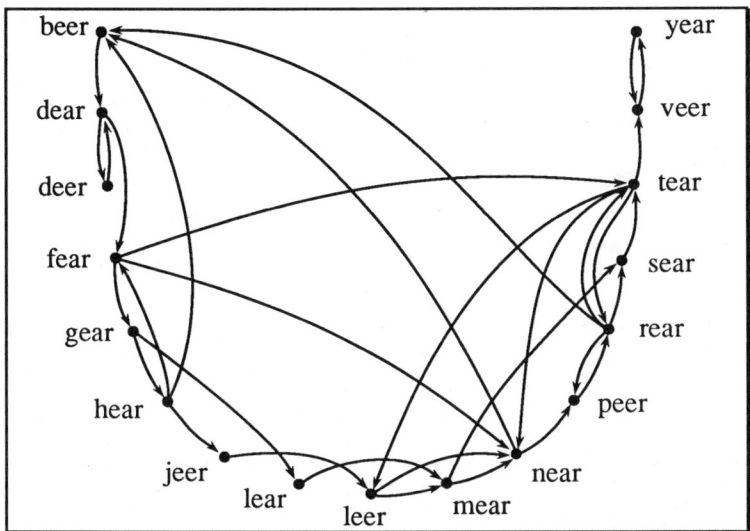

Figure 9. The Pathfinder solution for the rhyme category *ear* plotted to show the alphabetical organization of the words formed by a single letter followed by the *ear* rhyme ending.

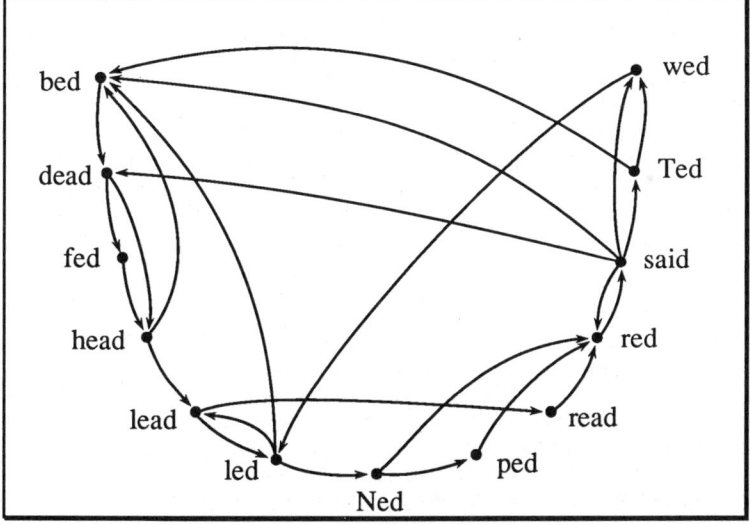

Figure 10. The Pathfinder solution for the rhyme category *ed* plotted to show the alphabetical organization of the words formed by a single letter followed by the *ed* rhyme ending.

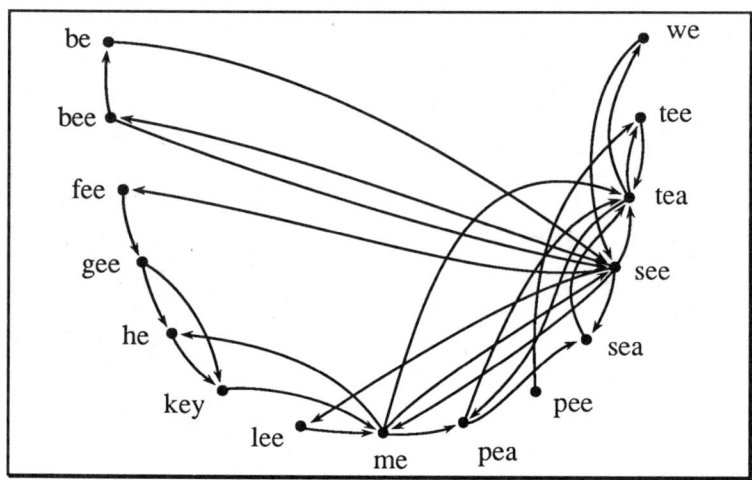

Figure 11. The Pathfinder solution for the rhyme category *ee* plotted to show the alphabetical organization of the words formed by a single letter followed by the *ee* rhyme ending.

Directed Graphs as Traces Through Memory

The search for the structure of rhyme categories demonstrates that Pathfinder offers many advantages for representing memory structures, as well as for uncovering retrieval strategies. The first major advantage of Pathfinder is that it allows associations among concepts to be asymmetric (i.e., the links among nodes to be directional). In a spatial analogy, such as MDS, the distance from point *a* to point *b* has to be the same as from point *b* to point *a*. In general, there is considerable evidence that this is not always the case for associations among concepts (Anisfeld & Knapp, 1968; Deese, 1965; Ekstrand, 1966; Rubin, 1980, 1983; Rubin & Friendly, 1986; Thorndike, 1932). In particular, for the rhyme data analyzed in this chapter, it is clear that a word is much more likely to lead to the word that follows it alphabetically than to the word that precedes it. The directional arrows in Figures 6, 7, 8, 9, 10, and 11 reveal regularities that the nondirectional distances in Figures 2 and 3 do not.

The second major advantage of Pathfinder is that it concentrates on local structure and does not try to maximize the fit for all possible links among all possible nodes. Semantic memory has a local as opposed to a global structure. Cluster size is very small (Gruenewald & Lockhead, 1980). There is great regularity in which item follows immediately after which in recall, but much less in what item follows five items later (Rubin & Olson, 1980). Few items will be recalled between *cat* and *dog*; the number of items recalled between *cat* and *cow*, however, is not as clear. That is why the next-to measure was used instead of a measure based on the number of intervening items (e.g. Friendly, 1977). Pathfinder makes use of this local structure by extracting the shortest links in paths and ignoring the longer, high-distance, low-similarity links. In contrast, MDS weights the discrepancy between the data and the fit equally for high and low distances (or similarities).

9 The Case of Rhyme

The emphasis on local as opposed to global structure may not be an advantage in all applications, but it is here because it mimics the structure in the data.

The third major advantage of Pathfinder is that its solutions are a natural analogy for search through memory. Figures 6, 7, 8, 9, 10, and 11 look like the directional trace of a path through memory. In fact, they are a kind of average of the order of output of all subjects. MDS solutions can be viewed as maps of the relationships among items; Pathfinder networks can be viewed as paths traversed through memory. Instead of showing us what could be seen as a static picture of a memory representation, Pathfinder shows us what could be seen as a record of the search process, that is, the record that memory traces.

Chapter 10

Proximities, Networks, and Schemata[*]
Roger W. Schvaneveldt

Over the past several years, my colleagues and I have been developing and evaluating empirical methods for eliciting and representing human knowledge. The network scaling technique known as Pathfinder (Dearholt, Schvaneveldt, & Durso, 1985; McDonald & Schvaneveldt, 1988; Schvaneveldt, Dearholt, & Durso, 1988; Schvaneveldt & Durso, 1981; Schvaneveldt, Durso, & Dearholt, 1985, 1987, 1989; Schvaneveldt, Durso, Goldsmith, Breen, Cooke, Tucker, & DeMaio, 1985) has resulted from these efforts. The essential idea behind Pathfinder is that proximities between entities should be represented as links in a Pathfinder network if the resulting links form the minimum weight paths in the complete (or nearly complete) network with proximity estimates as link weights.

Cognitive Structures and Network Structures

Recently, some of our efforts have been directed at identifying substructures, such as categories, schemata, and procedures in collections of concepts. Various standard tools, such as cluster analysis and sequence analysis, may be of value in this effort. In conjunction with our work on Pathfinder, we are also investigating various formal properties of graphs and networks, such as cliques, dominating node sets, and blocks, as methods of identifying substructures in networks (cf. Esposito, Chapter 6, this volume).

Another promising line of investigation stems from the extensive work in artificial intelligence and cognitive psychology on spreading activation in networks (Anderson, 1983; Collins & Loftus, 1975; Collins & Quillian, 1969; Meyer & Schvaneveldt, 1976; Quillian, 1967; Schvaneveldt & Meyer, 1973). Spreading activation reveals substructures in the activity levels of nodes in the network as a result of the spread of activation from selected nodes.

Networks and Schemata

Recent work presented under the banners of connectionism or parallel distributed processing (PDP) is based on the use of activation in networks to accomplish cognitive computation. The studies reported in the present chapter were originally inspired by the chapter on schemata in Volume 2 (Chapter 14) of the PDP books (Rumelhart, Smolensky, McClelland, & Hinton, 1986).

---

[*] Various versions of this chapter were presented at the Rocky Mountain Psychological Association meetings in Albuquerque, New Mexico in April, 1987, at a colloquium at the University of Oklahoma in October, 1987, at the 28th annual meetings of the Psychonomic Society in Seattle, Washington in November, 1987, and at the workshop on High-Level Connectionism in Las Cruces, New Mexico in April, 1988. Russell Branaghan and Steven Graves assisted in collecting data. Jim McDonald, Derek Partridge, and Jordan Pollack provided valuable criticism.

Rumelhart et al. presented an example of network representations of schemata using knowledge about different types of rooms. They selected 40 descriptors of rooms including such things as *ceiling, walls*, as well as things found in rooms, such as *oven, sofa, television* (31 of the descriptors are listed in Table 1). They constructed a network in which the 40 descriptors were represented as nodes and the weights on the links between nodes reflected the degree to which the two descriptors connected by a link tend to occur together. The actual values for the link weights were determined by asking people to imagine rooms of specific types (i.e., kitchen, living room, office, bedroom, and bathroom). For each room imagined, the subjects were asked to indicate whether each of the 40 descriptors applied to that room. By analyzing the co-occurrence of the descriptors in the imagined rooms, weights on the links were computed such that descriptors that tend to occur together are connected by positively weighted links, descriptors that tend to *not* occur together have negatively weighted links, and intermediate degrees of co-occurrence lead to intermediate weights with appropriate signs.

Dynamic instantiations of schemata are realized by a procedure in which particular nodes are activated continuously or "clamped on" (as if an external input is continuously signaling the presence of the item designated by the node), and the activation is spread throughout the network until a stable pattern of activation across all of the nodes is attained. The counterbalancing effects of excitatory (positive) links and inhibitory (negative) links is critical in producing stable activation states. In a stable state, some nodes will be active and some inactive.[1] The active nodes are taken to be the "completion" of the schema given the information that the items corresponding to the initially clamped nodes are present. For example, if the node corresponding to bed were clamped on, one might expect the nodes corresponding to other items found in the bedroom to become activated through the spread of activation. Table 1 shows the final states reached by the Rumelhart et al. network after clamping *sofa, bed, oven,* or *bathtub*.[2] The second line of the table shows which nodes were clamped on for the activation runs. The descriptor items (nodes) are listed in the far left column, and the dots in the columns indicate the nodes that are activated when the indicated node is clamped on. Inspection of the table shows that the network seems to lead to reasonable completions of schemata given central members of particular schemata as the starting point for activation. For example, clamping *bathtub* leads to a stable state in which *cupboard, toilet, sink, scale,* and *bathtub* are active. Note also that the individual items may belong to any number of schemata. For example, of the four cases presented in the table, *fireplace* is active in one, *carpet* is active in two, *clock* is active in three, and *ceiling* (not shown) is active in all four.

While these results are quite interesting and suggest that networks may provide a medium for representing schemata, there is a sense in which obtaining these results is not surprising given the method for establishing the network in the first place. People were told to imagine rooms in generating their judgments, and the method showed that their judgments revealed rooms. If one were looking for a way to identify schemata in a domain

---

[1] With the activation schemes used, nodes tend to be driven to an extreme state, either maximally active or minimally active. In practice, thresholds near zero and one (e.g., .001 and .999) are used to establish stability.

[2] Rumelhart et al. actually clamped on *ceiling* as well as the listed items. Since ceilings are found in all rooms, it doesn't contribute to differentiating rooms. Our own explorations suggest that clamping an item common to all of the schemata in a set doesn't make any difference. Representing still other schemata in the same set of units, however, may benefit from the presence of elements common to a subset of schemata. In general, we found that adding the additional 9 nodes that Rumelhart et al. used did not change the behavior of the derived networks with regard to the effects reported here.

that was not already familiar, it could prove to be difficult to properly instruct people to make the required judgments.

Table 1. Rumelhart et al. (1986) results.

Node	Clamped			
	sofa	bed	oven	bathtub
telephone	•	-	•	-
books	•	•	-	-
sofa	•	-	-	-
drapes	•	•	•	-
cupboard	-	-	•	•
toilet	-	-	-	•
bed	-	•	-	-
desk-chair	-	-	-	-
easy-chair	•	-	-	-
stove	-	-	•	-
sink	-	-	•	•
scale	-	-	-	•
typewriter	-	-	-	-
clock	•	•	•	-
coffee-cup	-	-	•	-
coffeepot	-	-	•	-
dresser	-	•	-	-
oven	-	-	•	-
bookshelf	•	•	-	-
picture	•	•	-	-
ashtray	-	-	-	-
refrigerator	-	-	•	-
television	•	•	-	-
computer	-	-	-	-
desk	-	-	-	-
carpet	•	•	-	-
floor-lamp	•	-	-	-
fireplace	•	-	-	-
toaster	-	-	•	-
bathtub	-	-	-	•
hanger	-	•	-	-

• = Active   - = Inactive

The critical aspect of the data for constructing a network appears to be information about co-occurrence of the items in schemata. Perhaps this co-occurrence information could be obtained by simply asking for direct judgments about co-occurrence. Such instructions would not presuppose particular schemata (such as asking people to imagine an instance of a particular schema, e.g., a kitchen), but the co-occurrence of items in people's experience may be sufficient to serve as the basis of schemata representations. Thus, one of the questions for the present study is: Can direct judgments of co-occurrence provide the basis for network representations of schemata?

Another issue of present interest derives from our recent work on using sparse networks to represent associational knowledge structures. The Pathfinder network generation algorithm takes proximity data as input and produces a network with a subset of the possible links. Frequency of co-occurrence can provide the proximity data required by the Pathfinder algorithm. In previous work with Pathfinder, the sparsest networks[3] contain about n links on n nodes in contrast to the $n(n-1)/2$ links in a completely connected undirected network. For the 31 items we included in the present investigation, this means that a Pathfinder network would consist of about 31 links in contrast to the 465 links in the completely connected network. This decrease in density of networks has implications both for the interpretability of the network structures and for the complexity of algorithms required to compute with the networks. Sparser networks are easier to interpret, and they lead to significant savings in the amount of memory they occupy. Provided that they include essential connections, sparse networks may also lead to more efficient computation. Thus sparse networks have both theoretical and practical value. So another question for the present study is: Will the sparse Pathfinder networks provide a basis for representing schemata, or are the complete networks of the connectionist variety essential?

In addition, this study also investigates the variation in network performance with different activation schemes. Since a variety of schemes have been used in various connectionist studies, it seems worthwhile to compare and contrast the results obtained with different schemes. Finally, we compare the results obtained with activation paradigms to some clustering methods including the familiar hierarchical cluster analysis as well as some methods for extracting local information from Pathfinder networks.

Method

Data Collection

A subset of the room descriptors used by Rumelhart et al. was used in the present study. The 31 descriptors selected are shown in the left-hand column of Table 1.

Two different datasets were collected. One set came from using the same method reported by Rumelhart et al. to allow a comparison with their results. For this method, each of 24 people was asked to imagine five specific instances of each of five room types (i.e., living room, kitchen, bedroom, bathroom, and home office). For each specific room, they were asked to indicate whether each of the 31 descriptors was present in that room. These are the *rooms* data.

A second dataset came from asking each of 24 people to judge the frequency of co-occurrence of the items in the 465 pairs of the 31 descriptors on a scale of 1 to 9. The scale represented a continuum from *never occur together* to *always occur together*. These are the *co-occurrence* data.

---

[3] The r and q parameters associated with generating Pathfinder networks lead to systematic variations in the density (number of links) of the network. The sparsest networks are obtained with $r = \infty$ and $q = n-1$, where n is the number of nodes in the network.

10 Proximities, Networks, and Schemata

Network Generation

Each of the two datasets was used to construct two different networks, a completely connected (connectionist) network and a sparse (Pathfinder) network. We consider each of the four cases.

Rooms Data - Connectionist Network. The rooms data were used to create a connectionist network following the procedures described by Rumelhart et al. (1986, p. 23). The method for establishing weights on the links between nodes essentially produces large positive weights for pairs of nodes that frequently occur together, large negative weights for nodes that never or infrequently occur together, and intermediate weights for the intermediate cases.

Rooms Data - Pathfinder Network. The Pathfinder network was generated from the rooms data by running the Pathfinder algorithm on the weights used in the connectionist network. Since the activation paradigms to be used require both positive and negative weights, the Pathfinder algorithm was run twice over the data. One run generated the positively weighted links, and by subtracting the weights from the maximum weight, the second run of Pathfinder generated the negatively weighted links. The Pathfinder network had 30 positive links and 31 negative links.

Co-occurrence Data - Connectionist Network. The co-occurrence data do not directly provide information about the probability of co-occurrence; they give relative information about co-occurrence. The weights for the connectionist network were obtained by averaging the ratings for each pair of items across the ratings given by the 24 subjects. The average ratings across all pairs were then converted to z scores with a mean of zero and a standard deviation of 1. The sign of the weights were such that positive weights corresponded to above average ratings of co-occurrence frequency and vice versa for the negative weights.

Co-occurrence Data - Pathfinder Network. Just as for the rooms data, the Pathfinder networks were generated from two applications of the Pathfinder algorithm to the average ratings of co-occurrence. Positively weighted links were determined from one run, and negatively weighted links from another run. The resulting network had 30 positive links and 30 negative links.

Network Activation

Various activation methods were used to investigate the nature of stable states in the four networks. With all methods, one node was clamped on throughout the activation procedure, and activation was passed through the networks until a stable state of node activation was reached. The active nodes in the stable state were taken as instantiations of the schema most associated with the clamped node. Different runs involved clamping different nodes. In this report we will focus on the results obtained from clamping either *sofa, bed, bathtub,* or *refrigerator*, which were taken as closely associated with living room, bedroom, bathroom, and kitchen, respectively. Thus, if the activation methods succeed in instantiating schemata, we would expect the items commonly found in each particular type of room to be activated when a typical node for that room is activated.

Psychometric Analyses

In addition to the activation analyses, we also examined the results of using various psychometric methods to analyze the data. Each of the datasets (rooms and co-occurrence) was represented as a proximity matrix which was analyzed by Kruskal's (1964) nonmetric multidimensional scaling (MDS) method, by Johnson's (1967) hierarchical cluster analysis (HCA), as well as by the Pathfinder network scaling method discussed earlier. The purpose of these analyses was to determine the extent to which the static results of psychometric methods reflected underlying schemata. We also compared the psychometric methods with one another and with the network activation methods.

Results and Discussion

A Pathfinder network for the rooms data is shown in Figure 1, and Figure 2 shows a Pathfinder network for the co-occurrence data. These two networks only depict the positively valued links. Figure 3 shows the negative links for the rooms data, and Figure 4 shows the negative links for the co-occurrence data.

The positive networks (Figures 1 and 2) are not particularly surprising. Items that co-occur most frequently are linked. The result is that items found in the same rooms tend to occur in sets of interlinked items. Of course, some method for determining where the clusters stop is required to isolate different rooms from one another because the whole set of terms is connected. If we consider the link weights, the single-link method of hierarchical clustering is embedded in the Pathfinder networks.[4] By successively adding links in order of the magnitudes of their weights and identifying the connected components of the resulting network, the clusters obtained in a hierarchical cluster analysis will correspond to the connected components of the network.

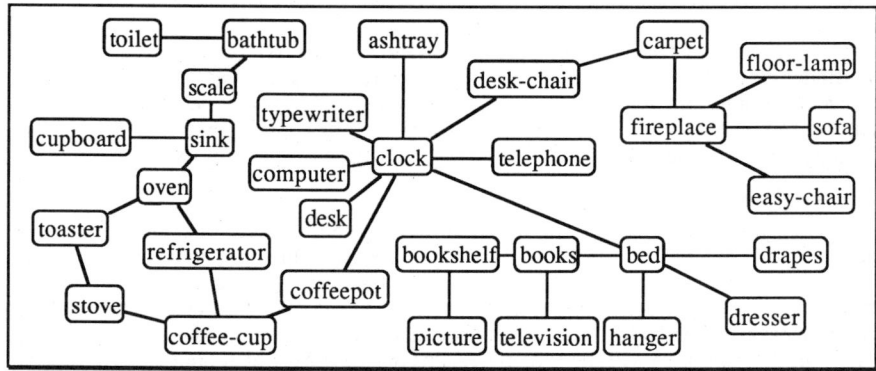

Figure 1. "Positive" Pathfinder network from Rooms data.

---

[4]This relation between Pathfinder networks and the single-link hierarchical clustering solution holds for networks generated with $r = \infty$ and $q = n-1$.

10 Proximities, Networks, and Schemata

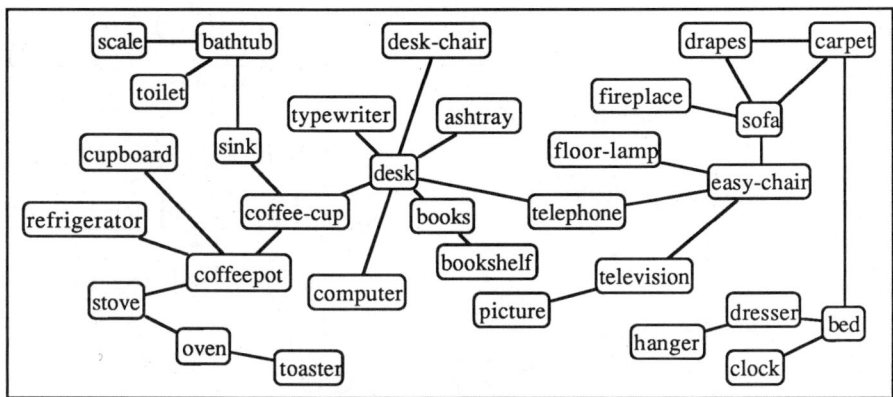

Figure 2. "Positive" Pathfinder network from Co-occurrence data.

The "negative" networks in Figures 3 and 4 show connections between items that co-occur least frequently. These negative links may help to identify items that do not belong to clusters of the positively linked items. The negative links are also necessary to implement the usual activation schemes used in connectionist networks. Without negative connections, positive activation simply spreads throughout the network until everything is maximally active. Negative links serve to dampen the positive activation and produce stable activation patterns across the nodes.

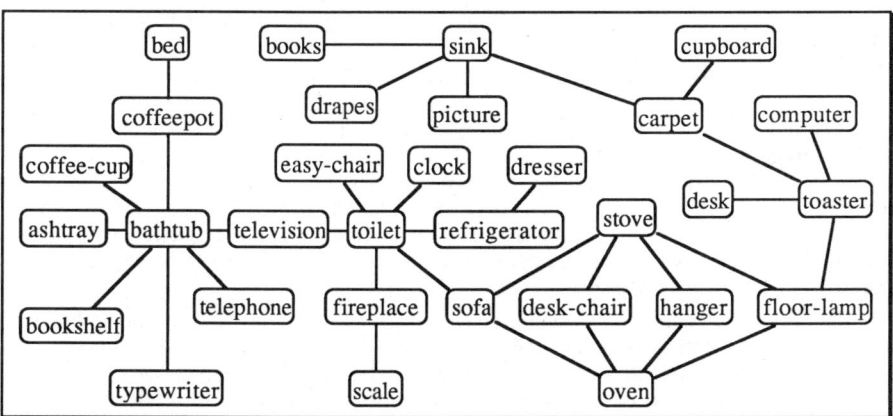

Figure 3. "Negative" Pathfinder network from Rooms data.

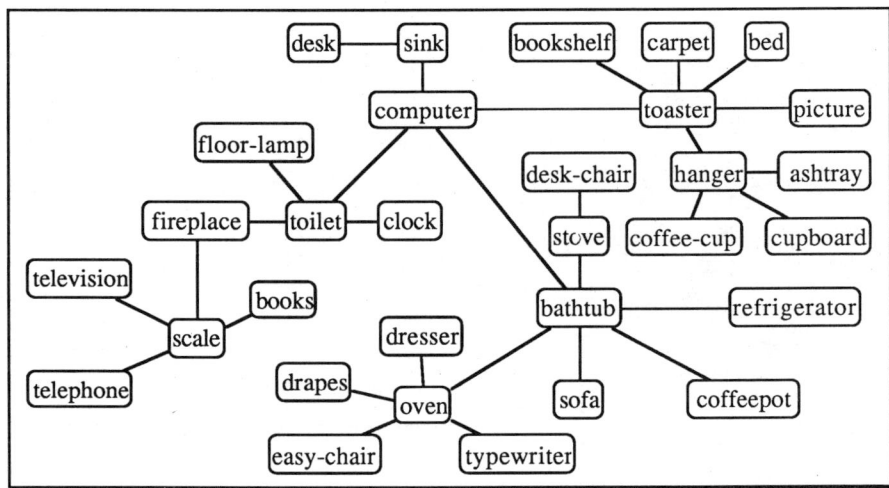

Figure 4. "Negative" Pathfinder network from Co-occurrence data.

Activation Experiments

Several different activation schemes were tried with the various networks we constructed. All of the methods yielded very similar results. A summary of these experiments is shown in Table 2. This table is similar to Table 1 except that results from four different networks and two datasets are shown. The second line shows which nodes were clamped on for the activation runs. The next lines show which dataset was used to construct the networks, and the next lines show whether the network constructed was a complete connectionist (CN) network or a sparse Pathfinder (PF) network. The CN network derived from the rooms data was based on the same method as that used by Rumelhart et al. (1986).

There are several aspects of these results that bear comment. First note that the activated items (large dots) tend to occur in pairs. This means that the complete networks and the Pathfinder networks are producing similar results. The one exception is that the Pathfinder network derived from the rooms data leads to activation of kitchen items when *bathtub* is clamped on. In this case, the positive connections between bathroom and kitchen descriptors, through *sink* (see Figure 1), are strong enough to overcome the negative connections between *toilet* and *refrigerator* and between *bathtub* and *coffee*-related items. Apparently a fine balance of positive and negative connections is required to achieve the desired selectivity in the activation process.

This lack of selectivity is seen in the activation patterns resulting from clamping *bed* and *sofa* for all of the networks. In general, the networks lead to activation of descriptors from living room, bedroom, and office whenever any of the descriptors typical of any of these rooms is clamped on. It is somewhat curious that these results do not replicate those of Rumelhart et al. (1986), but their footnote 7 on page 22 states that "Some slight modifications in the database were made in order to emphasize certain points in our example." Of course, some slight modifications in the networks used in the present examples would yield different results, but the interest here was in whether the activation schemes would reveal interesting patterns in the data. The answer to that question is negative.

A major lesson to be learned in this exercise is that activation schemes require a fine balance of positive and negative weights to yield interesting stable states. Collecting data in the two ways examined here does not necessarily result in such balance. A very careful selection of descriptors may help. What is needed are descriptors that are uniquely associated with particular rooms and negatively associated with other rooms. Such items are found in the present set for kitchens and bathrooms but not for living rooms, offices, and bedrooms. Of course a learning scheme could be used to adjust the weights to produce the desired results, but, again, that was not the goal of the present investigation. Perhaps some other ways of analyzing the datasets will be revealing.

Table 2. Results of Several Simulations

	Clamped															
	sofa				*bed*				*refrigerator*				*bathtub*			
	Rooms Data		*Co-oc Data*		*Rooms Data*		*Co-oc Data*		*Rooms Data*		*Co-oc Data*		*Rooms Data*		*Co-oc Data*	
Node	CN net	PF net	CN net	PF net	CN net	PF net	CN net	PF net	CN net	PF net	CN net	PF net	CN net	PF net	CN net	PF net
telephone	•	•	•	•	•	•	•	•	•	-	•	-	-	-	-	-
books	•	•	•	•	•	•	•	•	-	-	-	•	-	-	-	-
sofa	•	•	•	•	•	•	•	•	-	-	-	-	-	-	-	-
drapes	•	•	•	•	•	•	•	•	-	-	-	•	-	-	-	-
cupboard	-	-	-	-	-	-	-	-	•	•	•	•	•	•	-	-
toilet	-	-	-	-	-	-	-	-	-	-	-	-	•	•	•	•
bed	•	•	•	•	•	•	•	•	-	-	-	-	-	-	-	-
desk-chair	•	•	•	•	•	•	•	•	-	-	-	-	-	-	-	-
easy-chair	•	•	•	•	•	•	•	•	-	-	-	-	-	-	-	-
stove	-	-	-	-	-	-	-	-	•	•	•	•	-	-	-	-
sink	-	-	-	-	-	-	-	-	•	•	•	•	•	•	•	•
scale	-	-	-	-	-	-	-	-	•	-	-	-	•	•	•	•
typewriter	•	•	•	•	•	•	•	•	•	•	•	-	-	-	-	-
clock	•	•	•	•	•	•	•	•	•	-	-	-	-	-	-	-
coffee-cup	-	•	•	•	•	•	•	•	•	•	•	•	-	-	-	•
coffeepot	-	•	-	-	-	-	-	-	•	•	•	•	-	-	-	-
dresser	•	•	•	•	•	•	•	•	-	-	-	-	-	-	-	-
oven	-	-	-	-	-	-	-	-	•	•	•	•	-	-	-	-
bookshelf	•	•	•	•	•	•	•	•	-	-	-	-	-	-	-	-
picture	•	•	•	•	•	•	•	•	-	-	-	-	-	-	-	-
ashtray	•	•	•	•	•	•	•	•	•	-	-	-	-	-	-	-
refrigerator	-	-	-	-	-	-	-	-	•	•	•	•	-	-	-	-
television	•	•	•	•	•	•	•	•	-	-	-	-	-	-	-	-
computer	•	•	•	•	•	•	•	•	-	-	-	-	-	-	-	-
desk	•	•	•	•	•	•	•	•	-	-	-	•	-	-	-	-
carpet	•	•	•	•	•	•	•	•	-	-	-	-	-	-	-	-
floor-lamp	•	•	•	•	•	•	•	•	-	-	-	-	-	-	-	-
fireplace	•	•	•	•	•	•	•	•	-	-	-	-	-	-	-	-
toaster	-	-	-	-	-	-	-	-	•	•	•	•	-	•	-	-
bathtub	-	-	-	-	-	-	-	-	-	-	-	-	•	•	•	•
hanger	•	•	-	-	•	-	-	-	-	-	-	-	-	-	-	-

• = Active   - = Inactive

Multidimensional Scaling

Two-dimensional multidimensional scaling (MDS) solutions for the rooms data and the co-occurrence data are shown in Figures 5 and 6, respectively. While the fit of these two-dimensional solutions is marginal (stresses are 0.16 and 0.17 for the two solutions), the solutions are quite revealing. In particular, the solutions show a clear separation of bathroom and kitchen as well as a lack of differentiation of bedroom, office, and living room. This result is very similar to that obtained using activation techniques.

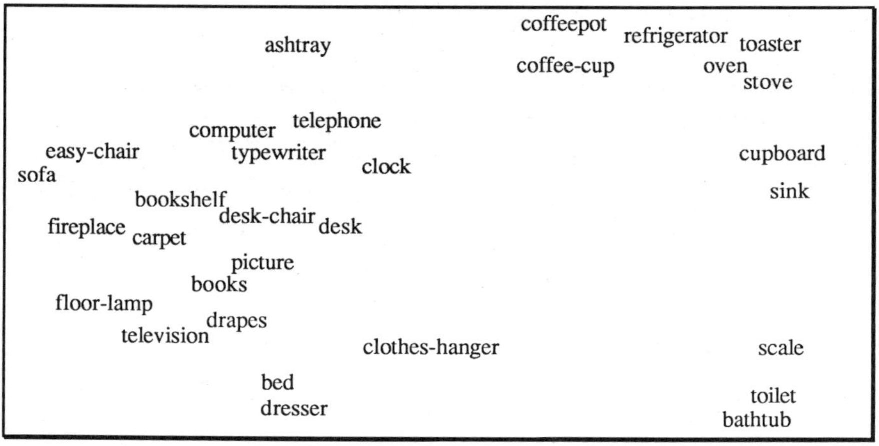

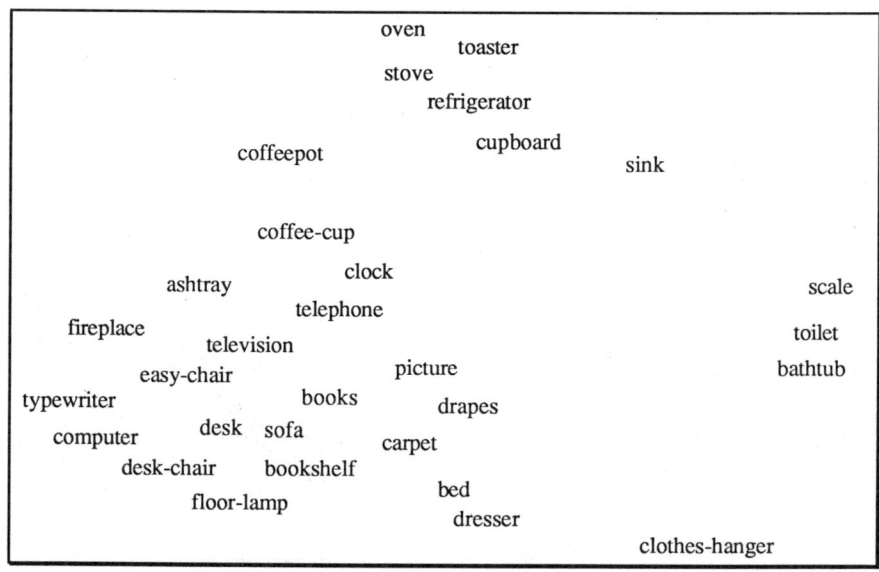

Figure 5. Two-dimensional MDS from Rooms data—Stress = 0.16.

Figure 6. Two-dimensional MDS from Co-occurrence data —Stress = 0.17.

10 Proximities, Networks, and Schemata

Cluster Analyses

Hierarchical cluster analysis is another common method for identifying groups of items on the basis of proximity data. Complete link hierarchical clustering solutions for the rooms data and the co-occurrence data are shown in Figures 7 and 8, respectively. Clusters are shown by the joining of lines emanating from the room descriptors. The figures may be interpreted as showing every descriptor in a separate cluster (at the left-hand side) that become a series of hierarchically organized clusters as one moves to the right in the figures. The figures do not show the final few steps which join the indicated clusters into successively larger clusters until all items are in a single cluster. At the point where the final clustering is shown, we see that there is a reasonable delineation of all five rooms.

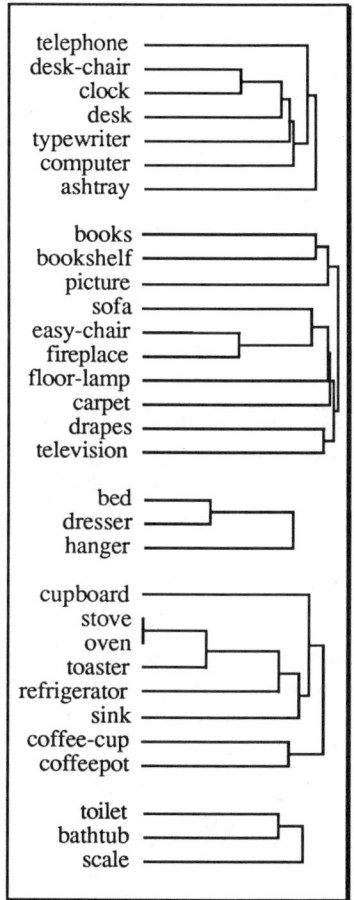

Figure 7. Complete-link cluster analysis from Rooms data.

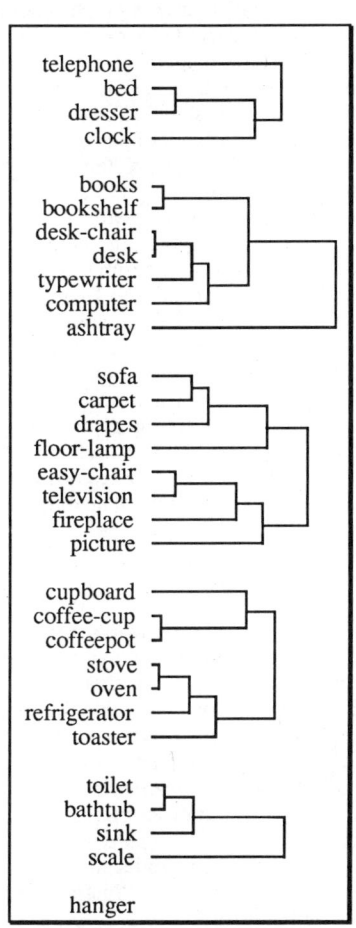

Figure 8. Complete-link cluster analysis from Co-occurrence data.

The cluster analysis is able to identify a separation of rooms where the activation methods and MDS could not. Of course, interpretation of the cluster solutions is required to decide that the clustering should be stopped at the point shown in the figures. These solutions are also not capable of revealing that some items (e.g., *carpet, clock, sink*) are commonly found in more than one room since the structure is strictly hierarchical. Nevertheless, the clustering solutions do capture some of the "room" structure in both datasets.

Connectivity in Pathfinder Networks

Still another approach to identifying substructures in the datasets is to examine patterns of connectivity in the networks generated by Pathfinder. Esposito (Chapter 6, this volume) has investigated various graph-theoretic structures in an attempt to identify "meaningful" clusters in Pathfinder networks. For illustration, a simpler approach is used here. Consider the nodes that can be reached within a certain number of links of a given node. This criterion would define clusters of terms that are likely to be "good" clusters if the number of links is small. Table 3 shows the result of using this criterion, collecting all nodes within one link of a particular node with the additional wrinkle that nodes that are only linked to a node defined by the first step are also included in the cluster. An "orphan" node can only join the node to which it is connected, so including the orphan node whenever the connected node is included seems reasonable.

With one exception, the items included in the groups starting with the nodes clamped on in the activation experiments constitute reasonable sets of items to accompany the starting node. The exception is found in the network derived from the Rooms data when *bed* is the starting node. Examination of this network (Figure 1) shows that the problem lies in the role *clock* plays in the network. Since several orphans are connected to *clock*, they will all be included whenever *clock* is included.

These groups of items are not necessarily hierarchical in that the same descriptor can occur in several different groups (notice the descriptor, *carpet*, in the network from the co-occurrence data). The general tendency, with this method of defining clusters, is to include too few items in the clusters. Still, unique clusters for every starting node can be generated, and including more than one starting node would generate still other unique clusters.

Conclusions

The Pathfinder networks appear to capture much of the information contained in the complete connectionist networks in as much as similar states are reached in the two types of networks as a result of activation. With the co-occurrence data, there were no substantial differences between the stable states reached with the complete networks and the Pathfinder networks. These similarities suggest that Pathfinder analysis could be a useful tool in simplifying the complex networks that are often found in connectionist models. For example, such simplifications may help to interpret the results of learning in such networks.

The particular descriptors chosen are critical to the patterns reached as stable states in activated networks. In particular, sufficient negative connections must be present to "damp out" the activation produced by positive connections. The undifferentiated living room, bedroom, and office are illustrative of this lack of sufficient dampening in the experiments reported here.

Table 3. Clusters from connectivity in Pathfinder networks.

Node:	sofa Rooms Net	sofa Co-oc Net	bed Rooms Net	bed Co-oc Net	refrigerator Rooms Net	refrigerator Co-oc Net	bathtub Rooms Net	bathtub Co-oc Net
telephone	-	-	•	-	-	-	-	-
books	-	-	•	-	-	-	-	-
sofa	•	•	-	-	-	-	-	-
drapes	-	•	•	-	-	-	-	-
cupboard	-	-	-	-	-	•	-	-
toilet	-	-	-	-	-	-	•	•
bed	-	-	•	•	-	-	-	-
desk-chair	-	-	-	-	-	-	-	-
easy-chair	•	•	-	-	-	-	-	-
stove	-	-	-	-	•	•	-	-
sink	-	-	-	-	-	-	-	•
scale	-	-	-	-	-	-	•	•
typewriter	-	-	•	-	-	-	-	-
clock	-	-	•	•	-	-	-	-
coffee-cup	-	-	-	-	•	-	-	-
coffeepot	-	-	-	-	-	-	-	-
dresser	-	-	•	•	-	-	-	-
oven	-	-	-	-	•	-	-	-
bookshelf	-	-	-	-	-	-	-	-
picture	-	-	-	-	-	-	-	-
ash tray	-	-	•	-	-	-	-	-
refrigerator	-	-	-	-	•	•	-	-
television	-	-	•	-	-	-	-	-
computer	-	-	•	-	-	-	-	-
desk	-	-	•	-	-	-	-	-
carpet	-	•	-	•	-	-	-	-
floor-lamp	•	•	-	-	-	-	-	-
fireplace	•	•	-	-	-	-	-	-
toaster	-	-	-	-	-	-	-	-
bathtub	-	-	-	-	-	-	•	•
hanger	-	-	•	•	-	-	-	-

(•) = Within one link plus orphans

The particular type of proximity data used does not appear to be critical. Similar results are obtained from simple co-occurrence judgments compared with the rooms-based judgments. In the two cases where there were differences, the co-occurrence data tended to produce somewhat better results (see Tables 2 and 3).

Cluster analysis and connectivity in Pathfinder networks reveal clusters of items belonging to particular rooms, but activation methods and MDS do not. Apparently, different information in the data is used by the activation methods and MDS in contrast with the hierarchical clustering method and the Pathfinder networks. Elsewhere (Schvaneveldt et al., 1989), we have discussed such differences between MDS and Pathfinder.

In a nutshell, MDS equally weights all of the data in determining a solution. There is a sense in which all of the weights are also used by iterations of activation spreading in a network, and to the extent that Pathfinder networks preserve the significant paths in more complex networks, activation should act similarly in Pathfinder networks. In contrast, the *link structure* of a Pathfinder network is more determined by the locations of the relatively small data values (high co-occurrence for the positive links or low occurrence for the negative links). Thus Pathfinder connectivity and hierarchical clustering tend to identify items that most frequently occur together. The constraints in MDS and activation methods are much more complex, and as we have seen, not always appropriate.

Chapter 11

Using Pathfinder to Extract Semantic Information from Text

James E. McDonald, Tony A. Plate, and Roger W. Schvaneveldt

How might the semantic information contained in existing textual material, such as dictionaries, be made more "tractable" for use by machines? In this chapter we discuss a technique for extracting relatedness information from text along with some potential uses for such information. The method is based on frequency of co-occurrence, that is, the number of times pairs of words occur together in selected units of text. We hypothesize that frequency of co-occurrence provides a reasonable basis for estimating relatedness among the objects, events, situations, states, and so forth, that "words" refer to, particularly for certain applications, and that it offers several advantages over the use of human judges to establish such estimates. We have been encouraged by the results of applying this method to the analysis of the *Longman Dictionary of Contemporary English* (LDOCE), and this effort has allowed us to identify several questions in need of additional research as well.

Although the focus of the present work is LDOCE, we believe that our method may be useful in other applications as well.[1] In the following sections we discuss (1) our objectives for this work, (2) co-occurrence data and relatedness functions derived from them, (3) the use of Pathfinder networks to simplify the representation of co-occurrence data, (4) some experiments aimed at validating the use of frequency of co-occurrence data to estimate the strength of relationships among words, and (5) two approaches to lexical sense selection, one which uses co-occurrence data directly and another which uses Pathfinder networks derived from co-occurrence data.

Our objectives for this work are primarily practical, although there are certainly theoretical implications for computational linguistics and cognitive psychology as well. Our immediate goal is to produce a modified version of LDOCE, one in which the words in the definitions have appropriate sense tags. The method will need to be refined, however, since our long-term objective is to make word-sense distinctions for unconstrained natural-language (general lexical sense selection). This is not meant to suggest that the statistical approach we propose is capable of solving all the problems of natural language understanding—or even lexical sense selection. Rather, we are investigating how a subsystem using co-occurrence can be built so that it will be useful as part of a natural-language understanding system.

---

[1] We have, for example, applied the same method to the UNIX online documentation system (the *man* system) as part of an effort to build a hypertext browser for UNIX.

The Nature of Co-occurrence Data

We assume that words co-occur in sentences because they are related to the idea being expressed by the sentence (the meaning of the sentence). The words are therefore semantically related in the context of the sentence. Our technique for lexical sense selection only relies on this claim being true in the aggregate—which is fortunate because individual sentences can certainly be constructed which violate this assumption. Strictly speaking, we contend that pairs of words co-occur frequently in *collections* of sentences because they are semantically related. In testing this claim we will discuss some experiments in which estimates of relatedness derived using our method are compared with human judgments of relatedness. The sense-selection experiments themselves also serve to evaluate this contention.

Co-occurrence in Text

In text, co-occurrence data record the frequencies of co-occurrence of pairs of words within some textual unit. The textual unit can be a phrase, a sentence, a paragraph, or any other identifiable unit. The co-occurrence data used in the work reported in this chapter were collected using the sense-definition as the textual unit.[2]

> The frequency of co-occurrence of two words is defined as the number of textual units in which some form of both of those words occurs. For two words, x and y, their co-occurrence frequency is designated f_{xy}.

The independent frequencies of occurrence of words in a textual unit are also important and are used in conjunction with frequencies of co-occurrence to calculate the values of various relatedness functions.

> The independent frequency of occurrence of a word is defined as the number of textual units in which some form of that word occurs. The frequency of occurrence of word x is designated f_x.

Extracting Co-occurrence Data from LDOCE

Some of the methods used to convey word meaning in dictionary definitions include giving examples of use in context (*illustrative definitions*), saying something directly about meaning (*descriptive definitions*), or simply providing other words with the same meaning (*synonym definitions*). In fact, all three methods are used in various combinations in different dictionaries. LDOCE, for example, relies primarily on descriptive and illustrative techniques, although cross-references (synonyms and related words) are often provided. However, the objective that "the definitions are always written using simpler terms than the words they describe," expressed in the introduction to LDOCE, limits the extent to which synonymy can be used in defining words (e.g., *copy* is used in the definition of one of the senses of *reproduce,* but not visa versa).

Although co-occurrence statistics could be collected on free text, and might prove useful, a dictionary such as LDOCE offers certain advantages. First, unique to LDOCE, the vocabulary used in defining word senses is limited (the LDOCE-controlled vocabulary contains approximately 2,187 words). A limited vocabulary makes the task of collecting

---

[2] A sense-definition is considered to be the entire definition of a sense of a word, including any examples. Sense definitions can be easily identified in LDOCE.

11 Semantic Information from Text

and storing frequencies of co-occurrence simpler, allowing the use of conventional computing techniques without requiring excessive resources.[3]

The second advantage of dictionaries for our purposes is that sense definitions provide small, coherent units of text centered around single ideas, natural units of co-occurrence. Other sources of text, such as thesauri and encyclopaedis, also provide coherent units of text, but free text is not as constrained. The point is that free text would not be as efficient at providing relatedness estimates among the words in the controlled vocabulary. This conjecture has been verified to some extent by comparing the co-occurrence information taken from the definitions in LDOCE with that obtained from the example sentences. In this comparison, estimates of relatedness derived from definitions correlated more highly with human subjects' ratings of relatedness than did estimates based on example sentences. This does not rule out the usefulness of free text as a source of relatedness information, but does support the claim that the definitions in LDOCE provide semantically focused units for co-occurrence analysis.

The last advantage of dictionaries over free text is that dictionaries provide definitions for all word senses. It is therefore possible to build a representation for every sense contained in the dictionary, not just those that occur frequently. Although it may be necessary to augment these representations through the use of other dictionaries or perhaps even free text, at least minimal representations can be obtained for even infrequent sense distinctions. Unfortunately, LDOCE doesn't contain very many words (about 1.2 million in total). Therefore, for many words in the controlled vocabulary the frequency of occurrence is quite low (280 occur 30 or fewer times). More importantly, not all of the senses of words are used in defining other words, limiting the accuracy of relatedness estimates that can be obtained from co-occurrence data.

Related Work

The automatic extraction of useful information from text has been a long-standing goal of several investigators. These efforts have ranged from work in artificial intelligence aimed at text understanding to systems for automatically indexing collections of documents. In the field of information retrieval, much of the work has focused on methods for determining the content of documents by examining the individual words contained in titles, abstracts, or entire documents (Belkin & Croft, 1987; Brooks, 1987; Dumais, 1988; Salton, 1986). Most of the attempts to develop automatic retrieval systems have relied, in various ways, on comparing the words that occur in queries and the words that occur in documents.

Many of the failings of indexing using individual words can be attributed to the complexities of the relations between words and meanings. Homography, polysemy, and synonomy all contribute to uncertainty about the similarities and differences in the meaning of words occurring in different documents. In various contexts, the same words can mean quite different things, and different words can have quite similar meanings. If the appropriate senses of homographs and polysemous words in a text could be determined, more precise comparisons could be made between the meanings occurring in different texts. One of the goals of our own work has been to develop methods of identifying the sense of a word in text. Others have had similar goals.

Lesk (1986) attempted to identify word senses by comparing the words in a sentence containing a target word with the words occurring in the sense definitions of the target word. He reports some success with the method (50-70% correct sense selections), and he

---

[3]For example, although the array of frequencies of co-occurrence for LDOCE requires approximately 4.7 megabytes of storage, it takes less than an hour to build on a minicomputer.

speculates that dictionaries with longer definitions would improve the performance. As we show later, better performance in identifying word senses is achieved by expanding the set of words to be considered by adding other words associated with the words in the context sentence. To the extent that expanding word sets produces the same result as longer definitions, this finding offers some support for Lesk's conjecture. Lesk (1987) uses a similar system based on overlap of words in dictionary definitions to identify words related to words in an information query. The use of this technique to identify related or associated words is very similar to our efforts to find relatedness estimates from the co-occurrence of words in LDOCE. Our work uses co-occurrences of words throughout the dictionary, in contrast to Lesk's method which looks for co-occurrences within the definitions of the words in question. We also go beyond the relatedness estimates derived from co-occurrence to create a network of interrelated words. The network can then be used to find sets of words related to any particular word.

The use of latent semantic indexing for information retrieval also attempts to identify a structure that represents the pairwise relatedness between words (Dumais, 1988; Dumais, Furnas, Landauer, Deerwester, & Harshman, 1988). This technique attempts to extract orthogonal factors from a large matrix of associations between terms and text objects. These factors can be used to define a semantic space in which both queries and text objects can be placed. Retrieval of text objects can then be based on their proximity to a query in this space. Fowler and Dearholt (Chapter 12, this volume) report a related approach to text and query representation using Pathfinder networks. Both of these efforts are attempt to go beyond matching words to uncovering some of the semantic structure that accompanies the use of words in documents.

Cohen and Kjeldsen (1987) developed a system to match grant proposals with funding agencies based on constrained spreading activation in semantic networks. A network of research topics is connected both to funding agencies and to a research proposal. Activation over these pathways identifies funding agencies who are likely to be interested in the proposal.

One approach to expanding word sets is to identify synonyms via a thesaurus (Furnas, Landauer, Gomez, & Dumais, 1987; Sparck Jones, 1986). Sparck Jones has conducted extensive studies of thesaurus generation.

Sparck Jones (1986) was concerned with finding thesaurus-like groups of words that could be used to resolve lexical ambiguity. She considered her work to be primarily practical, aimed ultimately at machine translation and discourse analysis. As a consequence, one of her objectives was the construction of a machine-tractable dictionary consisting of synonym definitions and organized according to "similarity."

At the core of the Sparck Jones approach is a method for precisely defining synonymy. This method assumes a model of language in which there are multiple signs for word senses and, possibly, multiple "similar" word senses for the same sign.[4] Unfortunately, it isn't clear from the description provided by Sparck Jones how much of the task of constructing the dictionary is to be accomplished by humans and how much by machine.[5] It

---

[4] Although clearly aware of homography and its impact, Sparck Jones seems to assume that it can be safely ignored. The model of language that she claims represents natural language, her Model 4, assumes that if two or more word uses have the same sign they are similar in meaning.

[5] Sparck Jones seems to believe that many of the definitions in dictionaries consist of synonyms or sets of synonyms, particularly in the *Oxford English Dictionary* (OED). However, in a footnote (Section 6.2) she states that "the thesaurus information in the OED is so inaccessible and so unsystematic that it can be said that it is hardly there at all" (1986, p. 259-260). Whichever is the case for the OED, definition by synonymy is not typically used in LDOCE.

11 Semantic Information from Text   153

seems likely, however, that considerable human participation would be required since only limited, hand-constructed examples were provided by Sparck Jones.[6]

In her thesis, Sparck Jones argued that it was essential to use a well-defined *linguistic* relationship as the basis for the analysis of natural language by machine. She selected synonymy for theoretical reasons, rejecting several alternatives, such as association and collocation, and went about operationally defining synonymy. She defined a "row" as a set of words (word signs) which can replace each other in a particular sentence without changing its meaning. Thus, the word signs in a row are, very precisely, synonyms and represent a particular word use (sense). Rows, each of which embodies the word use of the word-signs it contains, are *similar* to the extent that they have word signs in common. Although based on synonymy, the nature of the weaker "similarity" relation established by the Sparck Jones method is less clear, but it doesn't appear to be synonymy.

Sparck Jones believed that synonymy is fundamental to language. She rejected methods for establishing semantic relations based on co-occurrence on the grounds that the results may be due to extralinguistic factors. We agree that words can co-occur for many reasons, some pragmatic. However, we believe that this fact doesn't rule out the use of associations derived from co-occurrence for lexical ambiguity resolution. It is commonly recognized that natural-language processing systems will need to incorporate world knowledge in order to be successful. Such relationships can be established automatically using our method. Although the objective of being able to precisely label the relationships among word uses appears desirable, it may not be necessary. We define relatedness in terms of the operations used to establish it, much as synonymy is operationally defined in Sparck Jones' work. In a practical endeavor such as ours, the important questions seem to be "Can it be done?" and "Does it work?"

In spite of the apparent difference, however, there is a rather direct relationship between these two techniques. In her dissertation, Sparck Jones (1986) acknowledged that words can be classified on the basis of common rows, rather than classifying rows on the basis of common words. This is essentially the technique we employ, except that we use sense definitions rather than rows. Sparck Jones went on to speculate that this alternative approach might be more appropriate for machine translation, but that there may be some benefit in using both classification schemes.

Sparck Jones contended that because we don't know how to do machine translation it is difficult or impossible to evaluate the utility of semantic classification. We have taken a different position. We believe that the utility of the semantic classification approach can be assessed prior to completely solving the machine translation problem. Although there might be immediate gains to be had from incorporating syntactic analysis, the semantic and syntactic component are conceptually independent, and they can be evaluated independently.

Explorations of the Co-occurrence Data

We begin this section by emphasizing the magnitude of the task at hand. The co-occurrence data obtained for the LDOCE-controlled vocabulary consists of nearly two-and-a-half

---

[6]The decision that two words are substitutable in a particular sentence seems to require a very complex linguistic judgment. Such a decision could only be made by a system capable of understanding natural language. One objective of Sparck Jones was to accomplish limited natural-language processing, such as machine translation and discourse analysis. It is impractical to require a machine capable of understanding natural language in order to build one.

million frequencies of co-occurrence (the triangle of a 2187 × 2187 matrix). It is, therefore, impossible to examine these data in raw form. The information must be reduced in some way to be useful. The objective, of course, is to reduce the data while preserving interesting or useful information. In what follows we will discuss two techniques for getting at the important information in the LDOCE co-occurrence data. The first of these uses thresholds and relatedness functions derived from the LDOCE co-occurrence matrix directly. The second technique uses the Pathfinder network scaling algorithm (Schvaneveldt, Dearholt, & Durso, 1988; Schvaneveldt, Durso, & Dearholt, 1989) to determine relatedness and to identify important relationships.

Relatedness Functions

If related words are more likely to occur together than unrelated words, then statistics of co-occurrence provide some indication of relatedness. The problem is to find some function of co-occurrence that reflects the relatedness of pairs of words, that is, a function that will yield the relative strength of relatedness for various pairs of words. We will refer to such functions as *relatedness functions*.

A good relatedness function should have a number of characteristics. Ideally, it should be equally sensitive across the range of independent frequencies. In other words, estimates of relatedness should be independent of the base frequencies of the words involved. This is a particularly difficult characteristic to obtain in practice, and many of the relatedness functions discussed below produce more "accurate" estimates of relatedness when the words have approximately equal independent frequencies than when they differ greatly in frequency. The relatedness function should, of course, also produce valid results. In our evaluations, a good relatedness function will provide a measure that correlates with human judgments of relatedness and one that is successful in selecting appropriate senses of words in sentences.

We examined several relatedness functions with various characteristics (cf. Salton, 1968). Some of the relatedness functions we have considered are shown in Table 1, along with comments about their *bias, sensitivity*, and *symmetry*. Bias refers to the extent to which sensitivity varies with the independent frequency of words. Sensitivity refers to the extent to which a measure varies as dependencies in the occurrence of words in a pair vary from chance co-occurrence to maximum possible dependence. Since the maximum possible co-occurrence depends on base frequency, this definition of sensitivity makes sensitivity independent of frequency. Symmetric measures are the same for $f(x,y)$ and $f(y,x)$, whereas asymmetric measures may yield different estimates of relatedness for the two uses of the function.

As an example of the differences produced by applying the various relatedness functions, about 20 of the words most strongly related to *bank* for each of the relatedness functions are shown in Table 2.

Some of these functions include more closed-class words (especially determiners and very common prepositions) in the set of highly related words. Such words seem to provide very little semantic information. The dcp_{min} and *iou* functions yield the best sets of related words on intuitive grounds. Because the very common closed-class words do not provide much information about the meaning of other words, the most common of these were omitted from the sense selection experiments discussed below.[7]

---

[7]For the purposes of the experiments described in this section, the following words were omitted from the controlled vocabulary: *a, and, be, for, in, of, or, than, that, the, this, those, to, what, when, where, which, who, with*.

11 Semantic Information from Text

We examined the relatedness functions as well as raw frequency of co-occurrence (coc) as measures of relatedness by comparing them to human judgments of relatedness and by attempting to identify the senses of words in sentences using the relatedness functions.

Table 1. Relatedness functions.

Name	Value	Comments
$cp(x,y)$	$\dfrac{f_{xy}}{f_y}$ $(= Pr(x\mid y))$	Conditional *probability* of x given y. Asymmetric. Insensitive and heavily biased for all f_x and f_y, except low, equal values.
$dcp(x,y)$	$Pr(x\mid y) - Pr(x)$	(*deviation of cp*) Asymmetric. More sensitive than cp but still biased. An attempt to remove some of the bias of cp.
$dcp_{min}(x,y)$	$\min(dcp(x,y), dcp(y,x))$	Minimum of dcp in both directions. Symmetric. Sensitive if f_x and f_y are similar, but results in zero if they are considerably different.
$iou(x,y)$	$Pr(x \text{ and } y \mid x \text{ or } y)$	(*intersection over union*) Produced by dividing the number of units containing x and y by number of units containing at least one of them. More sensitive than dcp_{min} when f_x and f_y are different.
$dex(x,y)$	$\dfrac{f_{xy} - f_x \cdot f_y}{\min(f_x, f_y) - f_x \cdot f_y}$	(*dependency extraction*) Normalizes f_{xy} by mapping it to [0,1] according to its scaled position between its minimum and maximum possible values. Symmetric. Fully sensitive for all f_x and f_y.

Table 2. Words most strongly related to *bank* for each relatedness function.

cp	a account an and as be by for from have in money of on or river the to which
dcp	a account as at be by from have in keep money of on pay river rob the to water
dcp_{min}	account cheque criminal earn flood flow lake lend money pay prevent promise rate river rob rock safe sand sum thief
iou	account busy cheque criminal earn flood flow interest lake lend money overflow pay river rob safe sand thief wall
dex	a account be by cheque clerk dollar in messenger money of overflow participle pay river rob September the to

Human Judgments

Although there are many questions that might be asked regarding the application of the technique we propose, one of the first seems to be the extent to which the associations derived from text are like ratings of relatedness supplied by human judges. If it can be established that frequency of co-occurrence data, or some transformation of them, correspond to human judgments of relatedness, then we should be able to use this approach for applications that typically require such information. Furthermore, by comparing proximity estimates based on co-occurrence with those obtained from human judges, it should be possible to specify the ways in which these estimates differ and to compensate for such differences if necessary.

The general procedure used in each of the studies in this section was (1) to select a set of LDOCE-controlled vocabulary, (2) obtain judgments of relatedness from human subjects for the selected words, (3) compute estimates of relatedness from the LDOCE co-occurrence matrix, and (4) compare the obtained human judgments and LDOCE-based estimates. The basic measures of correspondence were correlations between relatedness estimates and human judgments.

The human-judgment data for the following comparisons were all obtained using the method of paired comparison. We correlated these data with raw co-occurrence counts (*coc*), average conditional probabilities (*cp*), deviations of conditional probabilities (*dcp*), minimum of *dcp* (dcp_{min}), intersection over union (*iou*), and dependency extraction (*dex*) derived from LDOCE.

The Natural Category Set. For our first comparison we selected the 25 natural category words (e.g., *animal, plant, dog, rose*) for which estimates of relatedness had been obtained from 24 introductory psychology students and 24 biology graduate students (cf., Schvaneveldt, Durso, & Dearholt, 1989). Sixteen of the words used in these rating studies were in the LDOCE-controlled vocabulary. This set of 16 words served as the basis for the following comparisons.

Table 3. Correlations of human judgments of relatedness and estimates of associations derived from LDOCE co-occurrence for 16 primitives.

	A	B	C	D	E	F	G	H
Psychology Students - A		.94	.50	.69	.68	.58	.60	.68
Biology Students - B	.94		.50	.68	.68	.58	.59	.68
LDOCE *coc* - C	.50	.50		.83	.82	.95	.95	.71
LDOCE *cp* - D	.69	.68	.83		1.00	.90	.93	.98
LDOCE *dcp* - E	.68	.68	.82	1.00		.89	.92	.98
LDOCE dcp_{min} - F	.58	.58	.95	.90	.89		.99	.79
LDOCE *iou* - G	.60	.59	.95	.93	.92	.99		.84
LDOCE *dex* - H	.68	.68	.71	.98	.98	.79	.84	

As can be seen from Table 3, the correlations of the LDOCE conditional probabilities with human judgments are quite high, but not as high as the correlation between the two groups of human judges. These results are promising in that we are able to account for a significant amount of the variability in human ratings (48%). The two relatedness

11 Semantic Information from Text

measures (*cp* and *dcp*) correlate about the same with human judgments, and for this set of words, the two measures correlate almost perfectly with one another.

The Bank Set. Unlike the previous comparison, the set of words related to *bank* was selected directly from LDOCE. The first step consisted of obtaining all of the words that were associated with *bank* above a relatedness threshold of .01. In turn, sets of words were obtained for each of these words, which resulted in a fairly large set of associated words. The 20 words with the highest number of co-references were selected for the rating experiment. Five Computing Research Laboratory researchers served as subjects in the rating study. The correlational analyses are shown in Tables 4 and 5.

Table 4. Correlations of human judgments of relatedness for the *bank* primitives.

	JM	TP	RS	CE	JB
JM		.80	.83	.70	.78
TP	.80		.78	.70	.75
RS	.83	.78		.73	.78
CE	.70	.70	.73		.82
JB	.78	.75	.78	.82	

Table 5. Correlations of relatedness estimates from human judges and LDOCE for the *bank* primitives.

	A	B	C	D	E	F	G
Mean Ratings - A		.48	.66	.65	.60	.64	.61
LDOCE *coc* - B	.48		.73	.71	.68	.76	.66
LDOCE *cp* - C	.66	.73		1.00	.78	.96	.98
LDOCE *dcp* - D	.65	.71	1.00		.79	.86	.98
LDOCE dcp_{min} - E	.60	.68	.78	.79		.98	.64
LDOCE *iou* - F	.64	.76	.86	.86	.98		.73
LDOCE *dex* - G	.61	.66	.98	.98	.64	.73	

The results from these comparisons are again promising. Intersubject correlations ranged from .70 to .83. Correlations between ratings and LDOCE measures are also quite high (except for the direct co-occurrence measure). This set of words produces differences among the different measures derived from LDOCE co-occurrences. Conditional probability is better than frequency of co-occurrence, and nothing is gained from the more complex measures.

Lexical Ambiguity Resolution

Our general method for identifying word senses is relatively straightforward. There are, however, numerous refinements to be considered, some of which will be discussed later in this section. The method requires determining related-word sets for individual words. The related-word sets are defined in different ways, but each set is essentially selected on the basis of some index of relatedness. All words in the controlled vocabulary satisfying some minimum threshold of relatedness with a particular word are included in a related-word set for that word. These related-word sets may also be expanded by including words that are related to the words in the related-word set, and so on. The basic methodology consists of the following steps:

1) A context sentence is selected and a test word is selected from it for sense tagging (lexical sense selection).

2) A context set c is formed for the context sentence by combining related-word sets for each word in the context sentence, *except for the test word itself*.

3) Separate definition sets $(d_1...d_n$ where n is the number of sense definitions for the test word) are formed for each of the sense definitions for the test word using the words in the sense definition, *except for the test word itself*.

4) The proportion of words in each of the definition sets contained in c is computed.

5) The sense definition with the largest overlap with words in c is judged the winner and the test word is tagged with that sense.

This process is, of course, not as simple as it sounds. We have already discussed the problem of measuring relatedness. There are also various ways to expand the context and definition sets. Identifying the winner can also be done in a variety of ways. These issues are discussed in more detail below.

The Direct Use of Co-occurrence Data

Using the overlap of context sets and definition sets to identify word senses is not simply a "keyword search" approach because the use of related-word sets makes the decision depend on more than the words that a sentence (the context) and the definitions have in common. Often sentences will not share any words with the appropriate definition. For example, the definition of sense 4.1 of *bank* is shown below, followed by an alphabetized list of the base forms of the controlled vocabulary words, excluding the list of ignored words.[8]

> *bank*[4.1]: A place in which money is kept and paid out on demand, and where related activities go on. (*activity demand go keep money on out pay place related*)

---

[8] We use the convention of numbering the Mth sense in the Nth entry (homograph) for a word as "sense N.M."

11 Semantic Information from Text

An example of the use of sense 4.1 of *bank* is:

Any of various kinds of bank accounts earning higher interest than a deposit account. (*account any earn high interest kind various*)

Notice that the phrasal context shown above and the sense-entry for *bank* 4.1 have no words in common in the controlled vocabulary. If we consider the words in parentheses to be sets of words, their intersection is empty. This is not at all unusual in LDOCE, given the small number of words used in sense definitions. As a consequence, however, the straightforward technique of looking for the sense-entry which has the maximum intersection with the context doesn't always work well. Our approach to this problem expands the contexts and/or sense-entries to included related words, thereby making the intersection technique more reliable.

In this first series of experiments, we represented expanded contexts and sense-definitions as vectors of weights over all of the LDOCE primitives. Each weight represents the "strength" of the association between a particular word and the context or sense-definition set to which it belongs. Expanded sets were created by adding to each set all of the LDOCE primitives that exceeded a relatedness threshold using one of the relatedness functions. For each expanded set, the weights were the number of words in that set to which each word in the set is related according to the relatedness threshold. These weights were intended to represent the centrality or importance of words in the context or sense-definitions in the sense that the more words in the set that are related to a particular word, the higher its weight. Of course, words that are not in the expanded set have weights of zero. Vectors of weights can be treated as sets by converting non-zero weights to one and zero weights to zero. This operation is expressed as $X > 0$ below, where X is a vector of weights.

Once the context and sense-definitions have been represented as vectors of weights, an estimate of their "similarity" is computed (i.e., the strength of the relationship between the two vectors). All of the functions used measure vector overlap in one way or another. Some of them consider weights, others are based only on set membership.

In the similarity functions, *SUM* sums all of the elements of a vector. The dot-product function "·" is the sum of the cross-products. The following similarity functions were used in various experiments.

The commonality or *COM* function treats context and sense-definition vectors as sets.

$$COM(V, W) = \frac{|(V>0) \cap (W>0)|}{|(V>0) \cup (W>0)|}$$

$HIT^{\rightarrow}(V,W)$ counts the "hits" of V in W (i.e., it sums the weights for the words in the intersection of V and W) and divides this value by the sum of the weights in W. The right-pointing arrow is used to indicate that this is an asymmetrical function [i.e., $HIT^{\rightarrow}(V,W)$ is not necessarily equal to $HIT^{\rightarrow}(W,V)$].

$$HIT^{\rightarrow}(V,W) = \frac{(V>0) \cdot W}{SUM(W)}$$

HIT^x takes the product of $HIT^\rightarrow$ to produce a symmetric result.

$$HIT^x(V,W) = HIT^\rightarrow(V,W)\, HIT^\rightarrow(W,V)$$

Finally, we have found it useful to compute the similarity between two vectors using the normalized dot-product (i.e., the cosine of the angle between the two vectors).

$$NDP(V,W) = \frac{(V \cdot W)}{\sqrt{V \cdot V + W \cdot W}}$$

Using the general lexical ambiguity resolution procedure already described, we attempted to select the correct sense of *bank* for the 197 sentences containing the word *bank* in LDOCE. The test sentences were first manually sense-tagged by the authors using the sense distinctions made in LDOCE for *bank*. This was not always a simple task because, in the judgment of the authors, some of the usages of *bank* cannot be classified using the sense distinctions for *bank* in LDOCE. More generally, there is some question as to whether or not all of the sense distinctions made in LDOCE are legitimate or, conversely, whether particular sense distinctions are missing. Nevertheless, the automatic method was judged correct if it chose the same sense as that selected by the authors beforehand.

The word *bank* was selected as a test case for a number of reasons. First, it has a "moderate" number of sense distinctions (13), at least as far as words in LDOCE go, yet the senses of *bank* can be easily divided into larger groups. The two main (homographic) sense "groups" contain *financial* senses and *earth* or *river* senses, respectively. These two groups account for 7 of the 13 senses, and, more importantly, nearly all of the usages of *bank* in LDOCE. Some of the finer sense distinctions within these two groups are semantic, whereas others are syntactic. For example, two of the three *financial* senses of *bank* are the transitive and intransitive verbal forms. Because our method does not directly consider syntactic information, we did not expect it to be able to correctly discriminate these uses. We also suspected that the method would have difficulty discriminating among the *earth* senses of *bank*, which differ by fine semantic distinctions. These considerations led us to construct groups of senses that contained gross rather than fine semantic distinctions. For this purpose, we assigned the 13 senses to 6 *sense-groups*, and the ability of the method to assign occurrences of *bank* to these larger groups was also assessed.[9]

Identifying the correct sense of *bank* proved to be a difficult task. In only 38 of 420 experiments was *bank* correctly sense-tagged 35% or more of the time. However, the probability of correctly tagging a particular usage of *bank* by chance is only 7.7%. As expected, selecting the correct sense-group was a far easier task. In 120 of the experiments, *bank* was assigned to the correct sense group 85% or more of the time (the probability is only 17% by chance). The experiments yielding the best performance are summarized in Table 6. The results of using the NDP similarity function using only the words in the context sentence and in the definitions are included for comparison.

---

[9] Various combinations of relatedness functions, vector similarity functions, and relatedness thresholds for choosing word sets were used, which resulted in a total of 350 experiments. The purpose of this thoroughness was to discover the most promising combinations of functions for future experiments. It was *not* an effort simply discover a combination that worked. In fact, most combinations performed reasonably well, compared to chance. If all the results of all the experiments were due to chance, the probability of all 350 experiments producing 30 or fewer correct sense-assignments is 0.96. In fact, only 145 of the 350 experiments produced 30 or fewer correct sense-assignments. Thus, successes cannot be attributed to simply capitalizing on chance.

11 Semantic Information from Text

Table 6. Rates of correct sense selection for the experiments with the best performance.

Relatedness Threshold	Relatedness Function	Vector Similarity Function	Assignment to Correct Sense	Assignment to Correct Group
none	none (*coc*)	*NDP*	23%	52%
0.1	*cp*	HIT^x	45%	79%
0.03	*dcp*	$HIT^{\rightarrow}(R^C, R^S)$	15%	97%

At its best, this method was able to correctly tag 45% of the test sentences. This is a reasonably good performance, given that correctly identifying the exact sense of *bank* in these sentences proved very difficult for the authors. Remember that this method completely ignores syntactic information, including morphology. It is unreasonable to expect any method that does not take syntax into account to reliably distinguish between words that have very similar nominal and verbal forms, such as *bank*. Furthermore, as mentioned above, *bank* has several senses that are very close in meaning.

The technique of expanding contexts and sense-entries to include related words (i.e., words judged to be related according to some relatedness function) proved beneficial. Without expansion, the correct sense assignment was made at best 23% of the time, whereas with expansion the highest rate of correct sense assignment was 45%. The example sentence shown above, which had no words in common with the appropriate sense definition, was generally sense-tagged correctly, demonstrating that the technique can work even under difficult circumstances.

Pathfinder Networks Derived from Co-occurrence Data

One of the problems with co-occurrence data is the sheer quantity of it. There are nearly 2.5 million frequencies of co-occurrence for the words in the LDOCE-controlled vocabulary. Such an enormous amount of data is difficult to use in raw form. However, any reduction must be accomplished without eliminating useful information.

Pathfinder analyses were performed on a relatedness matrix (using the IOU relatedness function) for the 2,177 controlled vocabulary that occur in less than 10% of the text units. The Pathfinder *r* parameter was always infinity in the analyses reported. Figure 1 shows a fragment of a Pathfinder network including the words within three links of the word *bank*.

The networks that resulted from the Pathfinder analyses were used to select the related sets of words. Related sets were formed by selecting words that were directly linked to a particular word in the network. The number of links connected to each word (the *degree* of the node) varies, depending on the extent to which other words consistently co-occur with it. This means that related-sets also vary in size.

Sense-definition sets were formed for the test word by combining the related-word sets for the words in each sense definition. The union of the related-word sets was used in the experiments described here, although weighting words differentially may be valuable for future investigations of this technique.

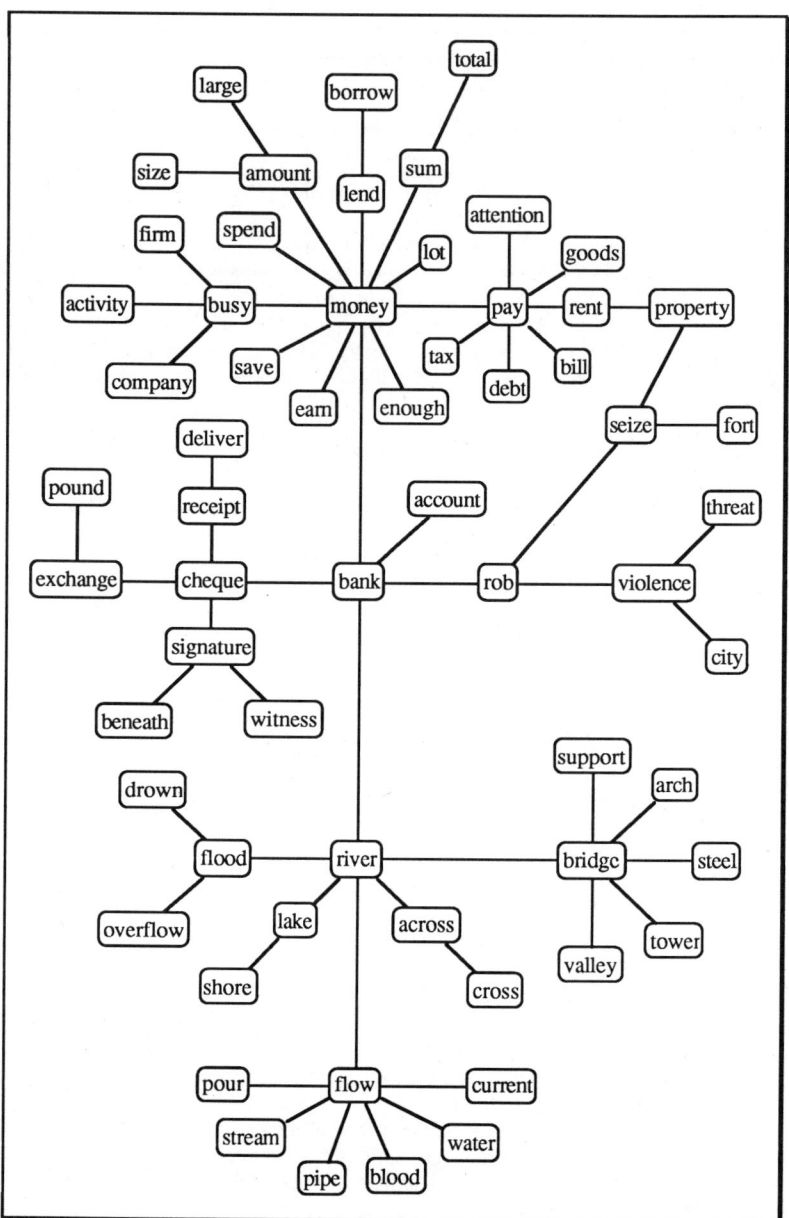

Figure 1. The fragment of the PFNET($r = \infty$, $q = 5$) showing the 61 words within 3 links of *bank*.

11 Semantic Information from Text

As with the sense-definition sets, context sets were formed by combining the related-word sets for all the words in the context sentence. However, our approach here has been to progressively expand the size of the context set by increasing the number of links, or network distance, used in determining relatedness. This is analogous to passing markers in the network starting with the words in the context sentence and continuing in steps passing markers over the links at each step. Finally, a measure of match was computed for each sense definition at each distance from the context set. Although several measures have been considered, the results of using the ratio of the number of words in both context and sense-definition sets to the number of words in either set are reported here.

We used three Pathfinder networks: PFNET(∞,2) with 16,955 links (Q2), PFNET(∞,5) with 3,136 links (Q5), and PFNET(∞,32) with 2,204 links (Q32). With each network, we attempted to identify the correct sense of the word *bank* in the 197 example sentences from LDOCE. In these tests, the sense-definition sets contained only the words in the definitions themselves (no related words, no weights). The context set was progressively expanded by adding the words directly connected to the words in the context set to obtain the Step 1 words, then the words connected to the Step 1 words were added in Step 2, and so forth. At each step, the COM evaluation function was used to compute a strength for each sense definition, and the sense definition with the greatest strength was taken to be the appropriate sense of the word *bank* for the particular context sentence.

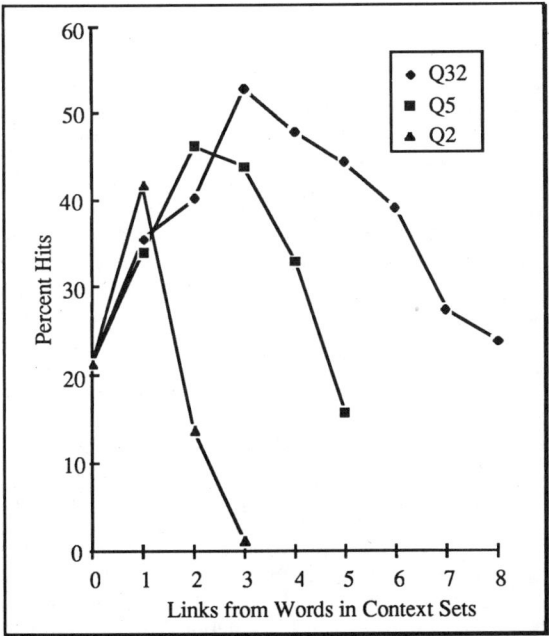

Figure 2. Percent hits using three Pathfinder networks to sense-tag 197 *bank* sentences.

The results of these sense-selection tests are shown in Figure 2. In terms of absolute performance, the network with the fewest links (PFNET(∞,32)) performed best, allowing *bank* to be correctly sense-tagged in 104 of the 197 example sentences (53%). Maximum performance occurred when the context set had been expanded to include items three links away (average context-set size = 102). Performance with the PFNET(∞,5) was next best (91 hits at Step 2; average context-set size = 81), and the PFNET(∞,2) was worst (82 hits at Step 1; average context-set size = 91). All of the networks improved on Step 0 performance which is similar to a keyword search using only the words in the context sentence.

The performance of the PFNET(∞,32) network is particularly surprising since it has the fewest links. Apparently, limiting the links to the relatively strongest relations available yields some advantage. It is possible that these links are more immune to the effects of incidental co-occurrences of words in LDOCE that have little to do with the inherent meanings of the words.

Although these results are promising, they may still be of limited practical value. However, the task of choosing the correct sense from a large set of highly similar senses (there are 13 senses for *bank*) may be too stringent a test. Therefore, we also examined performance with the PFNET(∞,32) when only the six sense groups for *bank* were considered. The hit-rate improved to 85% (167 out of 197), a far more usable result. At present, it appears that Pathfinder is capable of capturing the important relationships in the co-occurrence data without losing much of value, at least for our application.

Conclusions

The co-occurrences of words in the LDOCE-controlled vocabulary in the definitions in LDOCE appear to provide some useful information about the meanings of those words. Co-occurrence frequency correlates significantly with human judgments of relatedness, and the relatedness functions on co-occurrences yield even higher correlations. When the relatedness functions are used to derive Pathfinder networks on the LDOCE primitives, these networks serve to represent aspects of the intensional meaning of words. More specifically, one might say that the intensional meaning of a word is represented by the collection of words that are nearby in the network. The experiments on lexical sense selection suggest that the co-occurrence data and the networks derived from those data do capture some aspects of the meanings of words.

Lexical sense selection using co-occurrence data is promising but far from perfect. The Pathfinder networks dramatically reduce the amount of information that must be stored in order to do lexical sense selection. It would be useful to combine information from our method of sense selection with other methods for extracting information from text. Sampson (1986) presents a statistical technique for assigning part-of-speech labels, for example, which would be an excellent candidate for such combination.

At the beginning of this chapter we argued that LDOCE is a relatively good source of co-occurrence data, but that it isn't perfect. To reiterate, LDOCE contains short textual units, each of which is relatively focused. At the same time, the distribution of topics in LDOCE is relatively broad. Importantly, LDOCE is based on a controlled vocabulary, which makes co-occurrence data manageable. However, only a limited number of the senses of the words in the controlled vocabulary are used to define other words in the dictionary, and co-occurrence data cannot reflect relationships involving senses that aren't used. Another potential shortcoming is that LDOCE contains relatively few examples of "definition by synonymy" as compared to other dictionaries. One technique for improving our co-occurrence estimates would be to obtain additional co-occurrence data from other dictionaries and thesauri.

Chapter 12

Information Retrieval Using Pathfinder Networks*

Richard H. Fowler and Donald W. Dearholt

Collections of written documents are available electronically for a wide range of needs and interests. Commercial vendors now provide access to large collections of scientific, as well as popular documents, and the advent of low-cost optical storage technologies promises an even wider range of accessibility. Yet, widespread use of these collections will require solutions to long-standing problems of document indexing and query formulation that allow the user to express information needs in a style that makes full use of the electronic medium.

Pathtrieve is an information retrieval system that has served as a testbed for exploring the application of Pathfinder Networks (PFNETs) in solutions to these problems. The system interface is based on direct manipulation of graphic network representations of queries, indexing terms, and document texts. PFNETs are used to provide efficient, automatically derived graph representations for natural language text that retain some of the advantages of richer, but less tractable, knowledge representations. In addition, the different associational networks in the system provide the user multiple paths of access to information items.

Document Indexing

At present, document retrieval is the only practical means for accessing knowledge in large databases of natural language text. Information needs are met indirectly by providing documents or references to documents likely to contain relevant information. The basic retrieval strategy matches a query representation to representations of documents. In essence, this is a matching of conceptual representations, and the nature of these representations is central to the performance of a system. Most document retrieval system representations are based on the assignment of indexing terms to documents.

The goal of indexing is to provide a set of terms for each document that reflects the conceptual domain and can be used effectively in retrieving the document. Unfortunately, indexing documents with high reliability and validity is in practice not possible. By several measures, remarkably low consistency both between and within indexers has long been evident (Cooper, 1969). Furnas, Landauer, Gomez, and Dumais (1983) show that human object naming behavior is marked by a lack of agreement among individuals, even in restricted domains with persons of similar experiences. This lack of agreement has important implications for the design of information systems. Bates (1986) suggests that systems should provide a wide range of document descriptors that include associations beyond conventional thesaural relations to allow alternative paths of access to documents. One means

---

*This work was supported by Grant IST-850676 from the National Science Foundation.

for dealing with problems inherent in labeling is provided by associational structures that reflect not only semantic relations, but other types of relations as well. This is the approach taken in the Pathtrieve system.

User Knowledge

Information retrieval (IR) systems should be capable of assisting users with varying levels of expertise. For a subject area the user knows well, a principal function of the IR system is to facilitate the mapping of the user's knowledge structures and natural language vocabulary to the indexing terms used by the system for retrieval. However, when the user does not have well-developed knowledge structures in a subject domain, retrieval is problematic. This situation seems to present a paradox: The user's information need is to know the subject area, but in order to use the system to meet the information need, the subject area must be known.

One way to resolve this paradox is provided by iterative query techniques such as relevance feedback (Robertson & Sparck Jones, 1976). The user provides feedback to the system with respect to the relevance of documents retrieved by a particular query and the system automatically adjusts the query. Another approach is to develop an expert system to aid in formulating searches. The expert system automates the role of a human search intermediary who is familiar with the indexing terms and search capabilities of a particular IR system. By using information provided by the expert system about the knowledge structures and indexing of the database, the user can be guided to appropriate search strategies (Fidel, 1986).

Other approaches to assisting relatively naive users provide information for increasing the user's knowledge of indexing terms so that a more effective search is possible. The most widely used aid is a thesaurus of index terms that gives subject information by presenting the vocabulary of terms used in classifying documents in a domain. Most thesauri specify a limited number of relations (e.g., synonym, superordinate) among terms and group-related terms together to facilitate browsing and searches in a particular subject area. Conceptual analysis, in terms appropriate for searches using a particular system and database, can be facilitated by the thesaurus.

The Pathtrieve system provides a graphic associative thesaurus to assist the user in learning about terms and relationships of terms in a domain. The relations of the thesaurus are derived from statistical measures of term similarity and provide associations beyond those found in conventional thesauri. This network, based not only on linguistic relations, but also on other types of relations, provides a means for exploration of the term vocabulary to provide cues that are complementary to the user's cognitive structure. Information requests using terms from the thesaurus are formulated through exploration of associations. The associational structure also helps overcome some of the difficulties of indexing indeterminacy by providing multiple paths of access to terms used in query formulation. A broader, less structured approach to information description is possible, with the Pathtrieve system, that is perhaps better suited to the realities of document indexing than conventional thesaural relations alone.

Interface Styles

As online systems and optical storage have become more widely available, direct end-user searching has become the norm. However, elimination of the expert human intermediary requires an interface that facilitates conceptual interaction throughout the retrieval process. The Pathtrieve system addresses this concern by providing the user visual presentations of all conceptual structures used by the system. Work on the graphic presentation of problems suggests that a spatial, pictorial mode is useful in conveying certain types of knowledge (Billingsley, 1982; Bocker, Fischer, & Nieper, 1986), and graphic network displays are increasingly used to display knowledge structures (Fairchild, Poltrock, & Furnas, 1988; Mettrey, 1987). Additionally, viewing techniques for large data structures based on psychological considerations for display (Furnas, 1986) provide a promising approach for the very large databases often found in information systems. Though some IR systems have used graphic displays (Frei & Jauslin, 1983; Thiel & Hammwohner, 1987), this style of interface has received relatively little attention (Crouch, 1986). In particular, direct manipulation interfaces (Hutchins, Hollan, & Norman, 1986; Shneiderman, 1982) seem to provide an interaction style yet to be explored in IR systems, that is especially appropriate for the iterative, highly interactive processes of information seeking.

Pathfinder Networks in Information Retrieval

Graph-based representations have been a cornerstone of artificial intelligence (AI) research, but most cannot be used in large scale IR systems for two reasons. First, producing knowledge bases such as those used in AI applications is not done automatically. Virtually all AI knowledge bases are produced by extensive human encoding, that is, knowledge engineering. The size of the bodies of natural language text accessed by IR systems and the need to have documents made available in a timely fashion argue against deep manual encoding of documents. Second, the time complexity of procedures for manipulating these representations is not acceptable for the very large databases IR systems access. In IR systems even polynomial time procedures are to be avoided, and in most systems indexing is used extensively to allow efficient access. Though the development of tractable procedures for use with representations characteristic of AI research is a promising line of investigation, the direct transfer of procedural mechanisms from AI techniques to IR is not possible. Nonetheless, graph structure representations are increasingly found in IR systems (Belkin & Croft, 1987; Belkin, Oddy, & Brooks, 1982). Their use follows a general trend that systems designed to work on relatively small bodies of text incorporate more domain knowledge and apply AI knowledge representation techniques more directly than systems that are to be used with large bodies of text. Systems for very large databases rely at most on domain knowledge encoded as part of a term thesaurus and use shallow, statistical processing to automatically form representations of document texts.

Pathfinder provides an alternative procedure for automatically deriving network representations for IR systems. Statistically-based graph construction techniques used in IR and Pathfinder can both utilize proximity measures in deriving link structure, yet there are important differences in the resulting representations. One difference is that statistically-based graph representations used in IR have seldom been concerned with providing a psychologically salient representation, whereas the PFNET representation has been developed from the outset as a representational scheme for human conceptual structure. Only a few IR systems, such as Jones' (1986) Memory Extender system using spreading activation retrieval, have explicitly applied models of human cognition in the retrieval process. Another difference is that most statistically-based graph derivation techniques in IR systems are based on link membership determined by similarity thresholds between node pairs; links are included in a graph if the association between two nodes is above some criterion. In contrast, Pathfinder considers similarity over a wider range of nodes in deriving networks. The two approaches lead to quite different link structures.

The Pathtrieve System

The functions of the Pathtrieve system, shown in Figure 1, allow the user to access several PFNET structures during query formulation and document retrieval. The retrieval process is begun by deriving a PFNET from the natural language of a user's information request and serves as the initial query representation. This representation is displayed and can be graphically modified using terms from a second PFNET structure, the associational term thesaurus. A document collection can be automatically selected through matching of the query and the different database thesauri available to the system. One document retrieval mechanism matches the query representation to PFNETs derived from document abstracts. These document PFNETs are also available for use in graphic modification of the query. The final PFNET structure used in the system is based on the interrelationships of documents in a database. A second form of retrieval uses a selected document as an entry point into this structure and retrieves document clusters by traversing the network. The following sections describe derivation of natural language-based PFNETs, network displays and navigation tools, and document retrieval techniques.

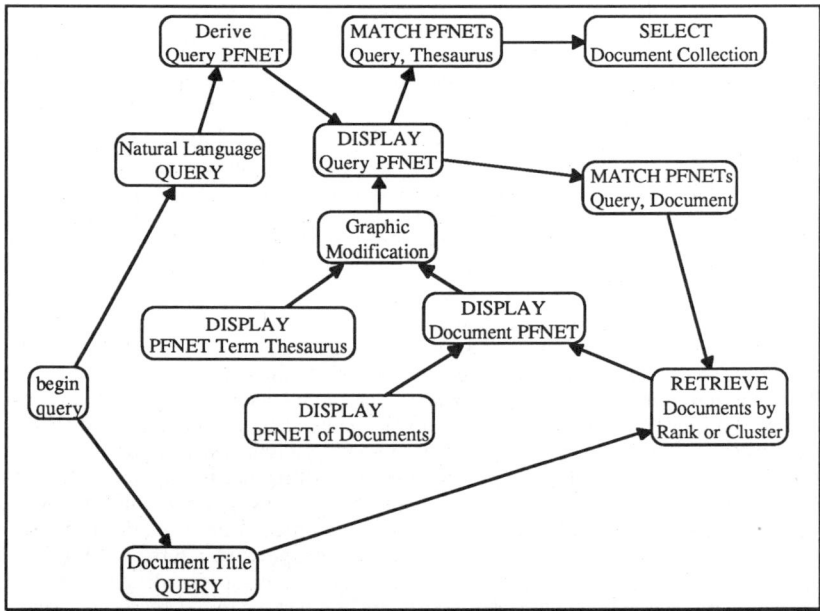

Figure 1. Overview of system functions showing PFNETs and graphic manipulations.

12 Information Retrieval

PFNETs from Natural Language

The term sets used in natural language-based PFNETs play a central role in the use and interpretation of the structures. The system allows access to multiple databases of document abstracts and for each database a unique set of terms is used. To construct a database's term set, a stop list is first used to eliminate function words from the texts of abstracts in a database. Stems are then derived from the remaining words and the most frequently occurring stems are used as the term set for that database. For some forms of retrieval this simple procedure suffers from the limitation that frequently occurring terms have relatively little value for discriminating among documents. However, with multiple document collections one of the functions of the term thesaurus is to provide a picture of all concepts in a document set. The most frequently occurring terms tend to be general terms that provide valuable information about the domain of the document collection and relations among concepts.

The statistical text analyses performed during PFNET derivations rely on recovering conceptual information from natural language by considering the form of text. This shallow analysis has advantages in that it can be performed sufficiently quickly for use in interactive query formulation, and term similarity measures reflect relatedness from a wide range of sources. The proximities used in constructing the different natural language PFNETs are derived from term co-occurrence in different textual units. The system uses PFNET representations of the user query, single document abstracts, and the term thesauri. Calculation of proximities for the term thesauri PFNETs uses all abstracts in a database and so provides similarity measures for all term pairs. In analyzing each body of text, stop list words are removed, the remaining words are stemmed, and stems are eliminated that are not in the database's term set. A weighted co-occurrence measure, similar to that used by Belkin and Kwasnik (1986), is computed using the remaining terms. For term pairs, similarity is increased by 5 for terms adjacent in a sentence, 3 for nonadjacent terms occurring in the same sentence, and 1 for terms occurring in the same document. PFNETs use $r = \infty$ and $q = n-1$, where n is the number of terms.

Query PFNET. A query is usually initiated by the user's entry of a natural language request for information. The request is transformed to a PFNET and the Pathfinder Graph (PFG) is displayed in the Query window, as shown in Figure 2. Two means of modifying the query can be used: further input of natural language text or graphic manipulation using other PFGs displayed by the system. When additional text is entered, it is appended to the existing text and reanalyzed to produce a new query PFG. The query graph can also be modified by graphically moving any term node displayed by the system (i.e., in document PFGs and the thesaurus PFG) into the Query window and connecting it to the query graph. Query nodes can also be deleted, and the link structure changed. The query representation can be made more elaborate and revised to better reflect information needs using knowledge gained through evaluation of search results and interaction with the thesaurus.

Document Abstract PFNETs. The representations of documents used in retrieval and graphic display are constructed from the texts of document abstracts. A typical Pathtrieve display is shown in Figure 3 with a document PFG as the central network display and the document abstract text in the lower right window. In all graphic displays, word forms of term stems are presented. In this example the 48-word abstract contains 14 unique term stems that provide the document PFG nodes. The upper windows display an abstracted overview of the term thesaurus, the natural language query PFG, and an ordered list of documents retrieved in a search based on the query PFNET. Any document title in view can be selected to display the abstract PFG or text.

In addition to providing information about term usage in a document collection, the document PFG also allows the user to rapidly preview a document. A document PFG can be scanned for terms or constellations of terms of interest. This ease of interacting with a document collection encourages exploration. In meeting information needs in subject areas not well-known to the user, exploration or browsing can be useful in directly retrieving documents, as well as in becoming acquainted with the domain.

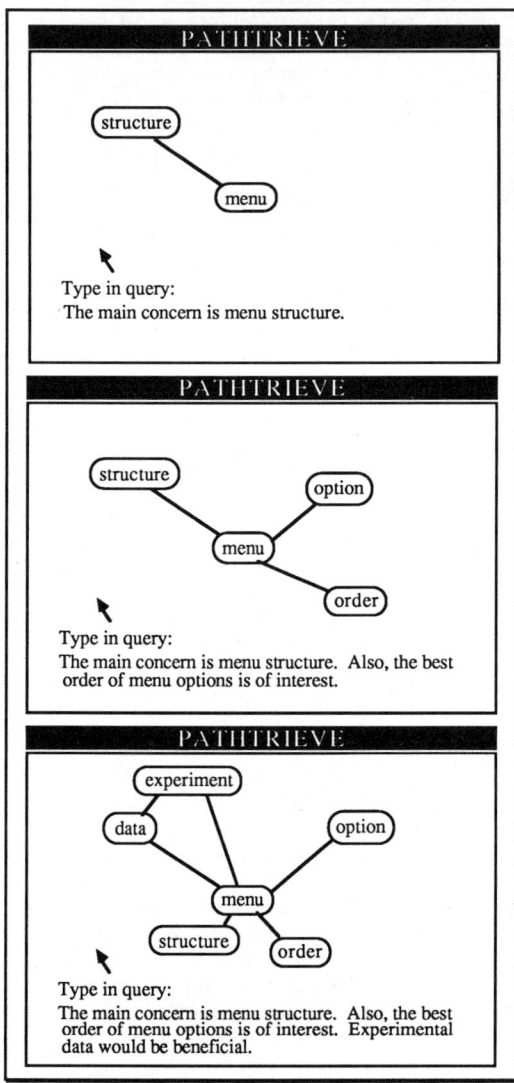

Figure 2. Query window displays of natural-language queries and PFGs.

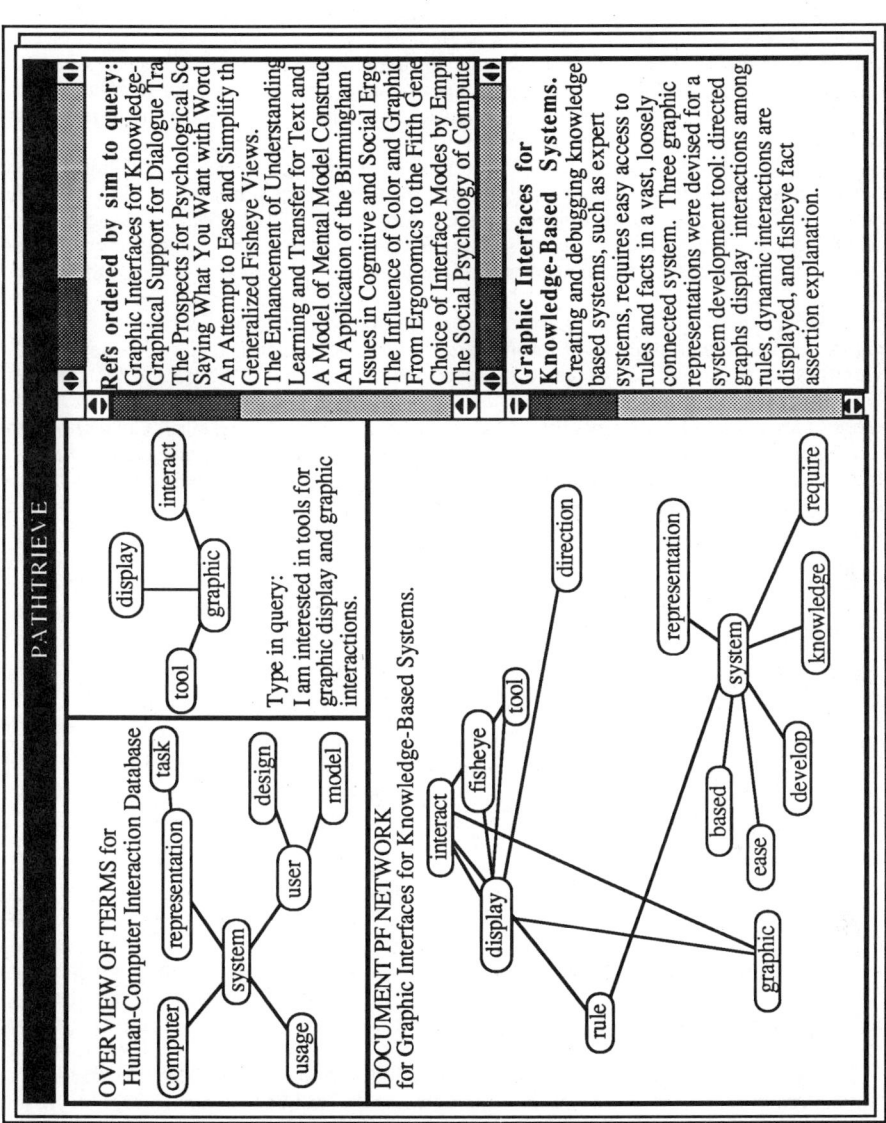

Figure 3. A typical Pathtrieve display. The upper windows display (1) an abstracted view of the term thesaurus, (2) the text of a natural language query with the PFG derived from it, and (3) an ordered list of documents retrieved in a search. The lower windows show (4) a document abstract PFG, and (5) the document abstract text.

Term Thesaurus PFNET. The term thesaurus provides information about indexing terms in a conceptual domain and should be flexible since users will think in different terms and perceive different relations as important. A thesaural organization that permits flexible navigation will be useful for both expert and novice users since the type of relation that makes a term relevant depends on the specific information need. Even in subject areas well-known to the user, an associational thesaurus is likely to provide cues that facilitate exploration. Both the need for an IR system to deal with different types of information needs and indexing indeterminacy suggest that mechanisms be provided to allow multiple paths of access to information objects. In the Pathtrieve system a PFNET constructed from all abstract texts in a database, using shallow text analysis to reflect a wide range of relations, provides an associational network thesaurus as one mechanism to address these concerns.

The PFG displays and navigation tools are designed to exploit the potential of user-controlled visual presentations of information. A graphic thesaurus display has several advantages as compared to textual display of the same material (Bertrand-Gastaldy & Davidson, 1986; Craven, 1984; Doyle, 1961). Network displays can group related terms more compactly and appropriately than the lists used in printed thesauri. Closely tied to this feature of graphic display is the ability to easily and quickly move or navigate through the terms by following thesaural relations. The associational thesaurus is one mechanism for conveying domain knowledge and information about term usage. The user is provided with more detailed term and domain knowledge through viewing the natural language of abstracts together with the term PFG for an abstract. The incorporation of a conventional thesaurus based on linguistic relations would provide an additional means for conveying knowledge about terms, thereby complementing the information conveyed by the associational PFNET thesaurus.

Network Displays and Navigation Tools

The development of procedures for spatial organization in graphic displays is a pervasive problem. Fortunately, some scaling techniques suggest a spatial form, for example, a tree for hierarchical clustering results. In knowledge engineering applications, display is often constrained by properties of the representation, such as temporal ordering for production systems or common slots in a schema-based representation. For PFNET displays, some constraints are needed to determine node layout.

In the Pathtrieve system PFGs are spatially organized using nonmetric multidimensional scaling (MDS). This solution to weighted network display has been applied to a variety of graph structures (Klovdahl & Brindle, 1987) and may provide additional information through the positioning of nodes. In determining the maximum graph size that can effectively be displayed using this technique, line (link) crossings are the limiting factor. As the MDS solution seeks to minimize a global criterion in deriving the spatial coordinates, and Pathfinder tends to emphasize local relationships in the link structure, limitations are readily apparent. However, with small (up to about 15 nodes), relatively sparse structures, MDS node layout is usually satisfactory. This includes most query and document PFNETs and determines a practical limit of the subgraph size that can be displayed effectively with this technique.

In accessing the PFNET thesaurus network, many more terms are available than can be presented on a screen, so only subgraphs of the complete thesaurus can be displayed in the viewing area. Two kinds of tools are available for subgraph viewing: an abstracted aerial view of the complete network providing orientation to the overall structure, and navigational functions allowing selection of subgraphs from the complete thesaurus for detailed viewing.

Aerial View. While a number of abstraction techniques have been explored for PFNETs (McDonald & Schvaneveldt, 1988), the abstracted, aerial view of the thesaurus PFNET used in the Pathtrieve system is simple yet useful in thesaurus navigation and orientation. The aerial view includes the nodes with highest degree in the term thesaurus PFNET. Links are displayed for nodes that are separated by paths not exceeding a criterion based on the size of the thesaurus. This sort of "bifocal display" has proven useful in providing orientation within large data structures when only parts of the complete structure can be displayed (Englebart & English, 1968; Ramsey & Grimes, 1983).

Subgraph Selection. The user can define a thesaurus entry point and initiate display of a thesaurus subgraph by selecting a focus node from one of several displays: aerial view, query PFG, document PFG, or the current thesaurus subgraph. A thesaurus subgraph is formed by following links from the focus node. If the subgraph size, defined by a default number of links (usually 3), exceeds a criterion set to reflect the number of nodes that can be satisfactorily displayed, shorter traversals are tried until the subgraph size meets the criterion. In the subgraph displayed in the system's network window, the focus node is inverted, and the size of nodes decreases with the number of links away from the focus node in order to convey a visual perspective centered on the focus node.

Orientation and Conceptual Elaboration within the Thesaurus. The different sources for defining an entry point into the thesaurus PFNET not only facilitate traversal of the thesaurus PFNET, but also provide different contexts and paths of access for a term. A goal of the system is to provide the user at least one structure that includes terms useful in formulating the information request. The PFG derived from the users natural language query supplies information about terms by showing, as the nodes of the PFG, which words of the query are indexing terms for a particular database. In addition, documents retrieved after a search can be inspected individually and provide another context and means of access to index terms. For a document, both PFG and abstract text can provide contexts for a term. Often other terms of interest will be found in documents retrieved in a search. Both of these structures can provide an initial context for terms before more information about a term's relation is learned through navigation within the thesaurus.

The abstracted, aerial view of the complete network is the principal means of orientation to the complete thesaurus. By incorporating the most frequently occurring terms, the most general terms tend to be selected. These terms are the most likely to be familiar to the user not acquainted with a domain and can provide a starting point for elaborating conceptual structures already available. For example, Figure 4 shows the system display produced by selecting the node labeled *user* in the Aerial View window. By viewing the thesaurus subgraph displayed in the central window, the user learns more about the range and relations of terms. Here, the user finds both general terms, such as *knowledge* and *software*, and less common terms, such as *UNIX* and *dialog*. Additionally, the information that all of these terms are associated with "user" is conveyed.

Selecting nodes from the aerial view to initiate subgraph display allows the user to quickly move around the thesaurus network and can provide an overview and orientation to the indexing terms used in a database. When changing the focus node within a subgraph display, the aerial view also provides orientation by tracking aerial nodes that come into view as part of a subgraph. Any aerial view node that is displayed as part of a thesaurus subgraph is inverted in the Aerial View window to show the position of the subgraph in the complete thesaurus.

In practice, thesaurus subgraphs usually contain an aerial view node due to the form that thesaurus PFNETs tend to have. Using term sets of the most frequently occurring terms, common, high frequency terms tend to co-occur with a relatively large number of

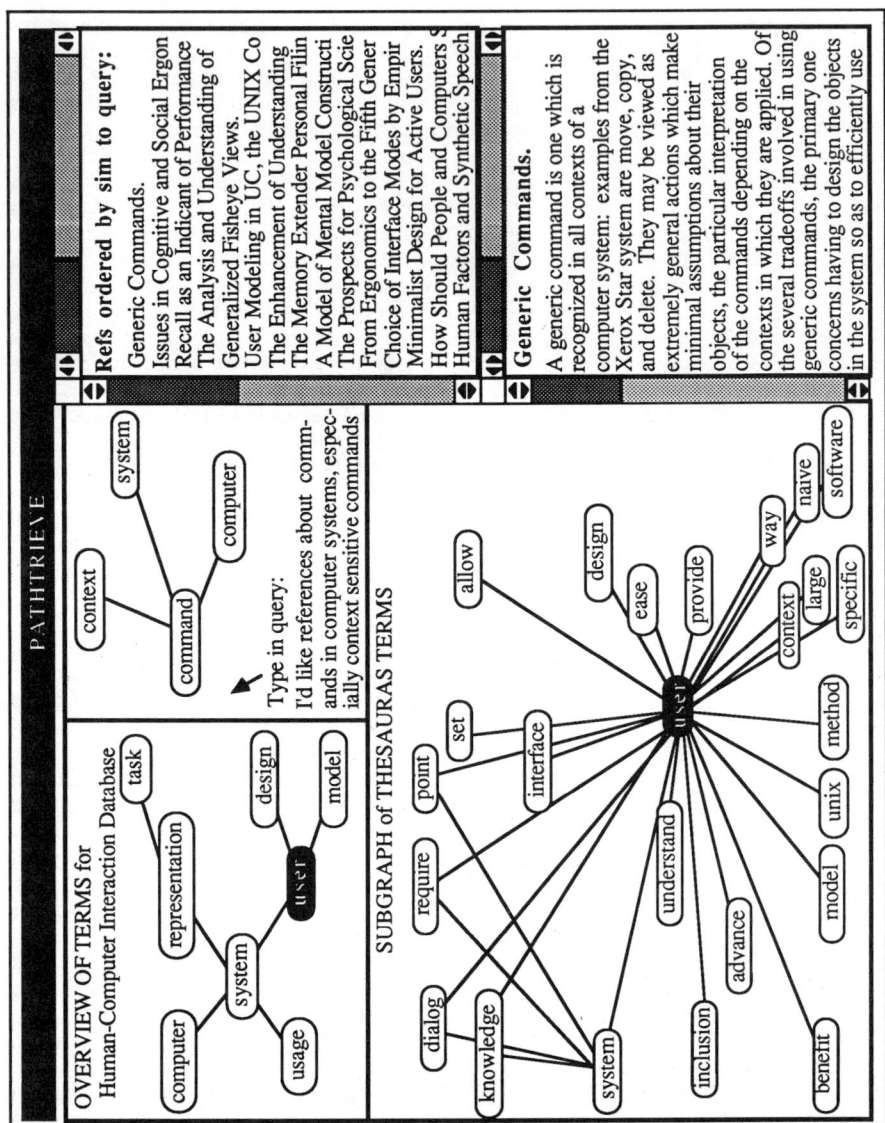

Figure 4. Pathtrieve display after selecting *user* node in Aerial View window. The thesaurus subgraph focused on *user* is displayed in the central Network window. The aerial view orients the subgraph to the complete thesaurus by tracking (inverting) aerial view nodes present in the subgraph.

infrequently occurring terms. Infrequently occurring terms co-occur less often with other low frequency terms than with the common terms. The resulting thesaurus PFNET has relatively few high-degree nodes (the frequently occurring terms) and many low-degree nodes (the infrequently occurring terms). These high-degree nodes are the nodes that are selected for the aerial view. In terms of path length the aerial view nodes are relatively close together. When traversing the complete thesaurus, an aerial view node is usually within the path length used for subgraph definition and so is tracked in the aerial view. This form of the thesaurus PFNET also allows traversal with relatively few changes of subgraph focus node.

Document Retrieval Techniques

Several different retrieval mechanisms might be usefully employed in an IR system to meet different types of information needs (Belkin & Croft, 1987; Jones & Furnas, 1987). Exact match Boolean searching, in which a query is formed by the combination of terms using the operators AND, OR, and NOT, is the retrieval technique most widely used in commercially available IR systems. Using this technique, query-document matching produces simply a set of retrieved documents. No use is made of relationships between query terms or relationships between individual documents, and the resulting unordered groups of documents are difficult for the user to evaluate. Because the match of query and document representation must be exact, the indexing vocabulary problem is particularly significant. Still, this approach is useful for information needs that can be articulated by the searcher in terms appropriate for the system. However, a precise specification of indexing terms is difficult or impossible when the subject area is not well-known to the user. Partial match techniques, such as term vector matching in which documents and queries are represented as vectors of terms, allow a range of match between query and document representations. The varying degree of match can be used to rank documents for presentation to the user with respect to how well each corresponds to the query.

IR systems use a wide range of techniques for representing the conceptual structures of queries and documents, from simple unweighted keyword lists to representations with unlimited expressive power, such as semantic networks. Our approach represents an intermediate position: PFNETs allow nodes to represent conceptual entities, but do not provide conceptually labeled relations as necessary for the expressive power of semantic networks. The choice of which retrieval techniques to include in a system depends on how all system components interact to meet those needs. PFNET-based components seem particularly useful when meeting information needs in subject areas not well-known to the user.

PFNET Similarity Measures. Providing similarity metrics for the PFNETs derived from documents and queries is an area rich with possibilities. We anticipate that Pathfinder network representations, which were designed with explicit consideration of psychological models of human memory, will lead to graph matching techniques useful in a wide range of document retrieval tasks. In comparing query and document PFNET representations, simply matching the node labels (stems) and making no use of the links is one form of query vector matching. A matching algorithm that takes into account both conceptual similarity, as might be measured by identifying common node labels, and structural similarity (Knoke & Kuklinski, 1982) should be more effective because link structure can be used to provide information about the relationship of terms.

The algorithm now employed derives metrics for two components of similarity. The first component is computed as the number of common term stems in the query and the document divided by the number of terms in the query. A second component provides a measure of structural similarity by considering the link structures of both query and document. The value of this component is increased when nodes closely connected in the query

PFNET are also closely connected in the document PFNET. For nodes common to both query and document, the measure of similarity of structure is increased by 2, if direct links exist in both query and document, between common node pairs; or by 1, if node pairs common to query and document are not directly linked in both, but have paths of 2 or less in both. This sum is divided by 2 times the number of links in the query so that the value is between 0 and 1, as is the measure of conceptual similarity. Conceptual and structural similarity measures are then weighted and summed to give the final similarity measure. The weighting of the different components provides one means for adjusting the algorithm's retrieval performance.

One major tractability concern in the system is the time required to apply a matching algorithm to document and query representations. Many forms of graph matching in the general case require a number of computations exponential in the number of nodes or links considered. Except for very small structures, time complexity of this order is unacceptable for the matching component of an IR system. Yet, in designing a matching algorithm for a specific application or problem domain, it is usually possible to exploit constraints so that a more tractable, but less general, algorithm can be found (Galil, 1986). Another consideration is that in IR the expected case complexity is often more important than worst case performance. Here, too, the nature of the domain plays a critical role in determining whether a particular algorithm is feasible for use.

Fortunately, well-known IR techniques can be applied to accomplish most aspects of the graph matching process efficiently. For example, in the matching scheme outlined above, the measure of conceptual similarity can be obtained using efficient vector matching techniques such as those provided by inverted file organizations (Salton & McGill, 1983). Determining similarity of link structures seems a more difficult task, but constraints of the problem common to many IR applications allow a tractable solution. In comparing query and document link structures in the above scheme, it is, of course, only necessary to consider documents with at least one query term. For the remaining set of documents only node pairs present in both query and document need be considered, thus typically eliminating most documents. In practice, PFNETs with $r = \infty$ and $q = n-1$, such as we have used tend to be quite sparse, usually the number of links is less than twice the number of nodes. Since the document PFNET link structure is static, it can be indexed to allow efficient access. The form of the indexing can be selected to be appropriate for the particular matching algorithm to be used.

For the user of an information system, the first step is often the difficult task of selecting a database likely to contain relevant information. The Pathtrieve system automates database selection using the natural language query. For each document collection database, a query PFNET is derived using the database's term set and the query network is compared to the thesaurus PFNET for the term set. The matching algorithm, slightly modified to provide a measure not normalized by the number of query concepts, provides a measure of similarity of the query network to each database term thesaurus PFNET, and the database with the highest similarity value is selected for use.

Document Collection PFNET: Browsing and Cluster Retrieval. Browsing among shelves in a library is exploration and search among documents guided by the classification system used in the library. Within an information system more search and organizational flexibility is possible. Access to documents can be based on a network of documents derived from interdocument relations. From some entry point the network can be traversed and documents selected. This technique has the advantage for some types of information needs of requiring little query formulation and knowledge of the subject area.

Finally, there is a PFNET reflecting relations among documents. In this PFNET, nodes are the documents in a database and the graphic labels are document titles. The proximities for deriving this network are obtained by applying the matching algorithm to all pairs of document abstract PFNETs in a database. The document collection PFNET is used for exploration in the network of documents similar to exploration in the term thesaurus. The PFG of documents can be displayed, as shown in Figure 5, at any time. As with the term thesaurus, an aerial view is provided as well as navigation tools for browsing and entry point selection. Selecting a title allows the user to view the document's abstract text and PFG. Functions using the PFG of documents provide an additional means of gaining domain knowledge and access to graphic network structures.

In addition to ordering documents by similarity to a query, a second form of retrieval is available in the system. Cluster-based retrieval uses the PFNET of documents and is based on traversing the network beginning at a particular document. The entry point can be directly provided by specifying a title, or by finding a document that best matches the query. Additional documents are then retrieved by following the links from the starting point. The sequence of retrieved documents is displayed to the user ordered by the number of links from the entry point document.

Summary

The Pathtrieve system has served as a testbed for the application of PFNETs in information retrieval and has allowed us to explore a number of uses of PFNETs. PFNETs have been used within well-understood IR techniques, such as retrieval based on document networks, as well as somewhat novel applications, such as direct manipulation of query network structures. The network representations are relatively shallow and, as such, contrast with current trends in IR for the use of deep representations more characteristic of AI techniques than traditional IR techniques. However, the associational representations may be appropriate to meet some of the more challenging user information needs by allowing multiple paths of access to information items. While our initial goal was to examine PFNET-based mechanisms solely as an adjunct to effective retrieval techniques in subject areas well-known to the user, our investigations suggest that graph matching retrieval techniques using PFNETs can provide a useful retrieval mechanism for a wider range of user information needs (Fowler & Dearholt, 1989).

The primary emphasis in development has been on mechanisms that allow the user to express an information need through the elicitation, elaboration, and revision of conceptual structures. In deriving the query PFNET from the user's natural language statement of the information need, an expression of the need using the system vocabulary terms and in the form used by the system for retrieval is automatically constructed. The interface provided for query revision and thesaurus use is designed to facilitate query elaboration and revision by providing mechanisms for graphic manipulation in order to reduce the perceived distance between user and conceptual structures of the system. The use of a uniform graphic PFNET representation is designed to reduce system complexity. These mechanisms are provided to exploit the potential of a visual, graphic interface for the display of knowledge during the highly interactive information retrieval process. The graphic PFNET structures provide representations that complement the user's conceptual structures. In many cases the Pathtrieve system guides the user in obtaining domain knowledge through exploration, rather than directly providing it. The graphic interface based on conceptual structure facilitates interaction for users at any level of expertise in a subject area and might extend information system use to a larger population.

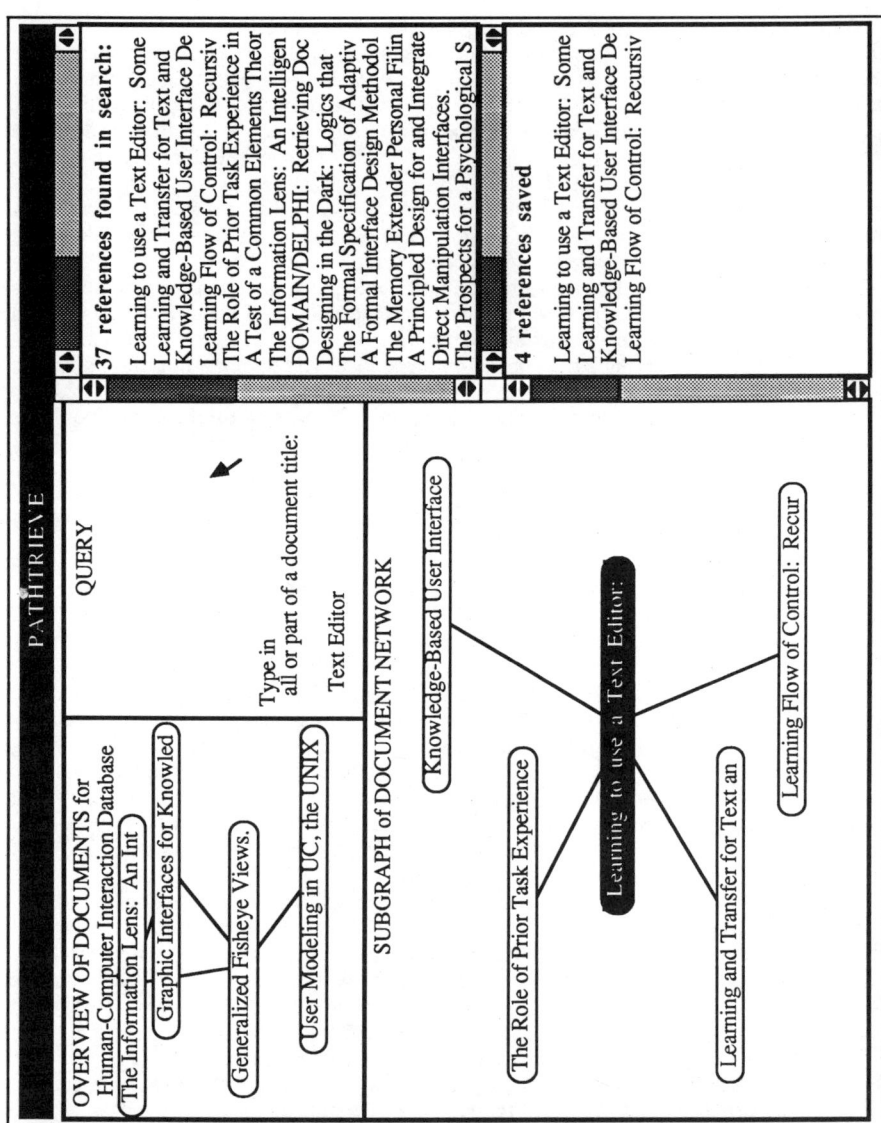

Figure 5. Pathtrieve display with PFG of documents subgraph in Network window and abstracted representation of the document PFNET in the Aerial View window. Node labels are document titles. The results of a cluster-based search are displayed in the upper-right window.

Chapter 13

Using Pathfinder to Evaluate User and System Models*

Wendy A. Kellogg and Timothy J. Breen

Mental Models and User Performance

The notion that a user's mental model of a software system has a critical impact on the user's ability to effectively use systems has gained widespread acceptance in the field of human-computer interaction (e.g., Carroll & Olson, 1988; Hammond, Morton, MacLean, & Barnard, 1983; Kieras & Bovair, 1984; Masson, Hill, Conner, & Guindon, 1988; McDonald & Schvaneveldt, 1988; Waern, 1987). The acceptance of the "mental model hypothesis" is motivated in part by the belief that faulty or incomplete representations (misconceptions) lead to errors, and that the kinds of errors users make can be understood once a model of their knowledge is derived (see, e.g., Masson et al., 1988). However, verifying this claim empirically and applying it in practice to system design or user training has met with mixed success (see Carroll & Olson, 1988, and Rouse & Morris, 1985, for critical reviews). Some studies have found performance benefits for users learning an appropriate model of a system or programming language (e.g., Kieras & Bovair, 1984; Linde, 1986; Mayer, 1987). Others have found little benefit or inconsistent benefits in giving learners device models (Halasz & Moran, 1983; Polson[1]).

In our view, there are two fundamental problems contributing to this state of affairs. The first is that the target body of knowledge represented by a software system has rarely been articulated. Without the definition of a system model, what the user *should* know and therefore what his mental model should contain are unknown. Defining an adequate model of a software system is made difficult by disagreement about what kind of knowledge it should encompass (e.g., "how to do it" vs. conceptual or "how it works" knowledge), and by our currently limited understanding of the relationship of different kinds of knowledge and user performance.

A second fundamental problem with improving the status of the mental model hypothesis is the difficulty of "capturing" the user's mental model, particularly in a way that can be systematically compared with a system model. The variety of techniques employed to date virtually spans the repertoire of psychological methods, and each method yields a different kind of mental model. Clearly, the way a researcher derives a mental model is critical to any assessment of whether the user's model is an important determinant of

---

*This work was carried out while the second author held a Predoctoral Internship 1985-1986 at the User Interface Institute, IBM Thomas J. Watson Research Center, Yorktown Heights, NY. We thank James McDonald and Roger Schvaneveldt for comments on an earlier version of this chapter. Portions of this research were presented at CHI+GI'87 and reported in Kellogg and Breen, 1987.

[1]Personal communication to W.A. Kellogg, 1987.

performance. In addition, for the purpose of applying assessment of user and system models to system design, pragmatic "cost-benefit" characteristics must be taken into account (e.g., whether the information gained is worth the effort to acquire it). The work reported here assumes the importance of the user's conceptual knowledge of a system. Our focus is evaluating the utility of *scaling techniques* as definitions of user and system conceptual models that can inform system design. Our interest in users' conceptual models of computing systems reflects an instantiation of the mental model hypothesis which asserts that the more congruent the user's structure of knowledge with that represented by the system, the more easily learned and usable the system should be. Our interest in scaling methods reflects both theoretical and pragmatic concerns, as elaborated below.

Defining Mental Models

Various methods have been employed in the study of mental models, including protocol analysis, production system modeling, and scaling techniques. The kind of model a researcher chooses to derive depends both on the kind of knowledge being represented (e.g., declarative or procedural) and the use to which the model will be put, since different methods have different strengths and weaknesses. Protocol techniques (real-time "think-aloud," post-task video confrontation or interviews, or inferring knowledge fragments or misconceptions from a protocol of user keystrokes) are particularly appropriate for relating user errors to misconceptions. However, they are hard to summarize and compare systematically. Analytic approaches, such as GOMS (Card, Moran, & Newell, 1983) or production system modeling (Kieras & Polson, 1985), may be useful for evaluating the efficiency and consistency of designed methods, but may be less useful in understanding the genesis of user errors, or in evaluating the difficulty for learners of the conceptual knowledge represented by a system. Scaling techniques have some of the advantages of both protocol and analytic methods. Like protocol methods, they explicitly represent conceptual knowledge and are empirically derived. Like analytic methods, once derived, models defined by scaling techniques can be systematically and quantitatively compared among users and between users and the system. But scaling techniques have their vulnerabilities as well: How do context-free judgments of the relatedness of system concepts bear on user performance, or on predictions of learnability and usability? The primary weakness of the scaling methodology is less a matter of the method itself than of its application—understanding what can and should be made of its results.

Pragmatic concerns may also influence the choice among methods, particularly if the focus is upon utilizing the information gained for system design purposes. Protocol techniques require a running system and are time-consuming to collect and analyze. Analytic modeling can be performed before a system is implemented, but requires considerable time and skill on the part of the analyst. Scaling techniques, in contrast, are startlingly simple to employ and can also be used before a prototype or running system is available. As such, they are an attractive candidate for evaluating the compatibility of users' knowledge of a task domain with a system or proposed system; the crux of the issue is whether they can yield valuable information to the design process.

Scaling analyses have been used in system design, particularly for organizing the presentation of information in the interface (McDonald & Schvaneveldt, 1988). However, the extent to which the use of scaling techniques can be extended to other interface issues is unknown. One goal of the current work was to extend the use of these techniques to the structure of a system's *functionality* per se. This was possible because of the nature of the system we studied, where the declarative structure of the commands as defined by the system model often determined when and where commands could be successfully used.

13 User and System Models

Using Scaling Techniques in System Design

Mental models defined by scaling techniques have the potential to be used to assess a system's usability and learnability in a way that depends on both the user and the system. In order to employ them in this way, it must be shown that the relationship between a user's model and the system model changes with experience in the predicted fashion. In particular, experience in using a system should be correlated with increases in the amount of overlap between the user and system models. This presupposes a methodology for extracting and expressing a model from users and the system in comparable forms. The use of mental models for assessing usability and learnability must also be empirically verified. The closeness of a user's model to the system model should predict performance on the system. Salient discrepancies between expert users' models and the system model should indicate modifications to the system that would improve its usability. Similarly, the distance between novices' models and the system model should correlate with the difficulty of learning a system. To verify these assumptions, a way of measuring the degree of agreement between user and system models is necessary. The work reported here attempts to verify the expected relationship between experience and the amount of overlap between user and system models defined by network scaling. This involved specifying a system model, deriving user models from groups of users with different amounts of experience with the system, and developing methods for assessing the degree of agreement between users' models and the system model. User networks were based on data representing subjective judgments about the structure of a system. A system model based on the system documentation was derived using the same scaling technique. To our knowledge, the work reported here is unique in specifying a system model from the documentation for direct comparison with empirically derived user models, and in its attempt to use the outcome of the comparison to suggest usability improvements.

Method

The Formatting System. The domain we studied was a command-driven text formatting language in which users format documents by labeling their components. The system is designed to take advantage of users' knowledge of typical document structure. Traditional formatting systems require specification of desired format from the user in terms of low-level components, such as spacing, line breaks, justification, and control characters. In contrast, the system under study defines more abstract components which entail a set of low-level formatting effects. These components (called *tags*) can be interpreted appropriately by different output devices. The user's task in formatting, then, is to label parts of the document with appropriate tags.

The structural diagram of the system from the documentation is shown in Figure 1. It divides the system into eight major categories: General Document, Headings, Basic Text, Displays, Lists, Index, Footnotes, and Process-Specific Controls. Within the categories appear the set of 51 tags which users might apply to parts of the document. All of the categories, except Basic Text and Process-Specific Controls, represent structural elements of typical documents.

The declarative structure of the tags is important to their appropriate use. In general, the system is hierarchically structured, and this structure has implications for where and when a tag can be used. In the General Document category, for example, Title Page elements (with the exception of Address and Address Line) can *only* be used within the scope of the Title Page tag. The system documentation explicitly lists Address and Address Line in both the Title Page group and the Basic Text group, since these tags can be used in the Body of the document as well as in the Title Page.

The implications of the system's view of document structure for obtaining desired formatting effects made the use of scaling techniques for evaluating learnability and usability as outlined above attractive. In particular, we hoped the document element networks derived with Pathfinder would show us several things: (1) the evolution of the user's view of document structure with experience using the system; (2) *what* restructuring of knowledge needed to occur for new users (an assessment of learnability); and (3) whether any disagreements with the system's view of document structure remained for experienced users (an assessment of usability). In addition, because of the central role of document structure in this system, and because nonusers should already have a good understanding of document structure in general, the use of scaling techniques represented a potentially strong test of the sensitivity of the method.

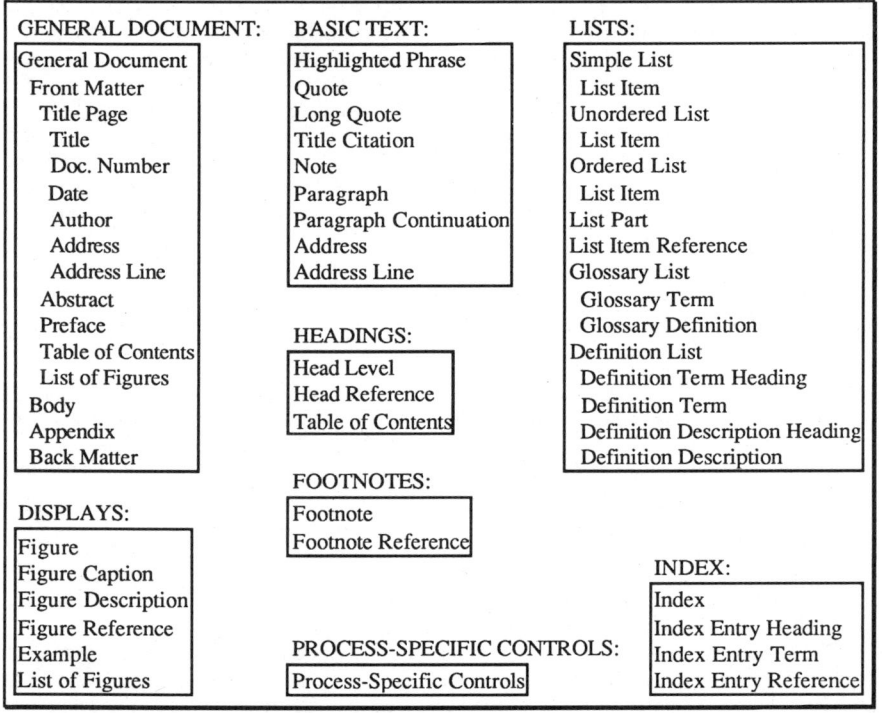

Figure 1. Structural diagram of system from documentation.

Participants. Fifteen experienced users and 15 nonusers volunteered to participate in the study. The experienced users had used the system in their work for a minimum of 7 months, with a mean of 3 years. These participants used the formatting system routinely in producing work documents ranging from research papers, to books, memos, and overhead projector foils. Most had some familiarity with formatting systems other than the one under study.

The nonuser participants had no experience with the system under study, and 4 out of 15 had no experience with formatting systems at all. One had not had any experience with computer systems. All except two worked in industrial research or academic settings.

Materials. The materials for the experiment consisted of a set of 51 index cards with one formatting command written on each card. The verbal labels for the commands were used, rather than the actual form of the command that would be used to markup a document.

Procedure. Participants read instructions describing the task. They were told that the set of cards represented parts of a formatting system, and that each card would do something to affect how a document would look when it was printed out (for experienced users, the system was explicitly identified). Participants were asked to look through the set of cards while thinking about what each might do, and then to sort the cards into any number of piles such that things that *seemed* related were in the same pile. They were also told they could use as many piles as they wished, and that they could change their arrangement until it seemed best. The experimenter clarified any questions; the participant then sorted the cards until satisfied with the arrangement. The number and content of the groups were recorded. Finally, participants were asked to rate each of the 51 commands for familiarity by checking any commands they had never used or with which they were unfamiliar.

Extraction of the System Model

The system model was based on a structural diagram created by one of the system designers for the system documentation. The diagram partitions the system into major components, each with hierarchical structure, as described above. Deriving the system model from the structural diagram was based not on the premise that the typical user would explicitly attempt to acquire a model of the system from it, but rather on the judgment that in this case it was the clearest and most objective representation of the system upon which to base a model.

Data for the system model were based on distances between pairs of concepts in the command set. Distances were determined from the structural diagram shown in Figure 1 as follows. Concepts within a subgroup were given a distance of 1. Concepts one level apart in the hierarchy were scored as distances of 2. Each additional level was scored by incrementing the distance count by 1. Concepts which were not direct descendants of a higher-level concept were given the distance score between levels plus 1. Thus, for example, the distance between Title and Front Matter was 3, whereas the distance between Title and Body was 4. Unrelated concepts were given an "infinite" distance of 5. Distances ranged from 1 to 5 for the entire set of concepts. The distances then served as input to the scaling analyses.

Results

The data from participants' card sorts were transformed into a measure of distances between concepts, which then served as input for the scaling analyses.

Individual agreement with the system data. Agreement between an individual participant and the system was determined by computing the mean correlation across the set of 51 concepts in the following way. For each participant, a vector of the 1275 possible pairwise combinations of the 51 concepts was constructed. Each value in the vector represented whether the participant had sorted those two concepts in the same pile or in different piles; the whole vector then represented the participant's entire card sort. The value for each possible combination was expressed as a linear transformation of the distance (0 for cards sorted together, 1 for cards sorted separately; transformed to 1's and 2's, respectively). The system data were represented in a similar fashion, with distances in the vector ranging from 1-5 as described above. The vector for each participant was then correlated with the system vector, resulting in a score for each participant representing the degree of agreement between each participant's card sort and the structure represented by the system diagram. The average correlation with the system for the experienced user group was .48; for the nonuser group, .30. The correlation for 10/15 of the experienced users was greater than the largest nonuser/system correlation. Participant scores from the two experience groups were compared with the Mann-Whitney U Test (Hays, 1973); the difference was significant: $z = 3.21$, $p < .003$, two-tailed.

Intragroup agreement. The vector for each individual was similarly correlated with all the other individuals in their experience group. In this analysis, 14/15 experienced users had higher correlations with other such users than the highest correlation among nonusers (for experienced users, mean = .37; for nonusers, mean = .20). The difference in intragroup correlation was significant by the Mann-Whitney U Test ($z = 4.33$, $p < .00007$, two-tailed). Replicating previous results (Cooke & Schvaneveldt, 1988), experienced users displayed higher agreement among themselves than did nonusers.

Functional grouping. Another finding reported previously (e.g., Kay & Black, 1984, 1985) is the tendency to structure knowledge more in terms of *function* than in terms of *surface features* with experience. Accordingly, we looked for evidence of functional grouping by individuals in each experience group. For four functional groups defined by the system (Basic Text, Front Matter, Title Page, and Displays), we set a criterion ranging from 67%-80% of the commands in the group for determining whether a user had sorted the commands functionally. For each functional group, we then looked at the number of users in each experience group who used that functional grouping, the mean percent of "correct" commands they included, and the mean percent of "extra" commands included in the card-sort group (relative to the size of the system-defined functional group).

The results of the Basic Text group are representative of the general pattern across the four functional groups. In addition, this group represents a particularly strong test of the presence of functional grouping since the commands it contains have few surface features in common. Using a criterion of 5/7 commands present, 9/15 experienced users had Basic Text groups. These groups contained 80% of the Basic Text commands, and 17% extraneous commands. Only 4/15 nonusers met the criterion for a Basic Text group, on the other hand, and while these contained 79% of the Basic Text commands, they also included 136% extraneous commands. The means across the four functional groups were: for experienced users, 10/15 had functional groups containing 85% correct concepts and 21% extraneous concepts; for nonusers, 4/15 had functional groups containing 84% correct concepts and 138% extraneous concepts. Experienced users showed strong evidence of functional grouping, with groups containing most of the correct commands and very few other commands. Nonusers were much less likely to group commands functionally, and when they did have an intact functional group, it was typically in the presence of many other extraneous commands.

In summary, analyses based on *individuals* indicated that experienced users were in closer agreement with the system than were nonusers, and were in closer agreement with each other than were nonusers. In addition, experienced users showed more evidence of sorting system commands on the basis of function than did nonusers. We then looked at the agreement of user groups as a whole with respect to system concepts based on the user and system models defined by Pathfinder.

Network Scaling

Comparison of User and System Networks

Pathfinder networks were derived from the distance data for both experience groups and the system with $r = \infty$ and $q = n-1$ (Schvaneveldt, Durso, & Dearholt, 1985). The resulting networks for the system, for experienced users and nonusers, are shown in the Appendix. In the user networks, bold lines are used to denote links *shared* with the system network, and thin lines are used to indicate links not present in the system network. In addition, on the user networks, "lassos" have been drawn around eleven subgroups defined by the system (General Document, Front Matter, Title Page, Basic Text, Displays, Headings, Lists, Definition List, Glossary List, Footnotes, and Index) to the extent that they are present. The system network contained 138 links, the experienced user network contained 82 links, and the nonuser network contained 73 links. Of the experienced users' 82 links, 69 (84%) were shared with the system network, 13 (16%) were not shared. For nonusers, 43 (59%) links were shared with the system network and 30 (41%) were not shared.

We examined the degree of agreement between a user network and the system network by computing for each of the 51 concepts the correlation of user-defined and system-defined links. This resulted in a set of scores for each user group, with each score representing the degree of agreement between a user groups' treatment of a concept (in terms of what it was linked to) and the system's treatment of the concept. Across the 51 concepts, experienced users correlated .640 with the system network, nonusers correlated .421 with the system network. This difference was tested by examining the difference (in direction and magnitude) between scores from each experience group for each concept with the Wilcoxon Test (Hays, 1973) and was significant: $z = 4.39$, $p < .00003$. Similarly, the correlation between experienced users' and nonusers' networks was .425; experienced users were more highly correlated with the system network than with the nonuser network ($z = 4.48$, $p < .00003$).

Our previous analysis of the extent of functional grouping in the raw data indicated that experienced users utilized functional grouping more than nonusers. Another aspect of this question is the role played by surface feature similarity in each experience group's network. We examined this question by constructing a three-way contingency table which classified each link in a user group's network as connecting nodes that were *functionally related* or *not functionally related* (nodes linked in the system network were considered to be functionally related) and which had *surface similarity* or *no surface similarity*. Surface similarity was defined as command names having one or more words in common (e.g., Front Matter and Back Matter). A test for the degree of association between functional relatedness and surface similarity for each experience group (Fienberg, 1977) was significant ($z = 3.03$, $p < .002$). Nonuser network links showed a greater dependency between functional relatedness and surface similarity ($\alpha = 3.99$) than experienced user network links ($\alpha = .291$), and the association was in opposite directions for each group. Nonuser links that were shared with the system model (i.e., were functionally related) tended to be those with surface similarity. Functional relatedness and surface similarity were much less strongly related for experienced users, on the other hand, and to the extent they were related at all, experienced user links tended to match the system network more closely for nodes *without* surface similarity.

The network derived by Pathfinder for the system (see Appendix) is regular and simple to describe. The system defines eight basic categories, three of which (General Document, Lists, and Displays) have hierarchical structure. Each basic category appears in the network as a *cluster* with members of the category fully linked (one category, Process-Specific Controls, does not appear in the network because it contained a solitary command). Certain commands *connect* different clusters (e.g., Title Page connects the Front Matter cluster with the Title Page cluster). Such commands reflect one of two characteristics of the system. The majority of connecting commands reflect hierarchical structure in the system. This can be seen clearly in the General Document cluster (containing Appendix, Back Matter, Body, and Front Matter), connected through the Front Matter tag to the Front Matter cluster (Abstract, Table of Contents, List of Figures, Preface, and Title Page), connected through the Title Page tag to the Title Page cluster (Address, Address Line, Title, Author, Date, and Document Number). In the three other cases where clusters are joined, the connecting commands belong to *two* categories in the system model (e.g., Table of Contents is contained in both the Headings and General Document categories). For two of these cases (Table of Contents and List of Figures), the system's double categorization reflects a functional relationship: for example, the Table of Contents tag *uses* Headings to construct the Table of Contents. The third case, Address and Address Line, categorized in both the Title Page and Basic Text categories, reflects the fact mentioned previously that these tags, unlike the other Title Page tags, can be used within the body of a document as well as in the title page. Thus, overall, the network produced by Pathfinder for the system cleanly reflects its categorical and hierarchical structure, and some of its functional aspects.

The user networks, in contrast, show categories (as evidenced by tightly interconnected clusters) and hierarchical structure (in the pattern of links) much less clearly. However, clustering is much more developed in the network of experienced users than in the nonuser network. Nonusers show rudimentary clustering for Lists and Title Page elements. Within the Lists category, some hierarchical structure can be seen in the cluster of Definition List tags, connected to the Lists cluster through the Definition List tag. Otherwise, the nonuser network tends to be linear (i.e., exhibits point-to-point connections between commands). Experienced users, on the other hand, show well-developed clusters for Title Page, Basic Text, Lists, and Displays, less-developed but discernible clusters for Definition List and General Document elements, and two clusters *not* defined by the system model—a References group and an Index/Glossary group.

The Reference tag group (List Item Reference, Index Entry Reference, Figure Reference, Footnote Reference, and Head Reference) serves as a bridge between the experienced users' Footnotes, Headings, Displays, Lists, and Index/Glossary clusters. The nonuser network also links four of the five Reference tags, connecting Lists, Figure, and Index commands. While a References cluster is *not* present in the system network (the system categorized each Reference tag within its category type), its presence in the user networks is not in *conflict* with the system model in two senses. First, its presence in the user network is similar to the kind of connecting nodes the system network displays for functional relatedness (e.g., the Table of Contents tag described above); from the user's point of view, the Reference tags share a similar function and are, in fact, the only semantic similarity that the "categories" they connect share. Second, the Reference bridge structure has no negative implications for *using* Reference tags in marking up a document; unlike some of the other tags, their use is not constrained by the overall document structure.

The presence of the Index/Glossary cluster in the experienced user network, on the other hand, is a deviation that *does* conflict with the system. The Index and Glossary List elements are connected through the Index command and in turn the Appendix command to

Back Matter. This pattern of network links suggests that experienced users think of Back Matter as a higher-order structure containing the Appendix, Index, and Glossary structures.[2] Examination of the system network, however, reveals a conflict, which in this case has implications for using the Appendix tag correctly. The system treats Appendix as a *major structural element*, along with Body, Front Matter, and Back Matter, under the higher-order structure of General Document. If a user tries to create an Appendix *within* the Back Matter of a document, it will not function correctly; in fact, the Back Matter tag acts as a delimiter for Appendices (i.e., indicating that the end of the Appendices section has been reached). The reason that Appendix was treated as a major document element, we found by contacting one of the system designers, was not one totally unfamiliar to aficionados of usability: The designer told us that Appendix became a major document element because it was an efficient means for implementing the automatic heading function associated with appendices. Interestingly, the pattern of links from Back Matter through Appendix to Index is also present in the nonuser network (although there Appendix is also mysteriously linked to Table of Contents and Preface).

In summary, overall comparison of the networks defined by Pathfinder revealed that the experienced user's view of document structure, as defined by patterns of shared and nonshared links, is closer to the system's view than is the nonuser's view, and is more like the system's view than like the nonuser's view. Examination of user networks, particularly in contrast to the cleanly structured "ideal" system network, reveals both the evolution in network structure with system experience and "failures to evolve," instances where discrepancies with the system structure persist and are indicated by the treatment of particular commands. In the next section, we describe a more systematic approach to the treatment of individual commands by experienced users and nonusers.

Command Definition

Following Cooke and Schvaneveldt (1988), we examined experienced users' and nonusers' understanding of system commands by categorizing each command with respect to the proportion of shared and extraneous links. For each command, the proportion of links shared with the system network links for that command and the proportion of extraneous links (relative to the number of links for that command in each experience group's network) was computed. Commands were then categorized as *well-defined, misdefined, overdefined*, or *underdefined*, based on a grand median split (over both experience groups) for shared and extraneous links. Well-defined concepts were above the median on shared links, below the median on extraneous links. The other categories reflected the other three possibilities: misdefined (low shared, high extraneous), overdefined (high shared, high extraneous), and underdefined (low shared, low extraneous).

In this analysis, experienced users had 49% well-defined concepts, 11.8% misdefined, 19.6% overdefined, and 19.6% underdefined concepts. Nonusers had 21.6% well-defined, 43.1% misdefined, 19.6% overdefined, and 15.7% underdefined concepts.

Consideration of the commands which fell into each of these categories, supported the analysis presented above based on inspection of the networks. First, the large difference between experienced users and nonusers in the proportion of well-defined concepts (49% vs. 21.6%) reflects both the overall amount of clustering in each experience groups' network and its degree of alignment with the system network. This is because the system network is so dominated by clusters: For a concept to be well-defined, it had to be linked to at least half of the other concepts in the system cluster (the grand median for shared links

---

[2]This view of Back Matter for experienced users was also confirmed by a hierarchical clustering analysis (see Kellogg & Breen, 1987).

was .50) and have *no* other links (the grand median for extraneous concepts was 0). This means that concepts could be well-defined in user networks *only* when they exhibited a fair degree of clustering (interconnectedness), and only with concepts included in the system cluster. In a complementary fashion, misdefined concepts reveal structural deviances in user networks. For example, the 22 commands misdefined by nonusers constituted substantial representation of seven of the eight basic categories defined by the system (the eighth category, Process-Specific Controls, was overdefined). The well-defined commands for nonusers suggest that only the Title Page and Definition List clusters were intact; the misdefined commands reflect the more general lack of correspondence between these participants' and the system's view of document structure. For experienced users, the well-defined concepts reflect intact clusters for five of the eight categories (General Document, Basic Text, Lists, Displays, and Process-Specific Controls) as well as the "subclusters" of Front Matter, Title Page, and Definition List. Two of the other categories (Headings and Footnotes) are congruent with the system definition, but were overdefined because of the links between the Reference tags. The misdefined concepts for experienced users did not reflect the kind of pervasive deviations from the system network shown in the nonuser network, with the exception of the Back Matter problem discussed above: Three of the six misdefined concepts were related to the Appendix and Index tags.

Finally, we analyzed users' familiarity ratings with respect to the definitional status of commands. Experienced users' ratings of commands were congruent with their well-definedness as defined by network links: Overall, .17 of these users rated well-defined concepts as unfamiliar or never used, .31 for misdefined, .30 for overdefined, and .28 for underdefined concepts, respectively. Nonusers showed less discrimination, as well as less of a tendency to rate commands as unfamiliar (none, of course, had ever been used). Overall, .12 of the nonusers rated well-defined concepts as unfamiliar, .10 for misdefined, .12 for overdefined, and .17 for underdefined concepts, respectively.

Discussion

User and System Models Defined by Pathfinder

Our intent in deriving users' mental models and an idealized system model with Pathfinder was to explore the utility of network scaling for evaluating aspects of a system's learnability and usability. The first step in assessing this utility is confirming that users' models grow closer to the system model with increasing experience. The models defined by Pathfinder confirm this assumption and reveal details of the evolution of user knowledge with experience using the system.

The nonuser network reveals a significant amount of disorganization in structure. Of the 11 "lassoed" groups, only four exist completely intact (Definition List, Glossary List, Footnotes, and Index), and *all* of the commands included in these groups have surface similarity as previously defined. Almost intact are Lists, Displays, and Title Page—but here, *discrepancies* related to surface features stand out: Nonusers linked "List of Figures" to the Lists group (it belongs to the Front Matter group), linked "Title Citation" to the Title Page group (part of Basic Text tags), and did *not* link "Example" to the Displays group (all of its other members had the word "figure" in the command name). The remaining four groups are not substantially represented.

What does the nonuser network suggest about the learnability of this system for naive users? First, despite the real-world familiarity of document structures (e.g., most naive users can be expected to be familiar with the structural layout and composition of books),

13 User and System Models

only one subgroup (title page elements) is well-developed for these users. The higher-level structure of major document elements (Front Matter, Body, and Back Matter) is not well expressed. This suggests that a strong emphasis be placed on the overall hierarchical document structure used by the system in user training and the interface itself.[3]

Second, it is clear from the network that the roles of Front Matter and Back Matter are not understood by the nonusers. This was confirmed by the familiarity ratings: 60% of the nonuser participants rated Front Matter and Back Matter as unfamiliar; in fact, these were the most frequent commands to be labeled unfamiliar. Finally, it is clear that nonusers will benefit from surface similarity in the naming of functionally related commands. This suggests that special attention be given to groups of functionally related commands that cannot be designed with surface features in common. In the system studied here, for example, commands in the Basic Text group did not share any surface features. Our results suggest that a redesign of these commands to reflect their common functionality in such features might enhance the comprehensibility of the system for learners.

Of course, it is an open question whether conceptual knowledge, such as the functional similarity among Basic Text commands, will have any real effect on learners. This question cannot be answered without performance data for learners of this system, which we did not collect. However, the present data *do* show that the comparison of system and nonuser Pathfinder networks can reveal the lack of appreciation of the functional similarity.

An example discussed by Carroll, Mack, and Kellogg (1988) suggests an analogous lack of understanding of functional similarity that did have performance consequences for learners. The example involved the task of creating new folders in the *Lisa*. All new documents were created in this system by "Tearing Off Stationery" from a paper pad icon. This method was applied to the creation of folders as well. However, the folder icon and other paper pad icons, which were functionally similar, did not share salient surface features (e.g., paper pads were all labeled "Paper" and had similar icons, but the folder icon looked different and was simply labeled "Folders"). Carroll and Mazur (1986) observed that learners had difficulty discovering how to create new folders with the system, though they were able to create other types of documents. In fact, subsequent versions of this desktop interface changed the new folder method by adding a special action for folder creation. In this case, of course, there is no Pathfinder data. But the present data strongly suggest that a Pathfinder network for naive *Lisa* users *would* reveal the lack of perceived similarity between folders and other paper pads.

The evolution of user knowledge toward the system model is shown by the network for experienced users. In their network, all 11 subgroups defined by the system are intact (see the "lassoed" groups in the Appendix). The experienced users' network also shows more developed clustering than the nonuser network and clearer hierarchical structure. Experienced users demonstrate an understanding of the major structural elements of the system model (with the exception of Appendix, discussed previously). They *use* this structure, particularly Front Matter and Back Matter, to organize the major document elements that appear from the front to the back of a document. Experienced users rarely marked Front Matter or Back Matter as "unfamiliar" or "never used."

On the other hand, there are also similarities remaining with the nonuser network. For example, the use of Reference tags as a bridge among different clusters can be seen in both user networks. Experienced users link the Basic Text group to the rest of the network through a "Footnote-Note" link which also occurs in the nonuser network. Thus, bridging

---

[3]Although the system we studied was command-driven, a menu-based interface to the system has been implemented and could emphasize the system view of document structure in a way the current system is unable to do.

links in the experienced users' network often reflects the kind of (a priori) semantic similarity that characterize nonuser networks in general.

The experienced users' view of document structure and system commands does not coincide perfectly with the system model. The two major discrepancies involve the linking of Reference tags and the grouping of Appendix, Index, and Glossary commands under Back Matter. As discussed previously, the linking of Reference tags is a discrepancy without usability implications. The organization of commands under Back Matter, however, does have performance implications: The experienced users' treatment of Back Matter, and particularly the Appendix tag, suggests that they would experience difficulty marking up a document containing an appendix, because they would place the appendix *within* the back matter, rather than *before* it as required by the system. The comparison of the experienced user and system models suggests that system usability could be improved if it was redesigned to allow placement of appendices within the back matter while still providing the automatic heading function. In the current arrangement, the system handles the headings, but at the cost of an unintuitive structuring of document elements.

The present results underscore the importance of defining a system model. Without representing the system model, the difference in experienced users' organization of major document elements would not have been found. The more typical comparison of expert and novice knowledge structures *can* show the evolution of the organization of user knowledge with experience, but for assessing usability in terms of "model congruence" as suggested here, a representation of the system's view of the task domain is essential. The more completely the system model incorporates the functional relations and organization of conceptual knowledge that users *ought* to have of the system, the more informative the comparison with user models will be. Alternative methods of defining a system model (e.g., obtaining judgments from system designers, using system specifications, examining the system directly) are possible and must be evaluated in terms of how well they represent the target knowledge for the system.

The system model we derived from the system documentation for this study, in retrospect, is a fairly good representation of the necessary conceptual knowledge for this system, but it could easily be improved. Our system model was incomplete in representing some of the functional relationships in the system: for example, it did not represent what the Back Matter is *supposed* to contain from the system's point of view (Index and Glossary List). Rather, Index tags were represented as an isolated cluster, and Glossary List tags were represented only within the Lists cluster.

Another aspect of functionality only partially represented in our system model has to do with the system's double categorization of the Table of Contents and List of Figures tags. The system linked Table of Contents to the Headings tags in addition to Front Matter tags because the Head Level tags are used to generate the table of contents. Similarly, List of Figures relies on Figure tags to construct itself. These functions only work properly if the user specifies a "twopass" command option when sending the marked-up document to an output device. The same is true of footnote references: They will only print properly with the "twopass" option. We might have been able to examine our users' understanding of the twopass mode had we included "twopass" as a concept to be sorted and allowed participants to place concepts in multiple piles (or had we insisted that duplicates of some of the concepts be included). Again, in retrospect, we might well have left out some of the more detailed commands (e.g., the subgroup of Definition List tags) which share obvious surface features (and thus are likely to be grouped together by all participants), and which are less interesting in terms of the functional relations embedded in the system. To gain the most information from the comparison of user and system models, and to keep the number of concepts to be judged within reasonable bounds, the analyst may have to select the most important functional and conceptual relationships to include in the system model and the comparison set.

Using Network Scaling in System Design

The results of the present study suggest that Pathfinder networks have much potential for providing information on the match between users' views and the system's organization of a task domain. In particular, it seems possible to extend the use of network scaling beyond issues of organizing information in the interface to questions about the inherent conceptual structure of the system and its functionality.

The comparison of user and system networks can yield interesting and potentially valuable information about a system's learnability and usability. However, discrepancies revealed by such a comparison must be evaluated in the larger context of the user's task domain if their import for usability in the system's real-usage context is to be correctly anticipated. For example, our results suggest that on the whole the system is congruent with experienced users' models of document structure. How significant the Appendix/Back Matter deviation is for the system's usability will depend on *how often* users will engage in the task of marking up appendices. Nevertheless, the ability of the Pathfinder analysis to reveal the discrepancy, and the relatively low cost of deriving user and system networks, suggests that scaling methods have an appropriate cost/benefit profile for use in system design.

A second issue is *how* the outcome of the Pathfinder analysis and comparison of user and system models can guide design. McDonald and Schvaneveldt (1988) recommend that systems be designed to conform to user models. The comparison of Pathfinder networks for users (or potential users) and a system can suggest specific ways the system might be restructured to be more compatible with users' models. We suggested this in the case of the treatment of appendices.

However, another use of Pathfinder analyses is possible. While we agree that advantages arise from structuring the system in congruence with the users' model of the task domain, we do not believe this is strictly necessary for a system to be learnable or usable. Discrepancies revealed by system and user networks can be viewed as imposing a *communicative burden* on the system and the system image: They indicate where designers must take extra measures to convey the system's (deviant) conceptual structure to the user.

By this we do not mean to suggest that discrepancies with the system model are the users' problem, nor to ordain that user models be based on system models. In fact, it seems likely that attempting to migrate the users' model to the system model through interface design, as opposed to redesigning the system to be in congruence with the users' model, is the more difficult alternative. Rather, we mean to emphasize that user/system congruence is the desired goal state, and there is more than one way to support it through system design. McDonald and Schvaneveldt (1988) offer several suggestions about how interface characteristics might be designed to effectively communicate a system's structure. The Pathfinder analysis can indicate where special attention should be given when using such techniques.

Pathfinder networks can provide a useful summary and representation of the conceptual structure of a system, from both the system's and the users' points of view. The more completely and accurately the analyst models the functional and conceptual structure of the system in the system model, the more the comparison with user models can reveal about the system's learnability and usability. Pathfinder networks do show the evolution of structure in user knowledge with system experience, and the treatment of individual commands and patterns of links are interpretable. While further work will be needed to use network scaling models as part of an empirical test of the mental model hypothesis (i.e., relating user/system discrepancies to performance outcomes), the results reported here validate the ability of network scaling analyses to reveal potential usability and learnability problems, and lay the foundation for empirical validation of the relationship between a user's declarative knowledge of an interactive system and performance.

13 User and System Models

Appendix 1 - System Network

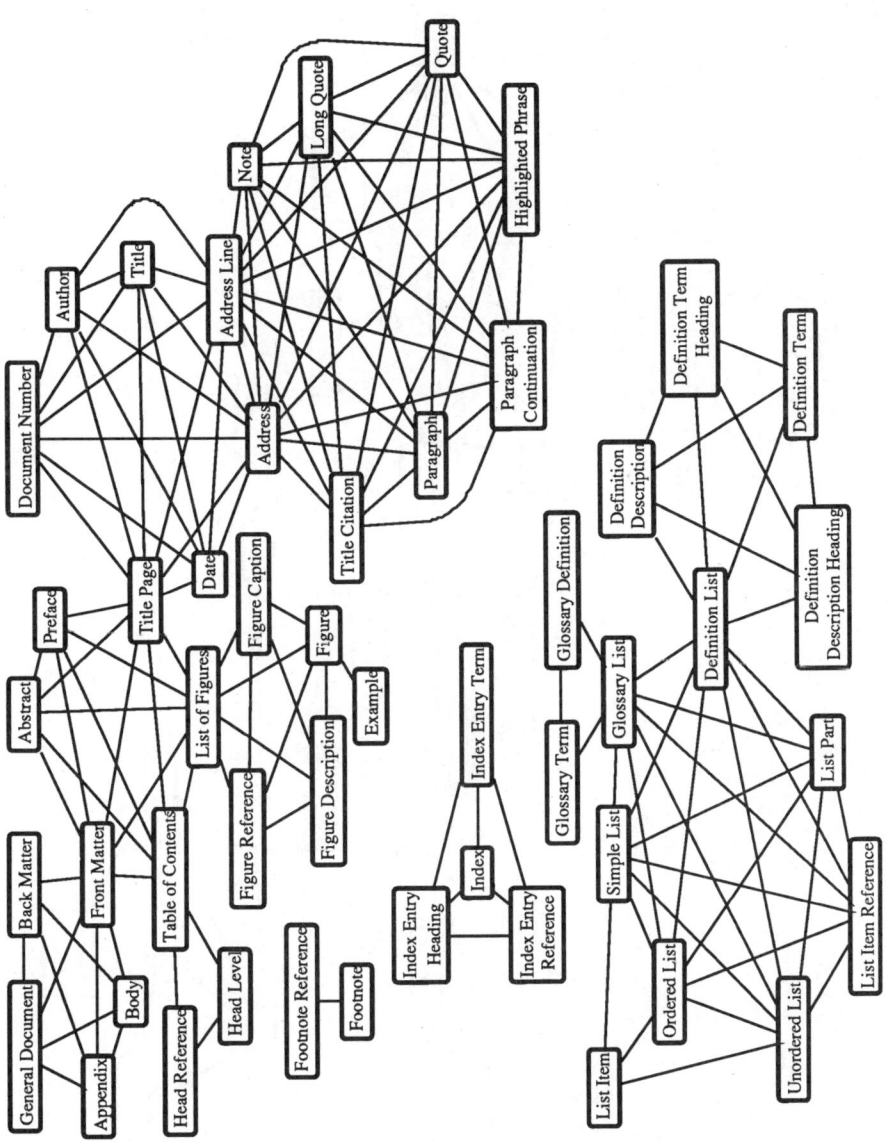

Appendix 2 - Experienced User Network

13 User and System Models

Appendix 3 - Nonuser Network

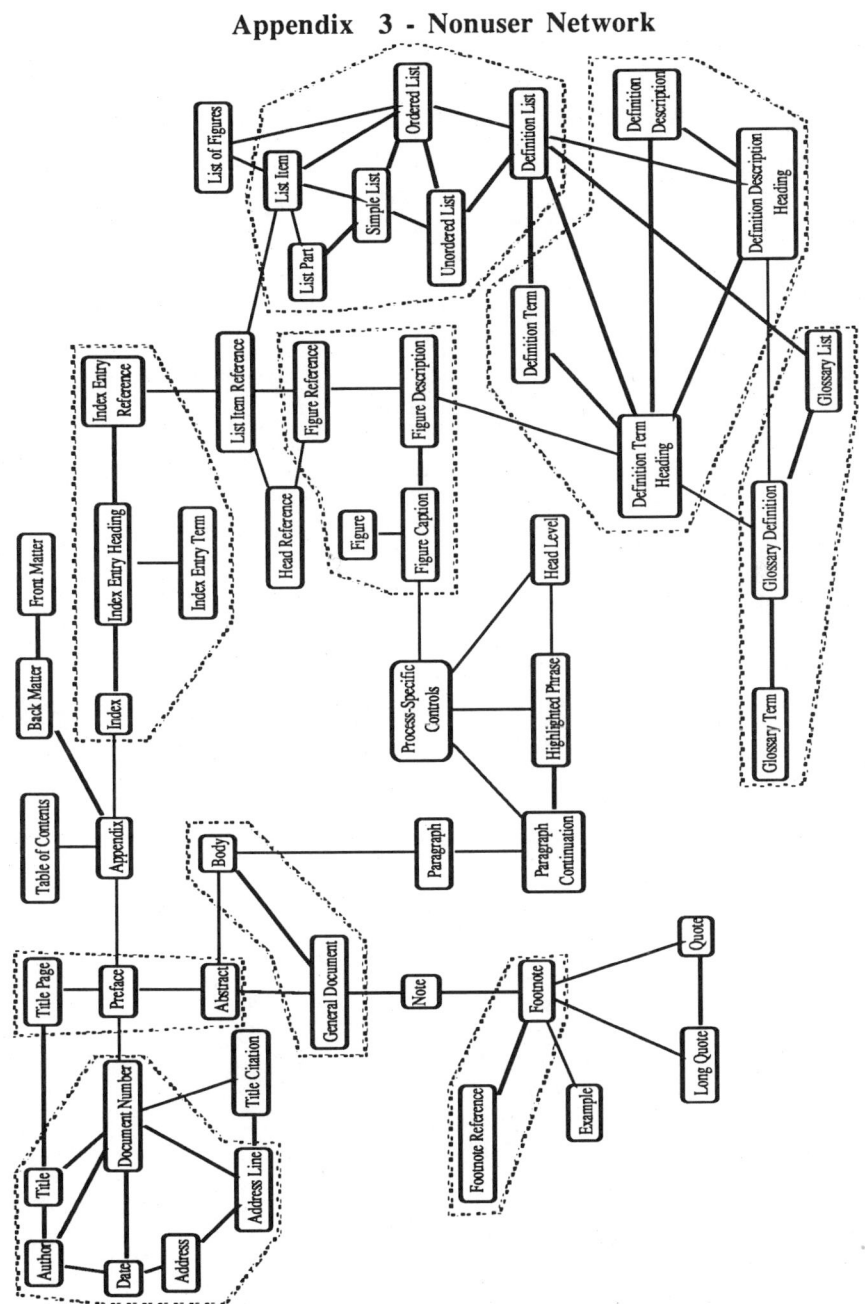

Chapter 14

Hypertext Perspectives:
Using Pathfinder to Build Hypertext Systems

James E. McDonald, Kenneth R. Paap, and Deborah R. McDonald

Hypertext is a "hot" topic. Interest in hypertext systems has burst upon the computing scene—which is a bit surprising given that the basic concepts have been around for more than forty years (Bush, 1945). Why this sudden interest in "nonlinear text?" In this chapter we examine some of the claims being made for hypertext and attempt to identify those that show promise of producing real gains in peoples' ability to acquire information, as distinct from purely technological embellishments. As with most technological advances, there seem to be costs associated with the benefits of hypertext and we will discuss how knowledge elicitation and representation techniques might be used to minimize the impact of some of these costs. In particular, we will show how Pathfinder network scaling can be used to construct hypertext systems in order to facilitate information access.

Hypertext

Vannevar Bush (1945), the father of hypertext, began his seminal work by identifying a problem that has become ever more acute: There is simply too much information for anyone to assimilate, and it's not readily available. Furthermore, even if all the information in the world were available, it would not all be relevant to any given task. Although technology has certainly facilitated the accumulation of knowledge, it has done little to make it more accessible. The trick, as Bush saw it, is to devise a system that allows users to access relevant information without forcing them to wade through a lot of irrelevant facts.

Bush proposed the *memex* as a way of mechanizing the scientific literature, placing it within reach of individual researchers. He envisioned this contraption as containing a very large library, along with personal notes, photographs, sketches, and so forth. Although not technically feasible in 1945 (many would question whether it is feasible even today), the design of the *memex* was inspired by Bush's belief in how the mind works:

> The human mind...operates by association. Man cannot hope fully to duplicate this mental process artificially, but he certainly ought to be able to learn from it. One cannot hope to equal the speed and flexibility with which the mind follows an associative trail, but it should be possible to beat the mind decisively in regard to the permanence and clarity of the items resurrected from storage.

Thus, researchers would use the *memex* to build *trails* by identifying links between items of interest. Trails could later be followed by the builder, or other users, to recreate episodes. Over time, networks of trails would organize knowledge in an efficient, useful way. The *memex* was intended to complement human abilities, not replace them. Bush was motivated by a perceived problem and formulated a solution based on his understanding of human strengths and weaknesses. Although a technological solution, the *memex;* was not inspired by technology.[1]

What is Hypertext?

One of the first tasks in identifying and evaluating the claims being made for hypertext should be to figure out exactly what it is. Unfortunately, perhaps because of the current hyperbole surrounding the topic, hypertext doesn't lend itself to such analysis. Although hypertext systems do seem to share a number of common characteristics, such as the ability to manage nonhierarchically organized information and to allow interactive branching among information "nodes," there doesn't seem to be any necessary or defining features. For example, the use of "windows" containing chunks of information is often viewed as important, as is the ability to provide users with a single, coherent interface. However, these are neither necessary nor sufficient criteria for classifying systems as hypertext.

In his excellent hypertext review article, Conklin (1987) identifies machine-supported links as an essential feature of hypertext. Without additional qualification, however, this requirement is not specific enough since a definition based on machine-supported links would encompass virtually all existing database management systems. To truly qualify as hypertext, Conklin contends that a system must support direct references from one information chunk to another and allow users to easily establish new relationships between chunks. In other words, it must be possible to link any information node to any other node without restrictions (e.g., the structure need not be hierarchical). Furthermore, if a user wishes to establish new links, a good hypertext system should have features that support their construction. Alternatively, if the user is simply engaged in information retrieval by browsing along a pathway of nodes, a true hypertext system should have features that support the selection of relevant nodes and minimize the likelihood of getting lost. Therefore, although windowing systems, outline processors (e.g., ThinkTank), text formatters (e.g., Troff, Scribe), database management systems, and even file systems each possess some features of hypertext, they all fail to qualify for various reasons.

A walk in the woods. In order to make the rather fuzzy concept of hypertext a bit more concrete, imagine a wilderness in which there are numerous crisscrossing and overlapping trails. Within this network of trails there are also many starting points and destinations for journeys through the wilderness, as well as numerous points of interest along the way. Travelers typically follow established trails, but may choose to blaze new ones depending on their knowledge and daring.

This analogy depicts the opportunities for exploration and discovery afforded by hypertext systems. It also suggests some of the potential problems, such as disorientation and confusion, are consequences of wandering around in hypertext without a good map and accurate trail signs. However, it fails to capture some important aspects of the information wilderness faced by users of hypertext systems as well. First, hypertext eliminates most of the limitations inherent in physical space. Specifically, hypertext removes the restrictions of traditional information systems consisting of printed books and journals, since information nodes are all potentially equidistant. Second, the ability to get from point A to point B efficiently is less important in hypertext than what's learned along the way. The content of information nodes and the order in which they are visited are what counts, not necessarily how quickly one arrives at the destination. Finally, hypertext systems often do a poor job of signaling established trails. In fact, the kinds of trails that Bush envisioned and that

---

[1]This is in contrast to the recent interest in hypertext, which often seems to be motivated more by technological feasibility than any perceived need.

exist in our hypothetical wilderness aren't typically represented in hypertext systems at all. Instead, users are presented with sign posts indicating nearby points of interest. Rather than trail signs such as "Rim Trail" or even "Trail #117," the user is more likely to encounter references such as "Dead Man's Gulch" or "Sourdough Hill." The point is that little indication of the relationship between trail segments and well-established trails is provided. A potential solution to this problem is to provide orienting aids, and the hypertext browser described later uses maps based on various perspectives for this purpose. Extending our walk in the woods analogy, one perspective may show trails relevant to botanical study while another map shows trails leading to the best fishing spots. Although segments of the two maps may overlap, confusion is reduced by selecting and viewing the map that is appropriate to the current intention or goal.

In spite of the difficulty of precisely defining hypertext, the excitement over this new technology continues, primarily because of the widespread belief that a linear organization of information is inadequate for many applications. Since information embedded in computers can be represented and used in qualitatively different ways than on paper, it seems reasonable to believe that these new representations will aid in the process of knowledge acquisition in unforeseen ways (Fischer, 1987). For many of us, however, the nagging question remains, "Why hypertext?"

Text Comprehension

Reading and Writing

Although a number of hypertext systems are now in use and interest in hypertext is intensifying, the psychological ramifications of hypertext as a tool for acquiring knowledge have not been empirically validated— even worse, we claim that the important psychological variables have yet to be identified. In order to make the topic more manageable, we will explicitly distinguish between features of hypertext systems that are important for reading (or trail following) and those that are more concerned with writing (or trail blazing).[2] We will begin with a discussion of reading and go on to suggest a methodology for building networks of trails appropriate for hypertext systems. One area of psychological research that is clearly relevant to acquiring information from hypertext systems is text comprehension.

According to the van Dijk and Kintsch (1983) model of text comprehension, there are three levels of representation: surface memory, text-based memory, and the situation model. The *surface memory* representation for a text is the verbatim encoding of words and phrases in the text. The surface memory provides the basis for the *text-based memory* representation. Text-based memory is for the propositional content and structure (micro- and macrostructure) of the text. The *situation model* not only involves memory for the meaning of the text and the unique situation described, but also incorporates the reader's goals and prior knowledge. Situation models are sometimes referred to as *conceptual* or *mental models*, and it is this level that is most relevant to our discussion of hypertext.

Considerable evidence has accumulated for the validity of distinguishing between textbases and situation models (Dellarosa, 1983; Fletcher, 1984; Perrig & Kintsch, 1985; Schmalhofer & Glavanov, 1986; Weaver & Kintsch, 1985). The consensus of these

---

[2]It should be noted that Bush (1945) did not distinguish between reading and writing and, in fact, felt that much of the power of the *memex* would come from the ability to combine new knowledge with existing information. Current authors often blur this distinction as well.

researchers is that people do not always remember the surface form or even the syntactic deep structure of a text when they remember what the text was about. Furthermore, the experiments in this area show that people's representations of a written text go beyond information directly presented in the text. For example, many pragmatic inferences are drawn as the propositional content of the text interacts with the reader's world knowledge. Similarly, reading nonlinear text might influence the development of conceptual models, although to our knowledge no empirical studies have examined such effects of hypertext.

Unlike traditional text, hypertext represents multiple relationships among various "chunks" of text simultaneously. Traditional linear text encourages the reader to follow a single sequence of paragraphs (although readers may choose to jump around), whereas hypertext provides alternative organizations for the same information, encouraging readers to wander. In terms of our walk in the woods analogy, this would be comparable to restricting one traveler to a given trail while allowing another to travel any route he chooses. The mental representations that these two travelers would form of the wilderness are likely to be quite different, even for a particular trail they both know. The second traveler, free to explore the wilderness, selecting points of interest to visit at will, is likely to posses a more elaborate mental model and, if the trails are carefully laid out, a more accurate model as well.

The Benefits of Multiple Perspectives

What benefits might we expect from providing readers with nonhierarchical, richly interconnected networks of information as opposed to linearly organized text? hypertext proponents seem to assume that the benefits are obvious, although the only benefit that is obvious to us is that having access to necessary information is better than not having access to it, a tautology. At the same time, these proponents acknowledge the problems of confusion and disorientation associated with hypertext systems (Charney, 1987; Conklin, 1987; van Dam, 1988), which makes even the promise of increased information availability suspect. Postponing a discussion of disorientation, there has been some recent interest in the effects of providing readers with more than one perspective on a given topic. In particular, researchers have been interested in the ability of readers to remember details and draw inferences, based on the way in which information is presented.

Mannes and Kintsch (1987) were interested in the notion that learning from text is different from memory for text. Using outlines as advance organizers, they showed that when subjects studied advance organizers whose structure differed from the succeeding text they created more richly elaborated mental representations of the textual materials. In a related study, McDonald (1987) found that requiring readers to form a more complex conceptual model by having them read two different versions of the same text enhanced their learning of the content, compared to readers who read the same version of the text twice.

In her study, McDonald devised two factual texts that both described the violin. These texts contained approximately the same information, but one text described the violin from the perspective of the historical development of the instrument (the *historical* version), whereas the other described the same aspects of the violin from a structural/functional perspective (the *categorical* version). Subjects were either in the Inconsistent condition and read both the *historical* and *categorical* texts, or they were in the Consistent condition and read the same text twice (either *historical* or *categorical*).

After reading the texts, subjects were given a true/false test containing both memory items (either verbatim excerpts or statements that had been paraphrased from the texts) and inference items (conclusions of either formal inductive or deductive arguments), as well as statements that were extraneous to the texts (foils). For inference items, the premises were presented in the body of each of the texts, but the conclusions were not. The subjects' task was to decide whether each statement was true or false according to the information they had read in the texts. The important results for the present discussion were that, as expected, subjects in the Consistent condition were better able to recognize verbatim and paraphrased items than subjects in the Inconsistent condition. However, subjects in the Inconsistent condition were better able to draw correct inferences than subjects in the Consistent condition.

The results of this study are intriguing. Earlier we suggested that separate maps (perspectives) should facilitate following the trail relevant to the current goal. But suppose our adventurer pursues flora on one outing and fish on the next. If McDonald's results transfer to the wilderness, we would expect the eclectic outdoorsman to have made more correct inferences about the local terrain than the dedicated botanist (or fisherperson) who travels the same trail every time.

There does appear to be some support for the notion that providing readers with several perspectives on the same topic is beneficial, at least for drawing inferences, implying that their conceptual models of the topic are more complete. However, rather than several discrete perspectives (or well-marked trails), hypertext systems typically provide users with tangled, poorly marked perspectives. Although Eileen Kintsch (1988) has speculated that forcing readers to work hard at uncovering structure may produce even more elaborate mental representations, neither the Mannes and Kintsch (1987) study nor the McDonald (1987) study investigated the consequences of flipping back and forth from one perspective to another. Subjects in the Inconsistent condition first read one coherent passage (or outline) and then read a different coherent passage. Unrestricted access and frequent switchbacks between perspectives may be harmful rather than beneficial. This is an empirical issue and one that McDonald is currently pursuing. We will now turn our attention to how hypertext networks might be built and represented in order to facilitate navigation and the development of useful conceptual models.

Empirically Derived Conceptual Models

A major theme of our work in the area of interface design has been that effective human-computer interaction depends on communicating system structure and organization to users via the interface (McDonald, Dearholt, Paap, & Schvaneveldt, 1986; McDonald & Schvaneveldt, 1988). The premise upon which this work is based is that users develop conceptual models of systems and that, in general, the more accurately these conceptual models correspond to the organization of the system the better. As a consequence, this line of research has focused on developing and evaluating methods for eliciting, representing, and communicating appropriate models of systems.

We believe that users' conceptual models can be usefully characterized as "schemata" in which parts of the system are associated because they are functionally related or because they have common features (Young, 1983). One of our objectives has been to develop an empirical methodology for eliciting and representing such conceptual models. The general methodology consists of (1) obtaining estimates of relatedness for all pairs of relevant system concepts (e.g., objects and actions), (2) analyzing the obtained proximity data using one of several scaling techniques, and (3) mapping the resulting representations onto interface organization (e.g., menu layout). A number of scaling techniques have been employed, such as hierarchical clustering (McDonald, Stone, Liebelt, & Karat, 1982; Tullis, 1985), multidimensional scaling (McDonald, Dayton, & McDonald, 1988), and simulated annealing (McDonald, Molander, & Noel, 1988). In this chapter, however, we will focus on the use of Pathfinder network scaling because we believe that it is most appropriate for hypertext systems.

Using Pathfinder to Blaze Trails

As we said, it is common for users to become disoriented when navigating through systems that require them to search for desired information by traversing machine-supported links. This may be due to the nature of the "trail signs," as suggested, or because of a more fundamental problem associated with the complexity of such systems. The type of disorientation common to hypertext systems is also found in many "deep" menu-based systems (cf., Paap & Roske-Hofstrand, 1988, for a discussion of the breadth-depth trade-off in menu design).

Roske-Hofstrand and Paap (1986a) compared several different prototypes for a Control-Display Unit (CDU), the primary interface for the complex flight management system being developed for NASA's advanced concepts simulator (Chappell & Sexton, 1985). The CDU is a fairly large menu-driven system that can be used to both acquire knowledge and take actions. In the prototypes evaluated, the pilot is shown a display which contains some text, sometimes a prompt for information that is to be entered, and a menu of options. The pilot selects options from the menu in order to display new panels of information. Pilots with limited CDU experience often have difficulty finding their way from where they are (the current panel) to where they want to go (the target panel). The goal of this study was to compare menu organizations based on Pathfinder networks to an organization based on the specifications of the original design team.

A subset of 34 panels from the original specifications for the CDU were selected. The information contained in each panel of the CDU was treated as a single, coherent concept. In the initial phase, four experienced pilots were familiarized with the panels, then rated the similarity between each of them. Various Pathfinder analyses were performed on the resulting proximity matrix and the Pathfinder networks (PFNETs) were mapped onto menu organization.

Sixteen experienced pilots participated in the validation test. The test was a training session that consisted of four blocks of 34 trials. On each trial the pilot read a scenario that described a set of current conditions, a general goal, and a specific question that could only be answered by accessing the appropriate CDU panel. The total task time for each trial was measured from the onset of the question to the pilot's response. Each of the 34 menu panels contained the target information for one trial in each block.

The results of this study support the use of PFNETs as a basis for interface design: Prototypes based on cognitive networks were consistently easier to learn than the prototype based on the original design specifications. In particular, this study shows how effective networks of hypertext trails might be constructed. Nevertheless, the CDU is relatively simple compared to most hypertext systems, and complex systems may require accurate maps as well as good route planning. A methodology based on user knowledge might be particularly appropriate for complex systems that are the products of evolution, rather than any systematic design process. In such cases, the "blueprints" which could be used to describe the systems to users simply don't exist. A classic example of this process of evolutionary growth, rather than controlled design, is the UNIX operating system and its associated documentation.

Using PFNETs to Build a Hypertext Browser

We have constructed a prototype of an interactive, graphic-based, hypertext Browser (HyBrow) for the UNIX online documentation system (the *man* system) using the general methodology outlined above. HyBrow enhances the user's ability to access documents, particularly in a nonlinear fashion, by providing machine-supported links among *man* entries. Our objectives for HyBrow are to facilitate the ability of UNIX users to locate appropriate information and to encourage the development of useful conceptual models of UNIX (commands and utilities). The design of the HyBrow system was based on data obtained from experienced UNIX users and information extracted from the UNIX documentation itself.

In line with the methodology described, we employed empirical techniques to elicit and represent the knowledge of experienced UNIX users. We began the HyBrow development project by obtaining estimates of relatedness among UNIX concepts using two different techniques. Estimates of functional relatedness were obtained by having experts sort commands into related categories, whereas estimates of procedural relatedness were derived from co-occurrence data, that is, the probability that particular commands will be used together (i.e., temporal proximity).

The Functional Perspective: Elicitation of Command Knowledge

The functional networks used in the HyBrow prototype were constructed from data obtained from 15 experienced UNIX users. Seven were classified as expert or intermediate-to-expert users, whereas eight were classified as intermediate users. The knowledge elicitation procedure used during this phase of the project was card sorting. An index card was prepared for each of the 219 UNIX *man* Section 1 functions (Berkeley UNIX version 4.2 running on SUN minicomputers). Subjects were instructed to sort commands according to functional relatedness. In an attempt to reduce the "hierarchical filtering" associated with sorting, subjects were encouraged to use duplicate cards when they felt a particular command belonged in more than one pile. They were also instructed not to sort commands they didn't know.[3]

Deriving the Distance Matrix. A proximity matrix was created for the subset of 152 UNIX commands that were known by at least half of the 15 experienced users. The matrix was constructed by subtracting the average conditional probabilities associated with each pair of commands from one. It should be noted that computing proximity in this manner tends to increase the relative distances between pairs in which both items are not sorted by all of the raters (see McDonald, Plate, & Schvaneveldt, Chapter 11, this volume, for a more complete discussion of relatedness functions).

Scaling: Hierarchical Clustering & Pathfinder Networks. A hierarchical cluster analysis of the 152 commands was performed using the minimum, or single-link, method (Johnson, 1967). The minimum method was selected because of its relationship to Pathfinder with $r = \infty$ and $q = n-1$, which facilitates mapping the resulting subnets onto clusters. The minimum method also tends to produce fewer distinct clusters, reducing the amount of effort required during the category labeling phase. The hierarchical cluster analysis produced a total of 83 distinct clusters. The Pathfinder analysis resulted in a network with 184 links.

Establishing the Category Structure. The design of HyBrow requires us to graphically represent "subnets" (connected portions of the Pathfinder network) that correspond to high-level concepts. Figure 1 shows one approach to representing subnets that involves the combined use of hierarchical clustering, MDS, and Pathfinder. First, clusters were obtained from the hierarchical cluster analysis. Then the appropriate subsets of distance estimates for each cluster were subjected to a two-dimensional, nonmetric MDS.

---

[3]Early experience with the initial set of 219 commands suggested that we had been too restrictive. Therefore, the command set was expanded from 219 to 373 by including new commands that were commonly used (see the Procedural Perspective).

Finally, the MDS coordinates were used to guide node (command) layout, and the links specified by the Pathfinder solution were added. Comparing the hierarchical and network representations of these clusters is informative. In Figure 1, for example, not only are the two clusters that consist of directory-level and file-level commands evident in the PFNET, but the "bridge" between these two clusters, from *rmdir* (remove directory) to *rm* (remove file), can be seen as well. It is tempting to speculate that this connection represents more than the simple fact that both of these commands remove things. Most UNIX users will recognize that these commands have another fundamental relationship as well; in order to remove a directory, all of the files in the directory must be removed first.

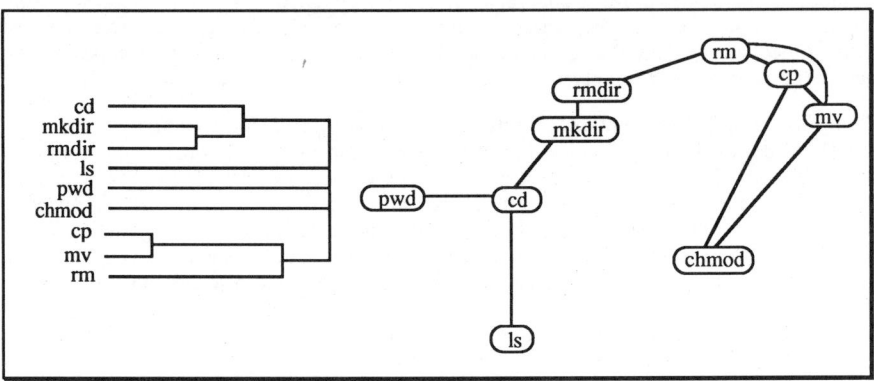

Figure 1. A comparison of hierarchical clustering (Minimum Method) and network analysis [PFNET($r = \infty$, $q = n-1$)] for UNIX directory commands.

A similar comparison of hierarchical cluster analysis and Pathfinder networks is shown in Figure 2. In this example, the subset of commands used in programming (e.g., compilers, interpreters, etc.) was selected. Rather than relying on an MDS analysis to help lay out the network, only the constraints inherent in the Pathfinder solution were used. This procedure is somewhat more subjective than the MDS approach, although many of the layout rules, such as minimizing link crossings, are relatively well-specified. For those familiar with programming, but unfamiliar with UNIX, the central role of the C compiler (*cc*) may seem a bit strange. An experienced programmer might reasonably expect the assembler (*as*) to occupy such a role, but probably not a particular language compiler. However, those familiar with UNIX will recognize the central importance of C to UNIX (e.g., UNIX is written in C). Other details in the Pathfinder representation can be observed in this comparison as well, such as the connections between the C utilities (*cb, cpp*, and *lint*) and the C compiler, and the particular way in which the Pascal functions (*pc, pi, pix*, and *px*) are linked. In both Figures 1 and 2 the PFNET provides more information than the corresponding hierarchical representation. Indeed, the minimum hierarchical cluster solution can be derived from the PFNET, but not vice versa (cf., Dearholt & Schvaneveldt, Chapter 1, this volume).

14 Hypertext Perspectives

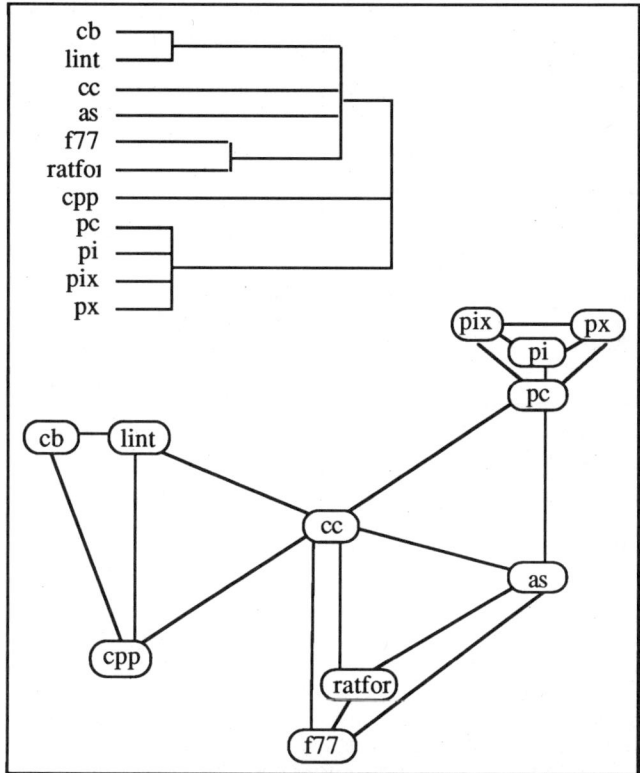

Figure 2. A comparison of hierarchical clustering (Minimum Method) and PFNET($r = \infty$, $q = n-1$) for UNIX programming functions.

Labeling Categories. The complete functional perspective requires not only valid categories, but appropriate category labels as well. Judgments of relatedness contain error, and the attempt to identify subnets, described above, is subject to the criticism that some of the clusters correspond to conceptual categories, whereas others are artifacts of the procedures employed. One approach to resolving such questions is to obtain more data from experts. Therefore, four of the most experienced UNIX users in the group were asked to judge the goodness of the clusters obtained from the hierarchical cluster analysis and to name them. Judges were shown all 83 clusters in a random sequence and asked to rate the goodness of each cluster on a five-point scale, ranging from 1 (Very Bad) to 5 (Very Good), and to provide appropriate names for all but the Very Bad clusters.

The results of the cluster-rating phase show a general tendency for the average cluster goodness to decrease as cluster size increased ($r = -.50$), with the ratings for clusters ranging in size from two to five commands averaging 4 or better (Good to Very Good). In the next step, the ratings and names supplied by the judges were used to eliminate artifactual clusters. As an example, Figure 3 shows the cluster corresponding to communications functions, with the average ratings of the original nine clusters on the left and the reduced version with four categories on the right. Reduction was accomplished by comparing the names given by the judges for each of the smallest clusters with those given for the clusters above them in the hierarchy, and collapsing smaller clusters into larger ones when the names were the same.

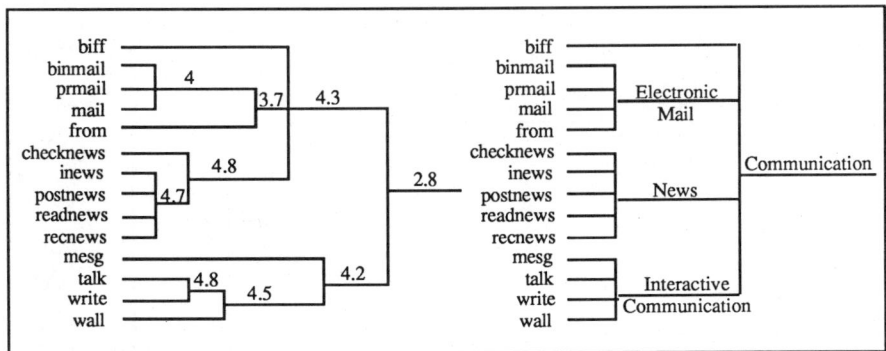

Figure 3. A hierarchical cluster analysis (Minimum Method) for UNIX communication commands with average "goodness" ratings and after reduction based on naming.

The final result of using Pathfinder to complement the hierarchical cluster analysis was 10 categories and 30 subcategories, with only 5 terms left in the miscellaneous category. It is our belief that this category structure resembles the schematic mental organization available to an experienced UNIX user. A common way of using HyBrow is to begin with the top-level category network and make successive selections first from the top-level category (see Figure 4) and then from any subcategories until the appropriate subnet of command nodes is revealed.

The Procedural Perspective: Obtaining Command-Event Protocols

Our approach to developing a procedural perspective was to obtain a number of protocols from nine experienced UNIX users. Users' *.login* files were modified to increase the size of their *history* records so that all of the events that occurred during a session were automatically captured and, at the end of each session, mailed to us. Event records were collected for a period of approximately six months and provided a database of over 75,000 command-line entries. A summary of the event-record data for the first of our two samples is shown in Table 1.

A total of 41,372 commands was obtained. Of these, approximately 58% came from section 1 of the *man* system. The number of unique commands used by individuals ranged from 42 to 133. Interestingly, as shown in Table 2, the top 10 commands accounted for 66% of all commands issued—the top three commands accounted for 32%! Although these statistics would undoubtedly change somewhat from one group of users to another (e.g., *rsh* is commonly used only when several computers are networked together), they are probably fairly typical of command usage.

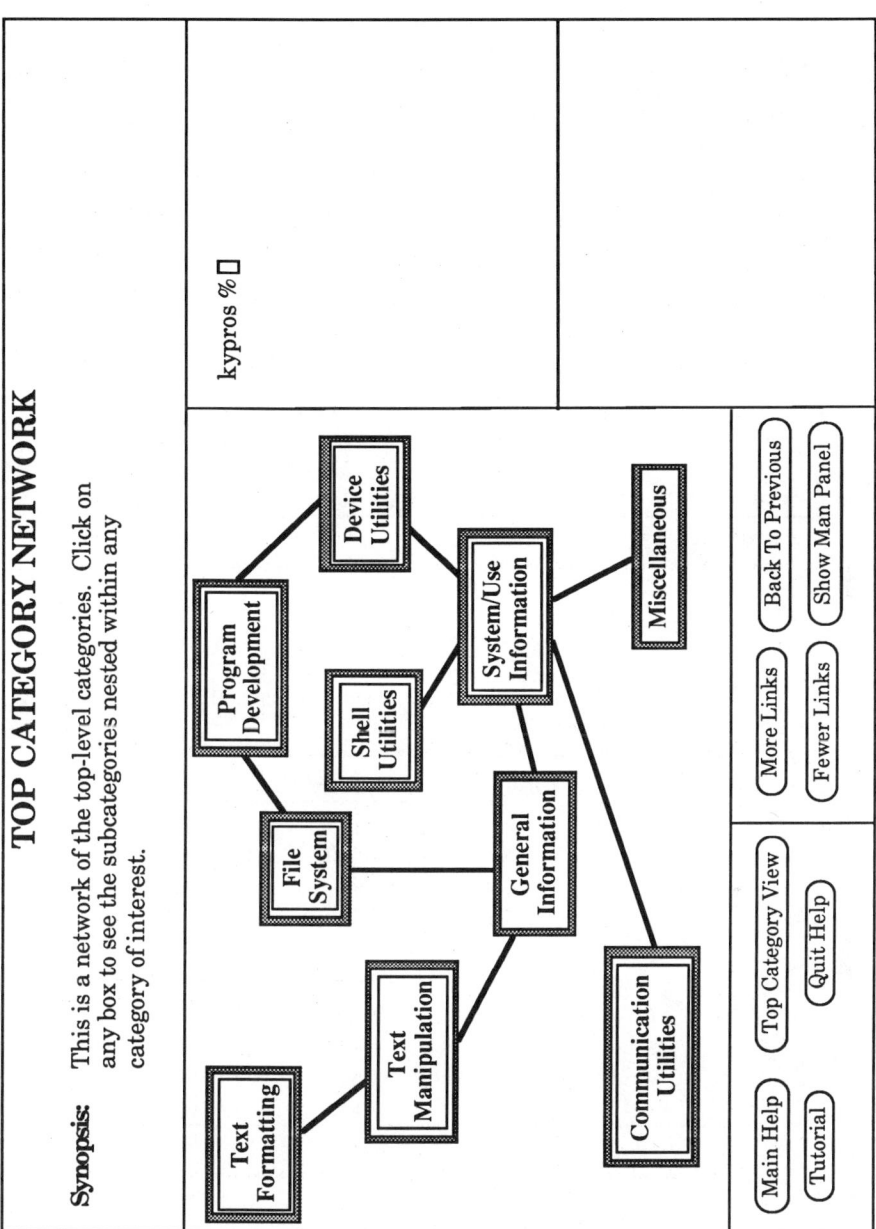

Figure 4. The top-level category network for HyBrow. Selecting any node will reveal the subcategories nested within any category of interest.

Table 1. A summary of the UNIX event record data showing the number of commands recorded for each user and the number and percentage of these commands that come from Section 1 of the *man* system.

User	All Commands		Section 1 Commands		
	Unique	Total	Unique	Total	Percent
1	58	2,124	39	1,197	56%
2	72	1,139	41	685	60%
3	47	1,370	32	814	59%
4	97	2,786	57	1,728	62%
5	42	863	26	592	69%
6	84	9,499	51	4,994	53%
7	122	15,894	75	11,225	71%
8	49	2,196	27	771	35%
9	133	5,501	84	3,178	58%
Average	78	4,597	48	2,798	58%

Table 2. The 10 most frequently used commands from the UNIX event record study.

Rank	Command	Frequency	Percent
1	ls	4,719	11.41%
2	cd	4,420	10.68%
3	more	3,961	9.57%
4		(pipe)	3,372	8.15%
5	mail	2,346	5.67%
6	emacs	2,078	5.02%
7	fg	2,060	4.98%
8	pwd	1,781	4.30%
9	rsh	1,588	3.84%
10	rn	815	1.97%
Total		27,140	65.59%

14 Hypertext Perspectives

Deriving a Proximity Matrix and a Procedural Network. The event records from all of the users were combined into a composite frequency of co-occurrence matrix by incrementing the appropriate cells each time a particular command followed another in an event record.[4] The data were then converted to conditional probabilities such that large conditional probabilities produced small proximities in the matrix, whereas small conditional probabilities produced large proximities.

After converting the frequency of co-occurrence data into conditional probabilities, a subset of 49 commands were submitted to Pathfinder analysis ($r = \infty$, $q = n-1$). The network of commands selected consisted of those that occurred in the event records of at least half of the users. The purpose of this analysis was to identify task *sequences*. Figure 5 shows a portion of this network that includes all commands within two links (arcs) of *kill*. As an example of the type of information contained in the network—and its potential utility—suppose several programs were executing "simultaneously." Further, suppose that one of these programs was a particularly time-consuming analysis that had to be terminated for some reason (e.g., to reduce processor load). This is one of the few cases where the name of the appropriate UNIX command is fairly easy to remember (i.e., *kill*). However, in order to *kill* a program, you have to know its job number, and the name of the command that returns job numbers is not so easy to remember. From the network in Figure 5 it is apparent that only one command, *ps*, frequently precedes *kill* (i.e., there is only one arc leading to *kill*). Indeed, *ps* (program status) is the command that returns job numbers, along with some other useful information.

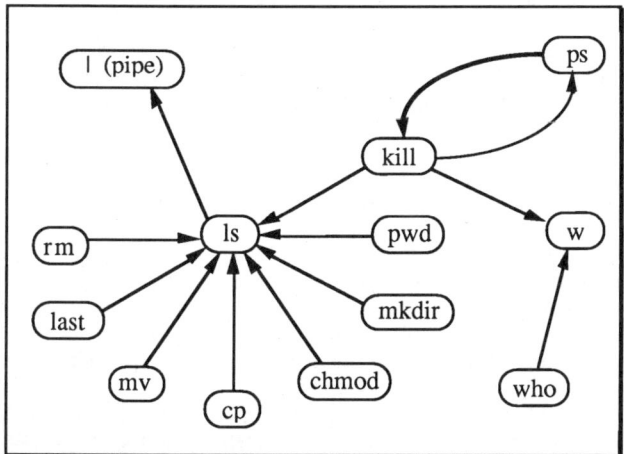

Figure 5. A Pathfinder analysis [PFNET($r = \infty$, $q = n-1$)] of UNIX event-record data, including all commands within two links (arcs) of *kill*.

---

[4]This is a relatively shallow method for extracting information from these data because only two-step sequences are considered.

The PFNETs derived from the two sets of proximity measures generate two different perspectives on the UNIX system. HyBrow attempts to provide users with a consistent documentation interface, relying on contextual cues provided by the network representation to facilitate the development of conceptual models. Either the functional or procedural perspective can be presented to the user in the form of networks with concepts as nodes and relationships among the concepts as links or arcs. In addition, there can be multiple levels of abstraction within perspectives, as described for the functional perspective. For example, Figure 4 shows the highest level of abstraction for the functional perspective.

A user can call upon HyBrow to display a network of commands that are functionally associated with any command of interest (see Figure 6). When a specific command is selected, a brief synopsis and description of the syntax for the command-in-focus appear in the window above the network. As shown in Figure 6, the user has selected the *cmp* command. The complete *man* panel for the selected command can be retrieved and displayed in the interactive window to the right of the network by clicking on the *show man panel* button. Technical terms that appear in the synopsis are highlighted, indicating Hypertext options. Selecting a highlighted term leads to the display of its glossary definition in the interactive window. This is illustrated in Figure 6 for a user who wants additional information about *standard input*.

The user can easily shift HyBrow's focus to any of the related commands in the network by clicking on a visible command node. The user has control over the network density through the use of the *more links* and *fewer links* buttons. HyBrow can also identify those commands that belong to the same category or the same supercategory. The Hypertext engine maintains a record of execution that permits the user to backtrack in stepwise fashion or to jump back to the beginning of the session.

A command subnet, like the one shown in Figure 6, can be accessed in two different ways. If the user wants help for a command he already knows, then the command subnet can be accessed directly by entering the command's name on any prompt line. Commands can also be accessed from a top-down search of the category network shown in Figure 4.

Conclusions

We are still a long way from Vannevar Bush's *memex*—but technology *has* provided us with the means to organize information in new and potentially useful ways. Considerably more research needs to be conducted into the utility of hypertext, both basic and applied, but the preliminary indications are promising: Providing readers with multiple perspectives facilitates their ability to draw inferences, and even superficially hierarchical systems may have naturally nonhierarchical representations (e.g., UNIX).

We have presented an interface design methodology that seems particularly suited to building hypertext networks, along with accurate "maps." It remains to be seen whether or not this methodology leads to better interfaces, but a number of advantages are already clear. First, the methodology is formal, meaning that it is well-specified enough that different designers following essentially the same set of procedures will arrive at essentially the same results. Second, it is empirically based, meaning that the method does not rely on the subjective impressions of designers but the mental representations of users.

In this chapter we have attempted to identify some of the potential and challenge associated with presenting information in hypertext systems. Although the current enthusiasm for hypertext may wain, there do appear to be legitimate reasons to believe that collections of information should be represented nonhierarchically.

14 Hypertext Perspectives

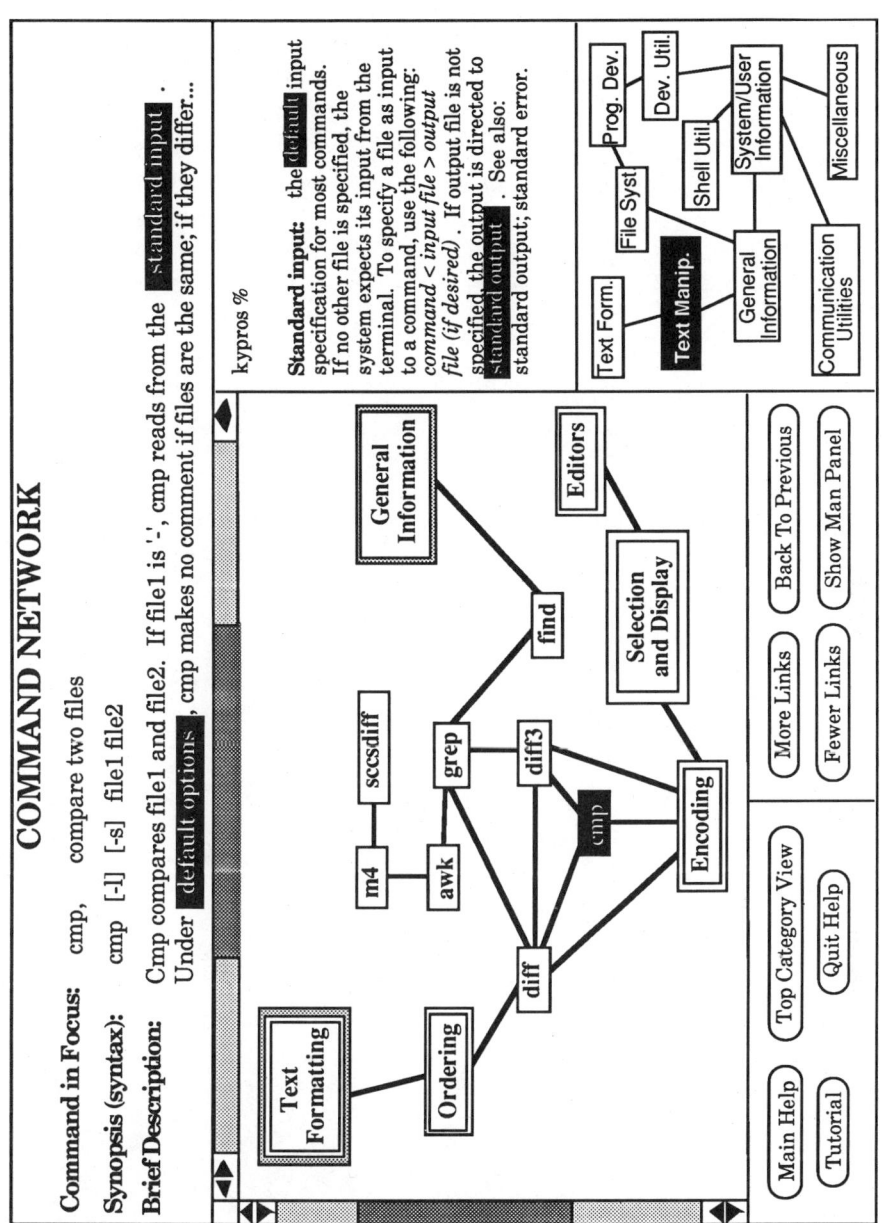

Figure 6. A bottom-level command subnet for HyBrow. Selecting any command node automatically retrieves a brief description of the command-in-focus and makes other information available.

Chapter 15

Expert Conceptual Structure:
The Stability of Pathfinder Representations*

John G. Gammack

This chapter describes the use of Pathfinder in modeling an expert's conceptual structure. The aim is to produce networks that represent domain knowledge underlying and informing expert classification decisions. To this end, several elicitation tasks are described, each of which produces a network relating domain concepts. These tasks have been developed in psychology and include relatedness estimates, repertory grid, and recall tasks. The techniques address distinctions among British steam locomotives, exemplifying the classificatory aspect of domain conception.

Practical application of these techniques in knowledge elicitation are described elsewhere (Gammack, 1987a, 1987b). Here the concern is with the role of Pathfinder within the approach and the stability of its representations under different elicitation conditions.

The elicitation techniques used in this study attempt to observe the results of information-based decisions without determining or otherwise influencing these results. Since only the results of cognition can be observed in this way, any conceptual structure in memory presumed to determine this may only be indirectly inferred. Such psychological tasks have traditionally been used to reveal pre-existing memory structures, on the assumption that they neither originate nor misrepresent that structure. Factual information is assumed largely to inhere in conceptual structures that are stored in memory and that reflect the actual relations of domain objects. Since each elicitation task provides a relationship network, these may be compared. By overlapping the (network) structure revealed by different techniques addressing the same stored information, it is assumed that spurious relationships due to experimental noise or error may be identified, and thus a core organization of stable relationships established.

Although the dictates of actual circumstance may modify the applicability of a given set of relationships, this constraint is minimized in an experimental laboratory setting. By abstracting the elicitation from any pragmatic context, any pre-existing structure in the mind is more likely to emerge uncompromised by workplace influences and expedient considerations.

These assumptions imply that a stable organization of related domain concepts will emerge. Two aspects of stability are considered here. The first concerns the stability of the Pathfinder representation of conceptual relationships under different elicitation methods. The second considers the stability of information in memory and of formal representations of its structure.

---

*I would like to thank Roger Schvaneveldt, Nancy Cooke, and Richard Young for help during preparation of this chapter.

The Elicitation Tasks

Five psychological tasks provided data for analysis by Pathfinder. The proximity matrices were also analyzed using cluster analysis (single-linkage, nearest neighbor) and multidimensional scaling (MDS), which respectively indicated prominent groupings and dimensions of conceptual variation. The five tasks are described more fully in Gammack (1987a, 1987b) but are introduced briefly here. Using the dominance metric ($r = \infty$) and omitting weights, the minimal Pathfinder graphs from each task are illustrated, and in the next section these will be formally compared.

Proximity Ratings

This task elicited estimates of the relatedness between pairs of locomotives to directly provide a proximity matrix. One locomotive was presented as a target and its relatedness to each of the others was considered in turn, thus providing the proximity estimates. Notwithstanding Tversky (1977), asymmetries were not considered to have psychological significance here, and targets were not replaced. To make the procedure easier, the expert presorted the remaining photographs into three piles according to degree of relatedness (strong, medium, or weak), and then assigned numeric values within each range in turn. This procedure was repeated until each locomotive had been used as a target, and the resulting half-matrix produced the graph shown in Figure 1.

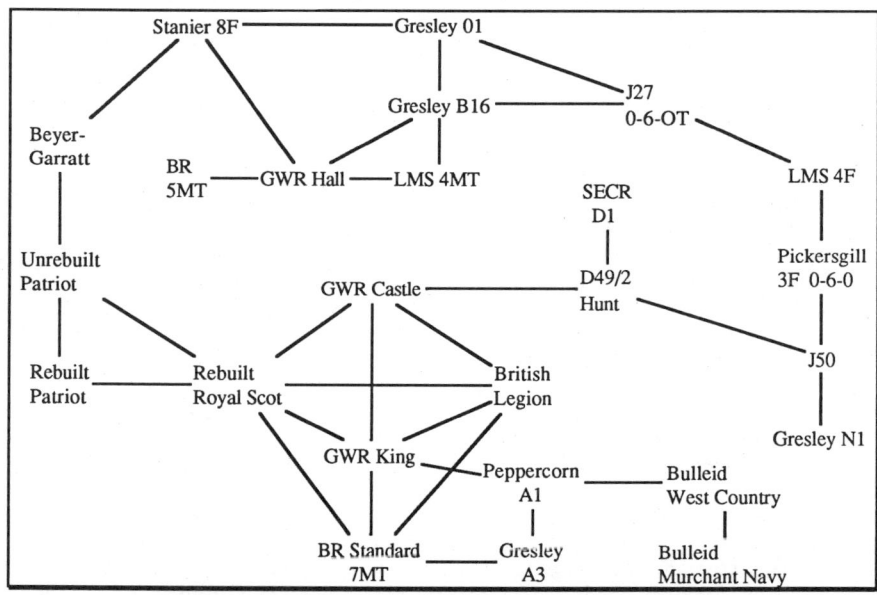

Figure 1. Pathfinder network from the proximity ratings.

Much domain information is implicit in the network, which reasons of space preclude detailing here; however, it is superficially clear that (for example) the two Bulleid engines are strongly related, as are the three Great Western Railway (GWR) engines. The expert later described the grounds for each link, and the detailed domain knowledge thereby emerging justified the relationships shown.

15 Stability of Representations

Repertory Grid

Relatedness estimates are not absolute values and are subject to contextual variation. Two locomotives highly related by geographical region might be unrelated in their design features. Such context will affect whether two locomotives are considered to be highly related or not.

In the absence of any context (other than the target locomotive itself) to provide criteria for relatedness ratings, it is probable that relatedness decisions will have been determined by consideration of the characteristic properties of engines, such as their age, size, and duty, and an overall estimate made on that informed basis.

To test this, and to establish the underlying information being used, the repertory grid technique was employed. This involved taking three locomotives at random from the set, and asking in what way two were alike and thereby different from the third. For example, two locomotives might be primarily involved in freight duty, but the third was rather more a passenger engine. The dimension (or construct) of freight/passenger thus elicited provided a bipolar scale along which the rest of the sample was rated from 1 (purely freight) to 7 (purely passenger). Repeating this procedure elicited a set of constructs, such as freight/passenger and older/modern, and provided a rectangular matrix profiling each engine in terms of these constructs. The city block distance through the multidimensional space implied by this grid was calculated for each pair to provide a full proximity matrix. This is illustrated in Figure 2, superimposed on the MDS plot and cluster analysis of the same data. Links in common with the previous solution are shown as bold lines, comprising about half the minimal network.

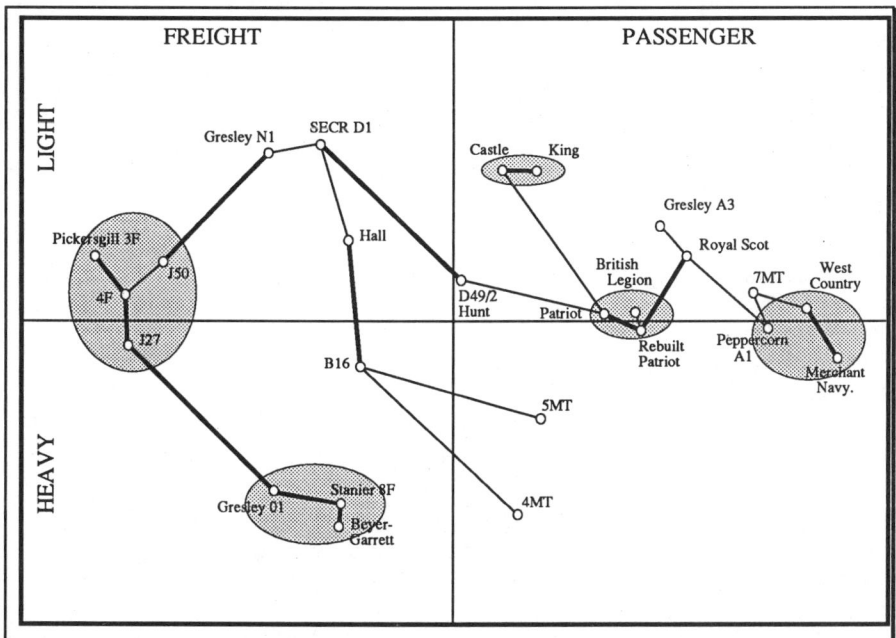

Figure 2. The analysis of the repertory grid data showing a Pathfinder network on an MDS layout with clusters of nodes.

Seeded Recall Task

This task follows the method of Reitman and Reuter (1980). In it the expert is given the name of one locomotive as a cue and is required to recall the rest of the set. This is repeated using the other items as cues (or seeds) and the output order is noted. The resulting data may then be converted to a proximity matrix by calculating the overall distance between each pair of items. The rationale behind the technique is that related items will be recalled consecutively, so that if two items are recalled consecutively on each of 25 trials, they will have a (very close) proximity of 25 units. The analyses are shown in Figure 3.

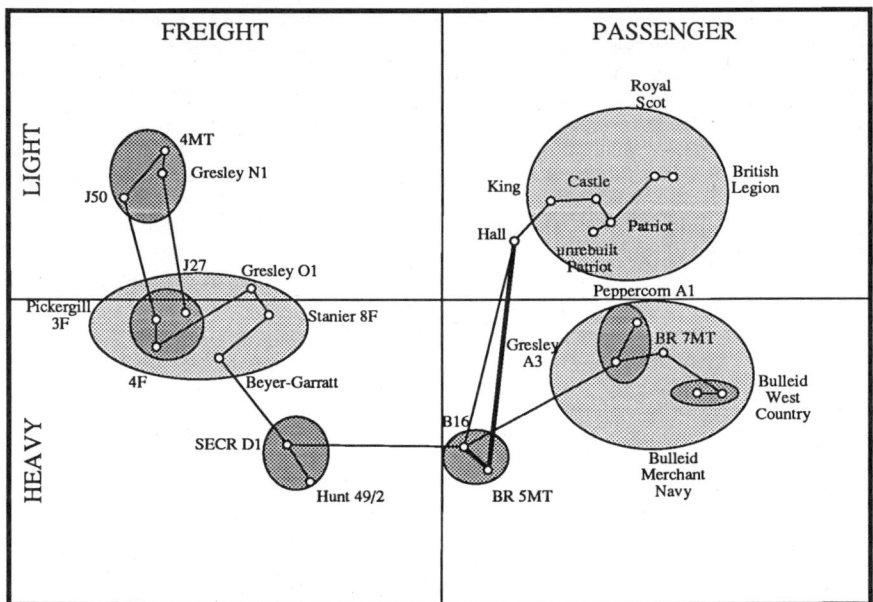

Figure 3. The analysis of the seeded-recall data showing a Pathfinder network on an MDS layout with clusters of nodes.

Bipolar Rating Scales

This is a simple consistency check on information given in the repertory grid and implicit in the MDS dimensions. Following a method reminiscent of the semantic differential (Osgood, Suci, & Tannenbaum, 1957), seven bipolar scales (e.g., freight/ passenger) were constructed. The expert marked the appropriate points on these for each locomotive, and this information was quantized as seven-point scales. These were again converted to a proximity matrix, using the same procedure as for the repertory grid, namely by taking the city block distance in multidimensional space. The analyses are shown in Figure 4.

15 Stability of Representations

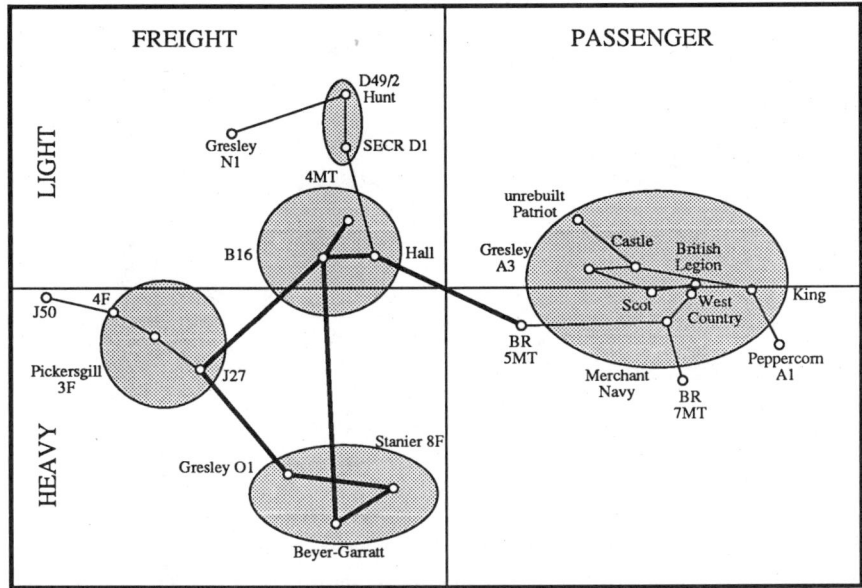

Figure 4. The analysis of the bipolar scale data showing a Pathfinder network on an MDS layout with clusters of nodes.

Twenty Questions

In this task, adapted from a parlor game, the experimenter chose one locomotive as a target, which the expert had to guess by asking questions phrased to be answered yes or no. For example, to the question "Are you thinking of a freight engine?" a "no" response would eliminate a whole class of items. In addition to being specifically informative, the elimination patterns were recorded. This allowed the similarity of locomotives across targets to be gauged to provide another proximity matrix. Using the rationale that locomotives with similar profiles would tend to be eliminated by the same criteria, this proximity matrix was calculated in two stages. First, a rectangular matrix was constructed, showing for each target item the question number at which each locomotive was eliminated. After this was done the city block distances for all pairs were taken as before, and the subsequent analyses are shown in Figure 5.

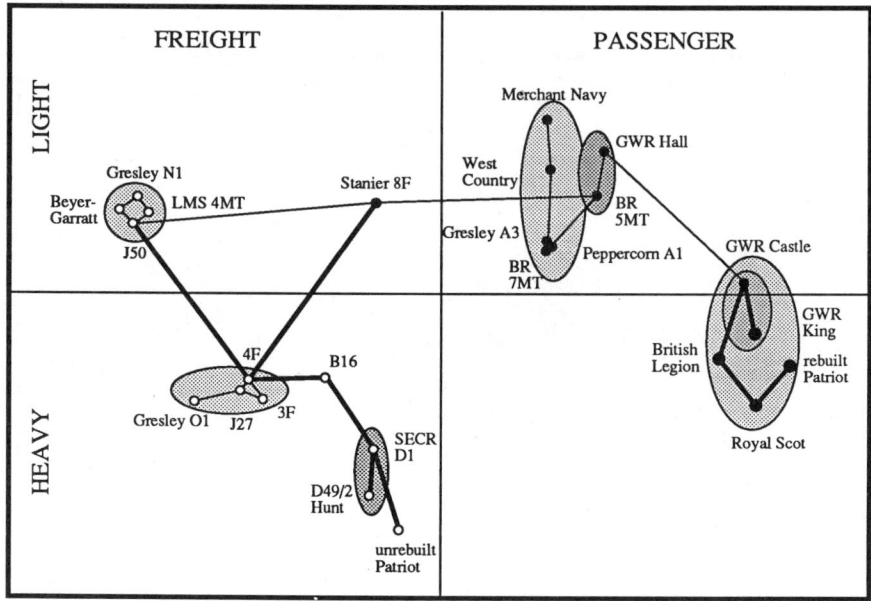

Figure 5. The analysis of the Twenty-questions data showing a Pathfinder network on an MDS layout with clusters of nodes.

Summary

These five tasks have provided five different networks showing relationships (in particular, similarity) among locomotives. Any or all of these may be (and have been) used to elicit the specific domain information they imply and concisely represent. Each network approximately fits the same two-dimensional space (e.g., Alscal's simple euclidean model in two dimensions explained virtually all the variance with little stress), but only a minority of specific relationships are preserved across tasks, and visually the networks differ considerably. In the next section some measures of the agreement between networks are introduced, prior to a more general discussion.

Analysis of Agreement Among Tasks

The previous section introduced the elicitation tasks and presented the minimal network for each. In this section some methods for comparing these networks are described.

Correlation

At first this method was not applied directly to the networks but to the original data matrices. Comparing these matrix pairs point for point using nonparametric statistics, such as Spearman's ρ or Kendall's τ, rarely produced correlation coefficients of greater than 0.2. Kendall's τ is appropriate to the comparison of interest since it attends to disarray in rank-ordered data. This test is as powerful as Spearman's ρ, but its conservative nature often

15 Stability of Representations

leads to lower values. Because of the high number of points involved, however, even such low values were statistically significant, suggesting that at least the datasets are related.

However, Schvaneveldt (personal communication) points out that a straightforward relationship between raw proximities and network distances is not captured by linear correlations of the raw data. Since noisy raw data may produce spuriously low correlations, it may be more meaningful to compare the minimal networks directly by correlating matrices reconstructed from the minimal network. This information is implicitly available from the Pathfinder program, although an extra procedure must be written to have it output. The distances through the minimal network can be computed for cells which would otherwise be empty in Pathfinder's output matrix. The full matrix, despite numerous ties, may then be compared with its counterpart from another elicitation task. Although to date the mathematical properties of this reconstructed matrix remain unexplored, the correlations were again disappointingly low for these datasets, appearing rather sensitive to properties of the raw data. For instance, an unusually extreme value would propagate through many paths to profoundly affect the overall pattern of rank orderings and lower the amount of correlation. Cooke (personal communication) suggests that the number of ties can typically be reduced by summing the weights along the shortest paths, rather than using the dominant link weight in each path. If psychologically justified, this procedure may be more appropriate.

Correlations were also performed on matrices of graph-theoretic distances taken from the illustrated minimal graphs. For each pair of locomotives, the number of links in the shortest path connecting them was computed and entered in a separate matrix for each of the five tasks. The matrices thus derived from each task were then correlated in pairs, comparing first the proximity ratings with each of the other four tasks and then the repertory grid with the other tasks. These compared favorably with the previous correlations; for instance, the correlation between the graph-theoretic distances for the proximity ratings and the repertory grid tasks was $\rho = 0.59$. The other correlations averaged about 0.3, and all correlations were significant or highly significant. This use of correlation is probably the most satisfactory one.

Hypergeometric Test

This simple statistic indicates the significance of the overlap between two networks. It views the networks as samples drawn from a known population, in this case the 300 possible pairwise links among 25 objects. The minimal network represents (typically) 24 of these links. Given a number of links common to two networks, the probability of this overlap due to chance alone can be calculated from the hypergeometric distribution. For example, the 12 links shown in bold in Figure 2 show the overlap between minimal networks from the proximity ratings and repertory grid tasks. This highly significant ($p < 0.001$) overlap is extremely unlikely to be due to chance, indicating the relatedness of the two solutions.

Two things, however, should be noted here. Firstly, links common to both tasks account for only about half the total number of links, so highly significant overlap can be achieved between rather different looking networks. Secondly, the hypergeometric test indicates significance given a relatively small number of common links. For instance, a mere 6 links in common between two 24-link networks is significant at $p < 0.01$. This means that the test will successfully indicate the presence of nonrandom overlap, but is not a strong indicator of agreement between networks.

It seems reasonable to expect that even minimal graphs over the same domain objects will overlap to a high degree. The significance of the statistic can often be increased by using lower values of the q-parameter, for example, using $q = n-1$ is conservative. However, in most of the networks illustrated here there was little difference between the number of links for different values of q, and neither the statistic nor the conclusions were significantly affected by adding more links.

Chi-squared Test of Association

The next test allowed different assumptions about the sensitivity of the data to be considered, and used the chi-squared statistic as a test of association between two datasets (in effect a sort of McNemar test). This test used information about the link weights rather than mere connectivity, but without being overly sensitive to properties of the raw data.

Each task has provided a matrix of proximity values for the relative relationship strength between each pair of locomotives. These ordinal data can be compared with their counterparts from other tasks by binning them into ranges, then noting the frequency of matches in a contingency table, as shown in Table 1.

Table 1. Frequency of matches for two elicitation tasks.

	Repertory Grid	
Proximity Ratings	*Strong*	*Weak*
Strong	129	18
Weak	39	114

Thus in the simplest case of two ranges, all cells with proximities below 50 (or the median) are designated "strong," and all those above 50 (or the median) "weak," and two such rescored binary matrices are then directly compared. The frequencies are noted in a 2×2 contingency table, showing the number of times a relationship was "strong" in both tasks, "weak" in both, and the number of mismatches in one direction or the other. The more consistently relationship strength is preserved across tasks, the larger the frequencies on the main diagonal, and the fewer the mismatches. Applying the chi-squared statistic indicates the significance of this consistency across tasks. In this case there was a highly significant agreement between the directly rated proximities and those derived from the repertory grid ($\chi^2 = 117.96$, $p < 0.0001$).

This basic procedure may be variously enhanced, for instance, by using more than two ranges or by treating values around the median (splitting criterion) more cautiously. Alternatively, distance matrices derived from Pathfinder networks can be used. While this test might benefit from more thorough investigation, the present data showed significant consistency across tasks using the median as the splitting criterion between strong and weak ranges.

Cluster Analysis

Having observed networks containing groups of items linked by small weights, distanced from other such groups by a single larger weight, the similarity to cluster analysis became obvious. It seemed likely that often the linkage structure of some subgraph comprised only one possible selection of relationships within that group of nodes, and in fact the whole group, as a group or clique, was stable. Comparing only partial selections within such a group would fail to indicate this overall stability. For example, the GWR King, the Castle, A1, A3, and Scot are similar express passenger engines, with different pairings of these linked in different networks. However, although the A1 is linked directly

15 Stability of Representations

to the King in two networks, and is never far away in others, measures sensitive only to common links will miss this.

By cutting the network successively at its longest links, such groups may be identified, resulting in a form of cluster analysis. The clusters emerging between tasks may then be compared directly at a chosen grain size of clusters. This is illustrated in Table 2.

Table 2. Common elements[a] in clusters from proximity ratings and clusters from seeded-recall.

Proximity Clusters	*Seeded Recall Clusters*						
	A P	L K H	B G M X Y	I J R Q S T U	C D E	O N	F V W
A P Q	A P			Q			
L K I		L K		I			
B M G X Y			B G M X Y				
J R S T U				J R S T U			
C D E					C D E		
H O N			H			O N	
F V W							F V W

[a]Individual engines are designated by individual letters.

The relation between the minimum network and single-linkage cluster analysis is formally homomorphic (Dearholt, Schvaneveldt, & Durso, 1985; Dearholt & Schvaneveldt, Chapter 1, this volume), but the network is more informative, since the cluster analysis can be derived from it, but not vice versa. However, the value of identifying the clusters as done in Table 2 is to show the grain size at which there is essential identity between two networks. Although the detailed networks of Figures 1 and 3 differ, at a slightly higher level of abstraction (grain size) they are barely different, as the main diagonal alignment of Table 2 shows.

Incidental information on unstably locatable items is also evident from the cluster comparison of Table 2. For instance, the GWR Hall (item Q) is justifiably associated with the mixed traffic engines A and P, but has a competing claim to be associated with the other GWR engines (items R and S) and their group of passenger engines. Comparing the cluster structure derived from the networks thus indicates agreement between tasks at another level of abstraction.

Summary

These measures are all candidates for establishing the agreement between pairs of Pathfinder networks derived from different tasks. Although the hypergeometric test and the correlation coefficient typically indicate that statistical agreement is significantly non-random, this may not be the comparison of interest. Two networks may have very little linkage structure in common and still be shown to be statistically related.

Chi-squared analysis and comparison of clusters aim to show that independently derived datasets are not significantly different from one another, more in terms of their implied content than of their structural details. While more satisfactory measures may be developed in the course of time, it may be sufficient for the moment to note that statistical agreement can be shown among these networks, despite surface differences in structure. In the next section the significance of these surface differences will be emphasized, as the stability of Pathfinder networks as a representation is discussed.

The Stability of Pathfinder Networks

This section asks whether differences between networks are important. Consideration of what a network actually represents is useful in this regard.

Firstly, each network is a statistical summary of an originally more informative set of data, and the algorithm selects the most salient relationships in that dataset. Unfortunately, which relationships are esteemed salient will vary across contexts, and relationships strong in one set of circumstances may not apply in another. Secondly, over longer paths, both cumulative distances and any implied conceptual relation may become meaningless, so that, as an overall picture the network's value is severely limited. To illustrate this, Tversky's (1977) example serves well:

$$\text{Jamaica} \xrightarrow{\quad 2 \quad} \text{Cuba} \xrightarrow{\quad 2 \quad} \text{USSR}$$

In a linked set of countries, Jamaica may be considered similar to Cuba with a strength of (say) 2 units, and Cuba may be considered similar to USSR, also with a strength of 2. Particularly under the dominance metric, Jamaica and USSR are very close in network terms, but such implied proximity is psychologically hardly very meaningful. For such multifaceted concepts, relatedness may be judged on numerous grounds, and this must be considered when interpreting a network. The criticism may not apply to all networks but is particularly relevant to those with semantic properties. De Groot (1983) found no evidence to suggest that semantic priming persists between indirectly linked concepts, such as Jamaica and the USSR in the above example. This may imply that values of $q > 2$ should be treated with caution.

As a representation of a dataset, a Pathfinder network is subject to noise and other experimental influences in those data. The actual values of certain cells assume critical importance when the algorithm is run and lead to the formation or rejection of direct links. Small changes to these critical cell values may dramatically affect the final structure and lead to apparent instability between comparable networks. In the final section, reasons for the evident instability of Pathfinder representations are discussed.

Some Reasons for Instability

Until closely examined, it seems disturbing that the networks are so structurally different, despite demonstrable statistical agreement. Whereas the effects of noise or task context presumably contribute to this phenomenon, there is nevertheless the assumption that the network represents only a slightly distorted form of a network structure actually existent in memory. Questioning this assumption suggests no reason to suppose direct correspondence between a network and a pre-existing memory structure. This implies that any apparent instability may be a fair reflection of psychological flexibility and not merely a

15 Stability of Representations 223

methodological bugbear. However, to make the point clearer, both methodological and psychological reasons for instability are examined.

Methodological Reasons

Trivially, it may be unsurprising that the networks differed from one another, after all, the elicitation tasks differed, and the proximity data were variously derived. If this is the explanation, it implies that task or experimental manipulations affect, if not determine, the relationships finally represented in a Pathfinder network. Far from these traditional procedures being passive measuring tools, they actively shape the resulting data. If this is extensive, the implications for methodology go beyond Pathfinder, since the source dataset is itself already compromised. For the moment it seems preferable to believe that the experimental task is not the major determinant of the data it produces.

Another possibility is that the variability results purely from noise, reflecting the finding that people are just not good at estimating numbers. While this may be part of the explanation, it is unlikely to be all of it. For instance, in the repertory grid and rating scales tasks, a locomotive's properties were considered individually, and rated on a scale of only 7 points. It seems unlikely that expert estimates could be seriously wrong under these conditions. Even if they were, the recall and twenty-questions tasks did not involve assigning numbers at all.

To say that people are not good at estimating numbers does not explain why estimates are unreliable in the first place. Probably here it is precisely because both relationship strength and the saliency of properties are relative to context, and there is no absolute value for relatedness judgments. The task demands of estimating relatedness are likely to be met by rough computations, guided largely by the expert's own experience and biases. For the data presented here, similarity of duty and size clearly contributed much to the consistency. However, relatedness judgments by an expert for whom design and vintage were important would produce a very different relationship pattern.

A third possibility is that all these networks are merely selections from a more comprehensive representation, for example, the total union of all defensible direct associations. It may be that no particular network is the only correct structure in memory, but that their complete combination is. Although a completely linked network might be a truer representation of connectivity, it must have differentially weighted links if it is not to be vacuous. For if everything is connected to everything else, and there are no intrinsic distinctions of weight, then the network in itself is utterly uninformative. However, in practical domains it is unlikely that all connections would be defensible, and for most intents and purposes it is sensible to establish the subset of useful ones. Pathfinder is a principled attempt at this.

However, there seems no reason to suppose that weighted links as such have any a priori existence in memory. Weightings that reflect the importance of relationships are likely to be computed online and in specific contexts (cf. Barsalou, 1987; Kahneman & Miller, 1986). Furthermore, the particular links formed between domain concepts are likely to be affected by factors extrinsic to properties of those concepts (Barsalou, 1985; Murphy & Medin, 1985; Schank, Collins, & Hunter, 1986). These and other psychological factors affecting observed structure are considered below in more detail. So even though regarding networks as selections from a totality at first seems promising it may lead to empirical vacuity (cf. Johnson-Laird, Herrmann, & Chaffin, 1984).

Psychological Reasons

A link-weighted network of concepts may be viewed as descriptive of semantic memory (in this case of an individual expert's category structure). However, the explicated network results from the interaction of pre-existing memory with various external demands. Recent research has identified various external factors which bear on the stability of conceptual representation and therefore may affect the resulting network. These include goals and needs (Gluck & Corter, 1985), context (Barsalou, 1987), and personal theories (Murphy & Medin, 1985).

Gluck and Corter (1985), for instance, note that traditional measures of category structure (such as cue validity) ignore the contexts and needs of people who create and use concepts. They propose a context-sensitive measure of the utility of categorizations. Schank et al. (1986) make similar points in their discussion of inductive category formation. They note that the function for which categories are formed helps to determine their structure, and this lies outwith the scope of purely inductive category formation systems, which operate on the intrinsic properties of concepts.

Murphy and Medin (1985) emphasize the importance of personal theories in shaping conceptual structure and give examples showing how such theories provide a coherent context in which concepts can meaningfully relate. Thus the seemingly disparate set of shrimps, moths, and grasshoppers becomes cohesively related given a deeper understanding of biological structure, in this case, a theory of the class of arthropods. Analogously, an individually held belief set gives a context of meaning within which conceptual relationships have cohesion. An expert's set of attitudes to a domain will affect its conceptualization, and some examples of this are demonstrated in Gammack (1988).

Barsalou (1982, 1985, 1987) has perhaps provided the most sustained arguments against the static definition of concepts. In Barsalou (1987), for example, his primary concern is with the instability of graded structure, that is, the extent to which exemplars typify a category. Barsalou shows how this varies with context and suggests that categories do not have an invariant representation that we should be trying to discover. Instead he proposes that concepts are temporary constructions in working memory, drawing upon (rather than retrieving intact) knowledge in long-term memory. This research suggests that long-term memory is relatively unstructured, supporting the flexible construction of concepts rather than containing knowledge in well-bounded packets of characterizing properties. This view is consistent with the present results.

Role of Pathfinder in Knowledge Elicitation

If the instability noted in this chapter signifies an inescapable issue not only for Pathfinder but for psychology in general, it would be easy to be pessimistic and conclude that a static description represented nothing more than the specific product of a few arbitrary conditions. However, this negative view ignores the many valuable gains to be made by considering instability seriously.

Supposing that the expert has a uniquely configured, stable network of 25 domain objects, which looks in retrospect, naive. Furthermore, the assumption that this knowledge structure would be transparently elicitable by objective methods having no influence on underlying structure also looks questionable. Since the tasks were different in nature, it may have been unreasonable to expect anything else. Clearly the tasks have had an effect in that they have elicited different structures, suggesting that they may be addressing different aspects of the same knowledge. Knowledge itself is surely stable, but representations of it are not, despite their specific value in particular contexts. Thus the bipolar scales and the repertory grid tasks act as mutual consistency checks on a constrained set of properties, whereas the proximity estimates permit other bases for relatedness than similarity of a small set of properties. The elicitation tasks do appear to have individual properties which make

15 Stability of Representations

them nonequivalent and thus uniquely valuable. The extent to which they help shape (if not determine) the form of elicited knowledge is a methodological issue which requires fuller investigation.

One argument against the existence in memory of an absolute weighted structure reflecting a "true state" is the effect of circumstantial changes on relatedness estimates. Although the relating criteria used here were freely chosen by the expert, an instruction to rate proximities in terms of vintage or design would have produced different patterns, as noted earlier. Exigencies arising in practical settings will have a similar effect. While some conceptual relations are likely to be more strongly held than others, and likely to emerge frequently across contexts (such as the Bulleid Merchant Navy and Bulleid West Country in this domain), there will be occasions when this relationship is relevant, and others when it is not. A weighted structure in itself describes something about an expert's domain knowledge, but one must recognize its range of applicability. Accordingly, when eliciting an expert's knowledge of the relatedness of domain concepts, the context in which two items are related must also be identified and acknowledged. It is likely that this will not be given merely by a linear combination of "intrinsic" properties, but by extrinsic considerations which give the concept its current definition. The meaning of a concept is not fully determined as a function of its analyzed parts, but is assigned within a context and background outwith a static representation (see Gammack & Anderson (submitted); Gregory, 1986; Winograd & Flores, 1986).

Theoretically, Pathfinder is likely to prove very useful in investigating and representing the stabilization of a systematic set of conceptual relationships. Instead of viewing Pathfinder as faithfully mapping a stable conceptual structure in memory, a narrow view in which personal interpretations, changing contexts, and methodological properties are problems to be reduced as far as possible, Pathfinder can become a technique in investigating those very areas.[1] For instance, a novice's domain theory, mapped out using Pathfinder and augmented with descriptions justifying the structure, can be compared with the same individual after training or with another expert to identify the effect of experience on domain conception. Whereas practicing experts are not limited to a single inflexible structure (see Murphy & Wright, 1984), such structures are likely to be particularly useful in formal teaching, routine tasks, and many other situations for which expert knowledge is elicited.

Given the existence of domains or parts of domains of expert knowledge that can be profitably systematized, Pathfinder has many advantages in producing a concise, visualizable, and informative representation. Several expert domains have already been modeled in which Pathfinder networks were an important part of the represented knowledge. In addition to domains described by Roger Schvaneveldt and his colleagues, these include choice of statistical analysis techniques and information scientists' sources (Gammack, 1988), central heating systems (Gammack, 1987a), and locomotive classification (Gammack, 1987b).

Being suitable for modeling the knowledge of experts as individuals or in groups, and having a variety of possible sources of input data, Pathfinder is a particularly flexible way of representing implied relationships among domain entities. The formal and quantitative aspects of Pathfinder networks are also attractive as a representation which knowledge engineers can use in producing expert system code.

---

[1] I am grateful to Nancy Cooke for reminding me of this.

Conclusion

In this chapter, an attempt was made to model an expert's conceptual structure using Pathfinder (and other) representations. However, across different elicitation conditions, no unique conceptual structure emerged when it might have been expected. Methodological and psychological reasons were considered as explanation for this instability, and with hindsight, it seems sensible to expect structural variation for a variety of reasons.

While the underlying knowledge may be relatively stable, structured representations of it are variable, context specific, and considered to be secondary phenomena of constructive processing, rather than homologues of any pre-existing memory structure. This implies that any given Pathfinder network is unlikely to represent a true state of memory, but a product of memory's interaction with task, contextual demands, and other factors extrinsic to similarity of defining properties. With a static description of relationships such as a network, one must also acknowledge the contexts in which domain objects relate coherently.

The instability of representation is neither caused by, nor unique to Pathfinder, which provides a tool for investigating the sources of instability in conceptual representation, such as individual theories and biases, and task or other demands. Despite some surface variation, Pathfinder descriptions were found to be particularly useful in the context of knowledge elicitation where concise and meaningful representations of expert domain conception were reliably produced.

Chapter 16

Using Pathfinder as a Knowledge Elicitation Tool: Link Interpretation

Nancy Jaworski Cooke

Over the last two decades cognitive psychologists have accumulated evidence that suggests that domain-specific knowledge is central to expertise (e.g., Chase & Simon, 1973; Chi, Feltovich, & Glaser, 1981). During this same period, the concept of "knowledge-based" systems has surfaced in the field of artificial intelligence. These systems incorporate and rely heavily on specific facts and rules about the problem-solving domain in which they specialize (Hayes-Roth, Waterman, & Lenat, 1983). This emphasis on specific knowledge can be contrasted to previous approaches to human and machine problem solving that emphasized general strategies that could be applied across various domains (e.g., Newell & Simon, 1972). Today most cognitive scientists do not deny that such general strategies exist, but they emphasize the importance of domain-specific knowledge.

This knowledge-intensive trend in cognitive science has also influenced several applied areas. Recent developments in intelligent tutoring systems (Sleeman & Brown, 1982) capitalize on differences between the student's knowledge (i.e., the student's model) and some ideal knowledge base that an expert might be expected to possess. The idea is that this information will provide clues to student misconceptions and direct the tutoring session accordingly. In the area of human-computer interaction, the knowledge that a computer user has about a system (i.e., the mental model) is thought to guide learning and interaction with the system (Norman, 1983). Therefore, it is beneficial for system designers to be aware of this mental model so that the system interface can be designed in accordance with user expectations. An example of this application is discussed by McDonald and Schvaneveldt (1988). Finally, in several domains such as medical diagnosis, computer configuration, and chemistry, expert or knowledge-based systems have been developed that perform some of the tasks of human experts (Waterman, 1986). These systems embody numerous facts and rules about the domain in question. The success of all of these applications depends greatly on knowing what the student, user, or expert knows. Unfortunately, the process of eliciting a person's knowledge is a difficult and time-consuming task.

The bulk of the recent work on knowledge elicitation has been done for expert system development by knowledge engineers who are saddled with the task of uncovering the facts and rules that a human-domain expert uses and the task of transferring this information to the knowledge base of the system. This process is not well-specified, but typically involves behavioral observations, interviews with experts, and the collection of thinking-aloud protocols (Ericsson & Simon, 1984) in which the experts are asked to verbalize their mental processes while performing a domain-related task. Although these techniques are successful at uncovering some basic facts and rules, they suffer from some serious limitations.

Of particular concern is the fact that experts have difficulty verbalizing their knowledge. Nisbett and Wilson (1977) reviewed a large amount of psychological literature that concluded that verbal reports are often inaccurate and incomplete. Experts in particular may have difficulties because, by virtue of their extensive experience, much of their knowledge is automatic (Shiffrin & Schneider, 1977) or compiled (Anderson, 1982). In addition to problems with introspection and verbal reports, these typical knowledge elicitation techniques are very subjective in that they require extensive interpretation on the part of the knowledge engineer in order to transform the verbalizations or observations into a computer-useable format. These techniques are also limited in the sense that they provide virtually no information regarding the appropriate organization of knowledge. Studies on human expertise have indicated that experts differ from novices not only in the facts and rules that they possess, but also in the way that those facts and rules are organized in memory (Adelson, 1981; Chi, Feltovich, & Glaser, 1981; Murphy & Wright, 1984). These limitations must be overcome in order to make advances in knowledge-intensive applications, such as expert systems.

Several investigators (Butler & Corter, 1986; Cooke & McDonald, 1986, 1987; Gammack & Young, 1985) have proposed that a formal knowledge elicitation methodology based on psychological scaling techniques like Pathfinder would overcome some of these limitations. It is assumed that the relatedness estimates that are required by such techniques reflect information about the subject's knowledge structure and in this sense, the scaling techniques are knowledge elicitation techniques. Because scaling techniques generate structural representations of a concept set, they address the knowledge organization issue. In addition, the procedures for collecting data and generating structural representations are more constrained than techniques such as protocol analysis. Most importantly, experts are required to make simple relatedness judgments about concepts rather than difficult verbalizations of the mental processes underlying such judgments.

Given the potential advantages of scaling techniques over traditional knowledge elicitation methods, advances could be made in knowledge-intensive applications by employing such techniques. However, there are several issues that arise concerning the application of these techniques to knowledge elicitation, such as selection of an appropriate method to obtain distance estimates and selection of a particular scaling technique (Cooke & McDonald, 1987). Furthermore, scaling methods empirically capture the structure of knowledge, but they do not capture the content of that knowledge; where do the concepts that are to be represented by these techniques come from? Cooke (1987) addressed this issue by comparing four concept elicitation techniques that could be used as an initial step in a scaling-based methodology. The techniques included: (1) concept listing, (2) step listing, (3) chapter listing, and (4) transcription of concepts from a 20-minute dialogue. Comparison of the concept elicitation techniques revealed that they differ in the number of concepts elicited as well as in the form of that knowledge (e.g., fact, rule, explanation). Cooke suggested that several techniques be used in combination to elicit a relatively complete set of concepts.

Another issue concerns the mapping of the scaling representations onto the application (e.g., student model, interface, knowledge base). It is usually necessary to interpret the representation in order to do this. For instance, multidimensional scaling arranges the concepts along dimensions, but does not identify the dimensions. The dimensions are typically interpreted by someone familiar with the domain. Similarly, in cluster analysis solutions, the cluster cutoff point needs to be identified along with category labels. Pathfinder also requires further interpretation because the links in the graph are not labeled or differentiated semantically. The focus of this chapter is on the interpretation of links in Pathfinder networks.

16 Link Interpretation

The link-labeling issue concerns using information obtained by Pathfinder in applications such as knowledge engineering. More specifically, how do networks of linked concepts relate to a knowledge base of facts and rules? In this respect, Pathfinder is limited in that the links in the networks are semantically impoverished. That is, the links are weighted with the relatedness estimate value, but they contain no labels[1] indicating the specific nature of the relation. In Pathfinder networks, the links are only indicative of a general association between the concepts represented by the linked nodes.

Without link labels, interpretation and comparison of links must be done cautiously. The fact that the same pair of concepts is linked in novice and expert networks does not necessarily mean that the novices understand the relation at the expert level, but instead, the two links might have quite different meanings. Likewise, in the knowledge engineering application it is not enough to know *that* two items are related, but it is also necessary to know *how* they are related. Currently there is no formal methodology for labeling links in Pathfinder networks. Typically the experimenter or domain expert assigns labels to the various links, but this strategy is subject to the same criticisms relevant to introspection and verbal reports. The remainder of this chapter describes a methodology for labeling the links in Pathfinder networks.

In summary, psychological scaling techniques can be used as knowledge elicitation tools, and thus have applications in education, training, interface design, and development of knowledge-based systems. These techniques also have advantages over traditional knowledge elicitation techniques, particularly in their reliance on data in the form of judgments rather than introspections. On the other hand, there are several aspects of the scaling methodology that pose difficulties for application to knowledge elicitation. This chapter addresses one of these issues that is particularly relevant to Pathfinder network scaling—the interpretation of links.

Interpreting Links in Pathfinder Networks

Attaching meaning to a link might simply involve labeling or naming the link, but the meaning is fully revealed only through all the inferences that are associated with it. For example, the inheritance properties associated with the typical *is-a* link give it meaning. This type of link implies that the subordinate concept inherits properties of the superordinate concept. Consequently, labels are useful to the extent that they suggest specific inferences; however, links associated with different inferences may be given the same name, or links associated with the same inference may be named differently. In short, labels are often ambiguous and imprecise and as a result they do not reliably differentiate among links. Therefore, the methodology discussed here approaches link interpretation first as a classification problem. Classification of links is achieved by asking subjects to sort linked items into groups according to type of relation. These data are then submitted to cluster analysis. The assumption is that the links in each cluster share a set of inferences. After classification, labels can be elicited for a cluster of linked items, instead of a single pair. This classification and labeling methodology should help differentiate links based on meaning that is deeper or more abstract than the meaning expressed in the labeling of individual links.

Chaffin and Herrmann (1984) also used a sorting procedure combined with cluster analysis to empirically derive a taxonomy of relations. Their subjects sorted pairs of items according to their relations. The pairs were selected to correspond to a taxonomy of

---

[1]"Label" is used in this chapter to refer to the meaning of the link, and "weight" refers to the numerical value associated with the strength of a particular link.

relations which was derived from an extensive literature review of natural language relations. The taxonomy contained five major categories: (1) contrast, (2) similars, (3) class inclusion, (4) case relations, and (5) part-wholes. Results of the sorting and cluster analysis supported this taxonomy. In addition, Chaffin and Herrmann's results suggested three properties or dimensions (contrasting/noncontrasting, pragmatic/logical, inclusion/noninclusion) that distinguished relations and also indicated that subjects could distinguish among many more relations than those typically used in the literature. Chaffin and Herrmann claimed to have captured most of the important families of relations, but they selected pairs of items so that they corresponded with the the taxonomy to be verified. It would be interesting to determine whether the relations identified in empirically generated networks would be accommodated by the taxonomy. Perhaps empirically derived networks could be used to extend the taxonomy. Alternatively, Chaffin and Herrmann's taxonomy could be used to distinguish between relatively common, easy to identify links and links that are either rare or spurious.

The link interpretation methodology discussed here was developed and evaluated using the set of common concepts presented in Table 1. Note that these concepts have a variety of relations. Common concepts were chosen instead of concepts in a specific domain of expertise for several reasons. Most importantly, the methodology could be best developed and evaluated using concepts with relations that are relatively obvious. Furthermore, subjects who know these concepts are common, whereas domain experts are difficult to locate, especially in the numbers necessary to develop a methodology. Finally, the methodology is best illustrated using concepts and relations that are understood by the general public.

Table 1. Common concept set.

blood	mammal	color	red
bird	raking	bats	leaves
flying	rabies	chicken	hair
brushing	barking	tree	injection
milk	dog	animal	egg

A Pathfinder network was generated for the 20 concepts in Table 1, from pairwise relatedness ratings obtained from 30 introductory psychology students at New Mexico State University. Subjects were told to assign ratings on the basis of their first impression of overall relatedness. The relatedness scale ranged from one (slightly related) to five (highly related). In addition, subjects had the option of responding with a "U" indicating that the pair was unrelated. These ratings were transformed to distances by subtraction from 6 ("U" was assigned a distance of 6). In order to account for individual differences using the rating scale, each subject's set of ratings was converted to z-scores. The datasets were then averaged across subjects and converted back to the original scale. These average distances were submitted to Pathfinder ($r = \infty$, $q = n-1 = 19$). The resulting network contains 20 links and is presented in Figure 1. For each of these links, the original ratings consist of no more than 10 percent judgments of unrelated ("U").

The link interpretation methodology was applied to the links in the Pathfinder network in Figure 1. The network consists of 20 links connecting the 20 concepts. It is a tree with the exception of a single link. The weights for the links *hair-dog* and *animal-blood* are tied at 19. Removal of either of these links would leave a tree. The two phases of the

16 Link Interpretation

methodology are link classification, from which clusters of link types are derived, and link labeling, in which the relation represented by each cluster is labeled.

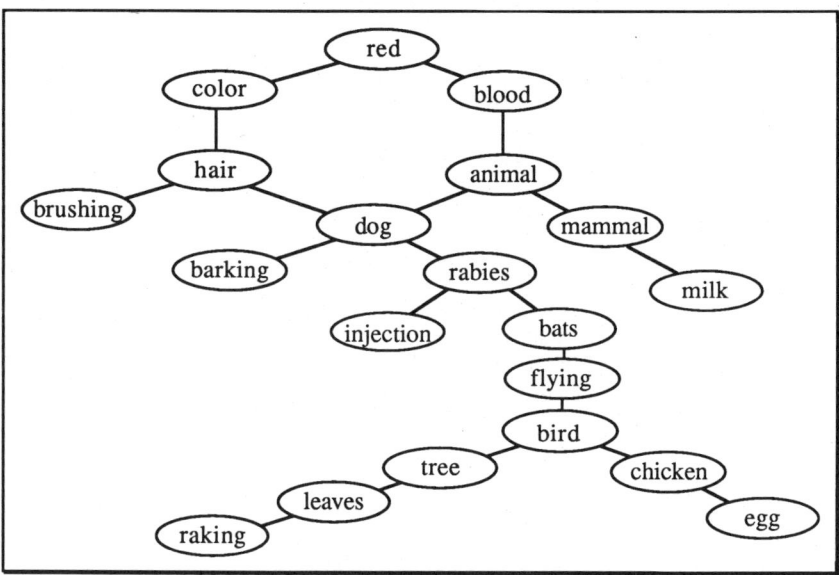

Figure 1. Pathfinder network of 20 common concepts: PFNET(∞, 19).

Link Classification

The purpose of this phase of the methodology was to classify links in the Pathfinder network according to the meaning of the relation associated with each link. Subjects sorted pairs of linked items according to the meanings of the relations, and cluster analysis was performed on the number of subjects who placed each pair in the same pile; frequency of co-occurrence was used as an estimate of distance between pairs of links. As a partial evaluation of the methodology, a control condition was included consisting of pairs that were not linked in the network. If the methodology is successful at classifying meaningful links, then these meaningless control pairs should be excluded from any major categories.

Method

Eight introductory psychology students from New Mexico State University participated in this study. Because pilot studies had indicated that the task (sorting pairs of linked items according to the meaning of the relation) was difficult, subjects who participated were required to correctly answer 15 to 20 analogy problems on a pretest.

Subjects individually sorted 58 index cards based on the meaning of the relation implicit in the pair of items on each card. Each pair was connected by an arrow indicating the direction of the relation (e.g., *robin → bird*). The 58 pairs included the 20 links in the network (see Figure 1) presented in both directions (e.g., *robin → bird, bird → robin*), because specific relations are dependent on order (e.g., *robin* is-a *bird,* but it is not the case that *bird* is-a *robin*). The remaining nine pairs (18 in both directions) were not linked in the network, but were included in order to determine how the methodology deals with spurious links. Spurious links can occur in a Pathfinder network because of noise in the relatedness estimates. Also, if the graph is to be completely connected, then some links may appear simply because a node that is not related to any other concept must be connected to the network. Fortunately, Pathfinder has an option that allows for unconnected graphs. Three of the nine unlinked pairs were linked in slightly denser networks (*blood-injection, chicken-animal, mammal-bats*) and three were linked in very dense networks (*bats-chicken, bats-tree, leaves-flying*). The three remaining pairs were not linked (*raking-rabies, brushing-egg, red-flying*) in even the most dense networks.

The sorting procedure was a variation on the method of controlled association discussed by Miyamoto, Oi, Abe, Katsuya, and Nakayama (1986). A card was randomly selected from the deck of 58 by the experimenter and the subject went through the remainder of the deck selecting any cards for which the relation was the same or nearly the same. Subjects were asked to concentrate on the relation for each pair, keeping in mind the direction of the relation. This procedure was repeated with the size of the deck decreasing on each successive sort until all of the cards had been sorted. After the cards had been sorted, subjects were allowed to make any changes to the piles that they wished. They were also permitted to use blank cards to duplicate a pair if they wished to place it in more than one pile.

Results and Discussion

The data for each subject took the form of a vector of 1163 zeros and ones, each corresponding to a different pair of linked items (58 taken two at a time). A one in this vector indicated that the subject placed both pairs of linked items in the same pile and a zero indicated that they were placed in separate piles. The vectors were summed across the eight subjects, resulting in frequencies of co-occurrence for each pair of linked items. These values were converted to distances by subtraction from nine. Thus, a value of one indicated that all eight subjects placed the pair in the same pile and a value of nine indicated that no subjects placed the pair in the same pile. These distances were submitted to a single-link cluster analysis procedure (Johnson, 1967) and the results are presented in Figure 2.

The analysis resulted in 36 clusters that ranged in size from 2 to 12 pairs. The cluster analysis discriminated among several very cohesive groups; 86% of the 36 clusters are based on the classification of five or more of the eight subjects. In most cases, clusters consisted of pairs ordered in one direction only, and each cluster tended to have a mirror-image cluster that contained the same pairs, but in the opposite direction.

It is interesting to note the fate of the nine unlinked pairs. The three pairs that would never be linked either never clustered with any other pair (e.g., *brushing* → *egg*) or clustered late with other unlinked pairs (e.g., *raking* → *rabies* and *flying* → *red*), or their inverses (e.g., *flying* → *red* and *red* → *flying*). Most of the other six unlinked pairs that would be linked in denser networks did cluster in the analysis and some clustered early (e.g., *chicken* → *animal*). It seems that unless the relation was bizarre, subjects were able to find a relation that fit the pair. However, it is reassuring that pairs that would never be linked by Pathfinder did not come together until later in the cluster analysis. On the other hand, some pairs that were linked in the network (e.g., *chicken* → *egg, hair* → *color*) also clustered late, suggesting that these links may also be spurious. Alternatively, late clustering might suggest that the link was unique and therefore unrelated to any other link in the graph.

16 Link Interpretation

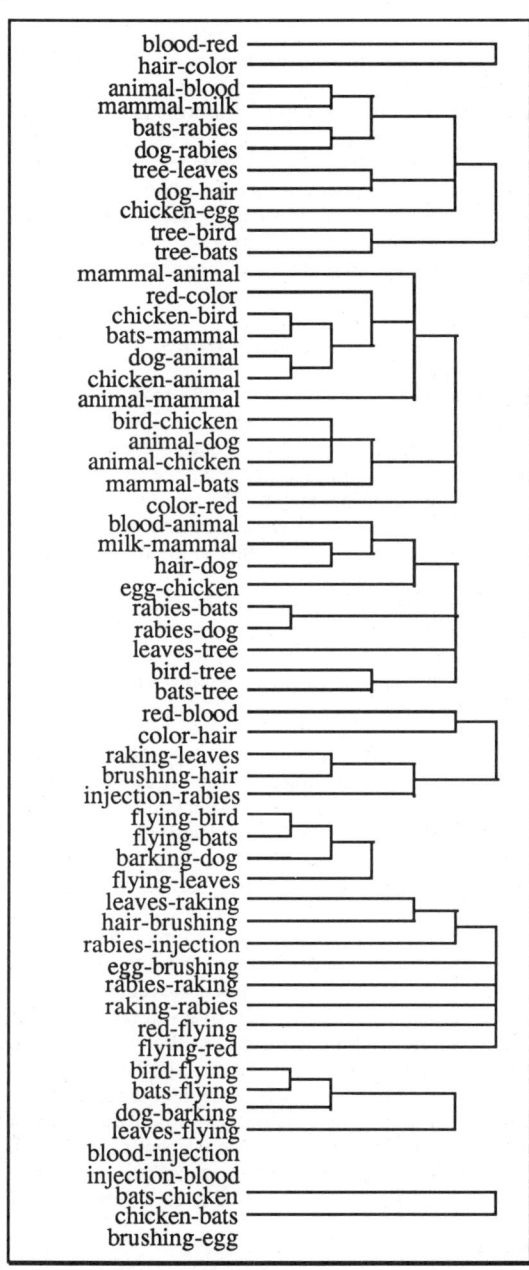

Figure 2. Cluster analysis (single-link method) based on link sorting.

The sorting and cluster analysis procedures generated intuitively reasonable clusters of linked pairs in the sense that there is a cluster that could be labeled *is-a* (*chicken → bird, bats → mammal*) and one that could be labeled *has* (*dog → hair, tree → leaves*). However, these clusters derived from the frequency data were not explicitly labeled by the subjects, but when subjects were asked informally to label the groups that they had sorted or to label individual links, they had difficulties. For instance, some relations were hard to verbalize (e.g., *bird → chicken*). It is possible that the clusters resulting from the frequency data would be easier to name than individual links because the clusters of links suggest more alternative labels—one or more of which may be "good." Therefore, in the following study subjects were asked to label the groups of links generated by the cluster analysis.

Link Labeling

The purpose of this phase of the methodology was to determine the nature of the relation represented by each cluster of linked items obtained in the classification phase of the methodology. Subjects were asked to assign a rating to each cluster indicating the goodness or cohesiveness of the cluster and to assign a label to cohesive clusters. The purpose of the ratings was to provide empirical support for the intuitive claim that the clusters of links generated in the classification phase were meaningful. Thus the ratings provided a means of evaluating the methodology. Additionally, the ratings and names should help to differentiate meaningful clusters from those that occur because of noise or because everything must ultimately come together in a cluster. The cluster diagram can be simplified by eliminating clusters that are not meaningful and by merging meaningful clusters that have the same label. Determining the appropriate cutoff level for a cluster is not peculiar to this methodology, but it is an issue relevant to cluster analysis in general.

Method

Twelve introductory psychology students from New Mexico State University participated in this study. As before, these subjects were required to pass an analogy pretest. Stimuli consisted of the 36 clusters of linked items obtained in the classification study, with an additional 15 groups of pairs that did not cluster until the very end. Unclustered groups included (*color → hair, blood → animal*) and (*egg → brushing, chicken → bats, tree → bats, chicken → animal*). These groups served as a control condition against which the clustered groups were compared.

Subjects were seated in front of a Sun Microsystems terminal and presented with instructions. They were told that for each trial two or more pairs of items would be displayed on the terminal screen. They were to first determine the relation for each pair in the group and then enter a rating (1, 2, 3, 4, or 5) by pressing the corresponding key on the keyboard. The ratings were to indicate the similarity of the relations present in the group of pairs with 1 indicating that the relations were not very similar and 5 indicating that the relations were very similar. The rating was meant to reflect the goodness or cohesiveness of each cluster of linked items. In addition, subjects were informed that after the rating was entered they might be required to enter a name that best described the relations in that group. Subjects were asked to name a group if they had given it a similarity rating of 3 or greater. The groups of linked items were presented to each subject in random order.

Results and Discussion

Average ratings for clustered and unclustered groups were calculated for each subject. The mean rating across all subjects was 3.42 for the clustered groups and 1.52 for the unclustered ones. This difference was significant, $t(22) = 8.33$, $SE = .229$, $p < .001$, supporting the claim that the clusters produced by the cluster analysis were meaningful. Groups of linked items that were clustered in the analysis were judged more cohesive than groups that did not come together until the end.

The average cohesiveness ratings, along with a cluster label (the name most frequently assigned to the cluster by the 12 subjects), are presented for each cluster in Figure 3. At least half of the subjects agreed on the same or nearly the same label for each of the clusters that had average ratings greater than three. The judgments about what constituted an "identical" label were subjective, but the decisions seemed straightforward and were easily made. The cohesiveness ratings were low for clusters that contained spurious pairs (i.e., those never linked in a Pathfinder network). Also, subjects did not agree on labels for these clusters. These results were expected, given that the pairs in these groups were judged as being unrelated.

The average ratings and the labels provide an indication of clusters that are real, as opposed to those formed as a result of noise, or because all items must eventually come together. For instance, the cluster that contains *blood → red* and *hair → color* received an average rating of 2.67, suggesting that this cluster is not very meaningful. Also, in several cases there were two or more clusters that were assigned the same label (e.g., the *is-a* clusters). The fact that these clusters were not differentiated by labels may indicate that the smaller clusters should be merged into a single cluster. Thus, low average cluster ratings and redundant labels were used to simplify the cluster solution, clusters were dropped (unclustered) if they had an average rating of less than three, and finally clusters that shared the same label were merged. The resulting simplified solution is presented in Figure 4. These simplification procedures reduced the number of clusters from 36 to 17.

An inspection of Figure 4 reveals four major clusters of relation: *is-a/superordinate*, *has-a/comes from, controls/is controlled by,* and *does/is done by.* Each of these four basic relations can be placed in the Chaffin and Herrmann (1984) taxonomy, but some of the specific relations that were nested within one of the four basic types were not as easily classified within their taxonomy. The *is-a/superordinate* (or *category*) relations fit into the CLASS INCLUSION group of the taxonomy. More specifically, these relations fall under the subheading of PERCEPTUAL SUBORDINATES or "objects that are principally characterized by their visible, physical properties" (Chaffin & Herrmann, 1984, p. 135) as opposed to functional properties. The *has a/comes from* relation is similar to the PART/WHOLE relation in the taxonomy. However, the taxonomy subtypes for this category do not seem to match the aspects of *has* that were captured in this study (i.e., *contains, is covered by, carry, is home of*). Both the *controls/is controlled by* and the *does/is done by* relations fall under the taxonomy heading of CASE relations. The *controls* relation can be considered an instantiation of an ACTION-INSTRUMENT relation and the *does* relation is an instantiation of an AGENT-ACTION relation. Thus, although the basic types of relations identified in this study could be placed in the taxonomy, there were some specific distinctions made by subjects within these categories that could be used to extend the taxonomy.

In summary, the results of this study indicated that groups of clustered links were more meaningful than groups of unclustered links. Low ratings and redundant labels were used to eliminate the less meaningful clusters, thus greatly simplifying the solution. Clusters of relations that were revealed in the analysis fit nicely into the Chaffin and Herrmann (1984) taxonomy of relations, although there were some fine distinctions within categories that were not compatible with the taxonomy. The sorting and cluster procedures, in combination with the rating and naming tasks, comprise a valid and useful approach to interpreting links in Pathfinder networks.

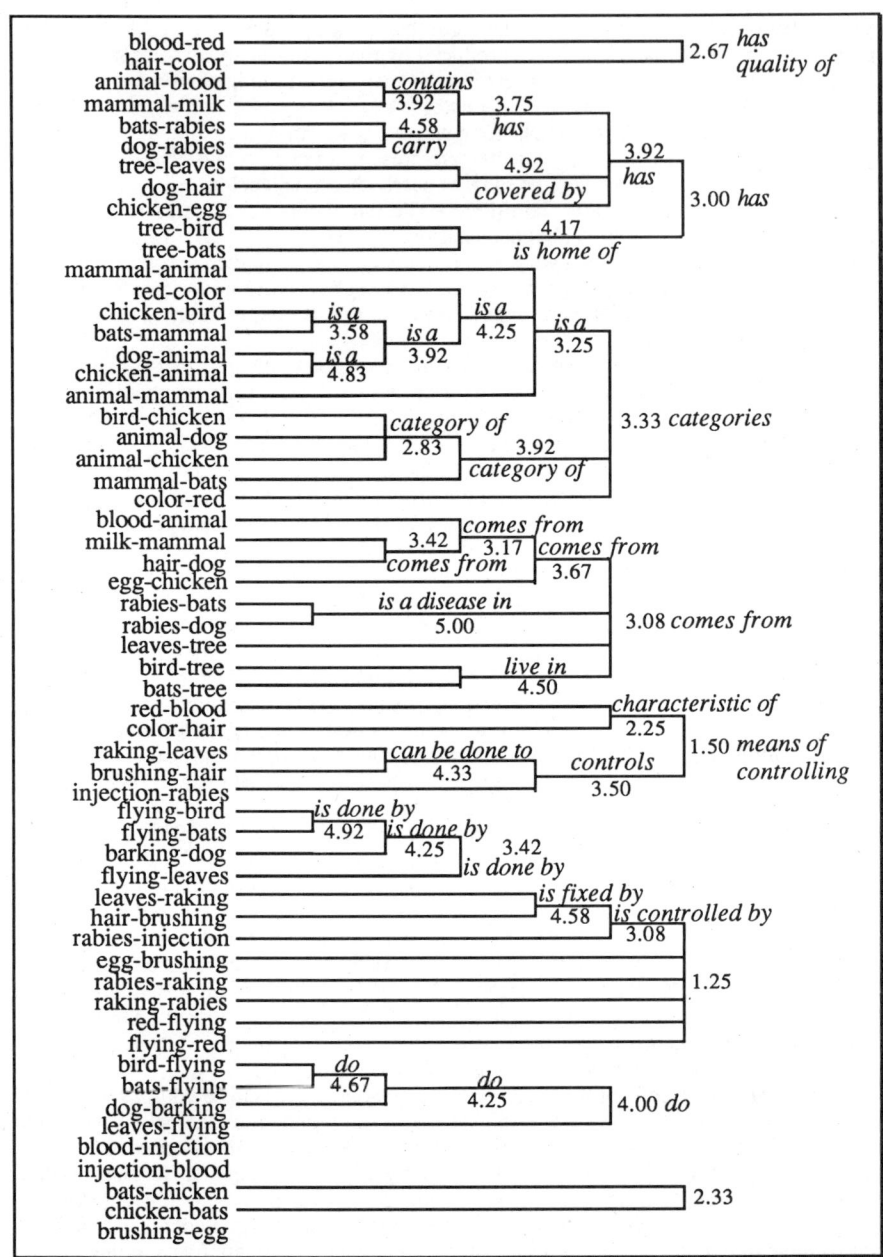

Figure 3. Cluster analysis for link sorting with average ratings and labels.

16 Link Interpretation

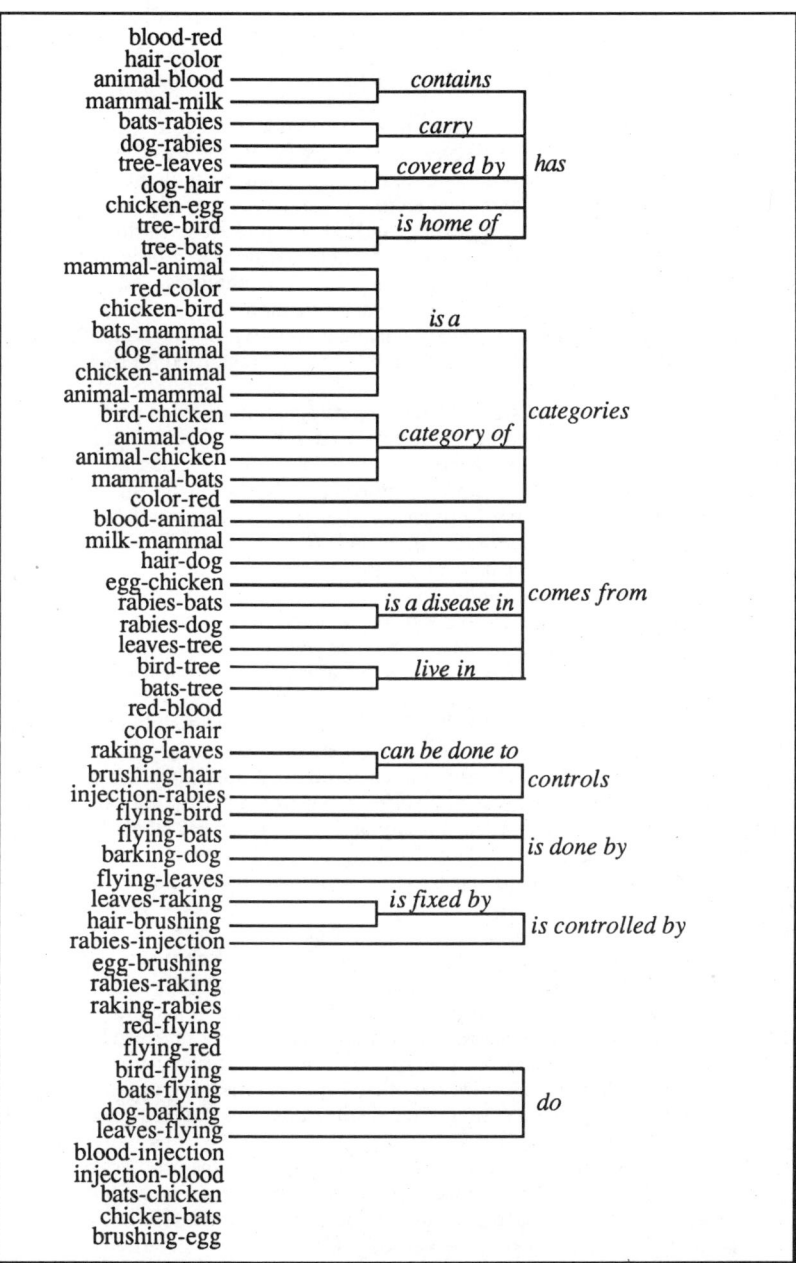

Figure 4. Simplified cluster analysis for link sorting.

Conclusions

Using the methodology described, links were interpreted in a Pathfinder network. This information enhanced the Pathfinder representation, and consequently adds to its effectiveness as a knowledge elicitation tool. Link interpretation can be thought of as part of the third stage of a knowledge elicitation methodology that consists of (1) concept elicitation (see Cooke, 1987), (2) psychological scaling, and (3) interpretation. In this third stage the scaling representations are interpreted by identifying dimensions in multidimensional scaling solutions, identifying cluster cutoffs and labels in cluster analyses, and identifying link labels in Pathfinder networks. The link interpretation methodology presented here consists of two separate phases: (1) link classification using a sorting procedure and cluster analysis, and (2) cluster rating and labeling in which relations are identified for meaningful clusters of links.

One possible criticism of the study described here is that it involved pairs of items with obvious relations. That is, the common concepts used in the study were not representative of concepts in specific domains of expertise, such as physics, computer programming, or flight maneuvers. However, this methodology has also been successfully applied to the domain of computer programming in order to label links in a network of abstract programming terms, such as *algorithm, subroutine,* and *character data* (Cooke, 1987). Cooke found that 24 links relating various pairs of programming terms were each classified as one of six bidirectional relations: (1) *can be done to/is operated on*, (2) *is part of/contains*, (3) *interchangeable/interchangeable*, (4) *is a type of/superset*, (5) *is a type of/could be*, and (6) *can be done to/is used with*. Again, these link types were consistent with the Chaffin and Herrmann (1984) taxonomy; however, there were some subtleties that could be used to expand the taxonomy.

It is informative to compare the results of the link labeling methodology described here with the more straightforward approach of asking experts to simply label links in a network. Cooke and McDonald (1987) attempted to identify link labels for the same programming network that was used by Cooke (1987), but by simply asking experienced programmers to label the individual links. Twenty of the 24 links were labeled identically by three or more of five programmers. However, there were 11 different labels assigned to the same 20 links, as opposed to the 6 labels generated using the classify and label methodology.

The sorting and clustering methodology could be characterized as a generalization mechanism to the extent that it reduces the number of distinct link types. Unfortunately, subtle differences among links could be masked by this methodology. It is important to constrain the number of link types for the sake of efficiency, but it is equally important to capture the distinctions that people are able to make. The rating task used in the second phase of the methodology provides one check on overgeneralization by eliminating clusters of pairs that are given low ratings. On the other hand, the merging of clusters based on similar labels encourages generalization. Fine distinctions between clusters may exist, yet may be hard to verbalize.

The issue of the level of abstraction for link labels is tied to the problem of identifying an appropriate cutoff level in a cluster analysis solution. There are some problems raised by the rating and naming approach used here to determine cutoff. Names are subject to variation and therefore require experimenter judgment in order to determine whether two names are equivalent. Also, the naming task is difficult and subject to the usual problems with introspection and verbal reports. Ratings might overcome these problems by enabling subjects to express subtle differences among clusters in that they are unable to convey through cluster names. Unfortunately, ratings are also plagued by variance. For instance, does an average rating of 3.2 differ significantly from an average rating of 3.1? Variance should be taken into account in cases in which fine distinctions need to be made using ratings.

In regard to the task itself, it is encouraging that the experienced programmers used by Cooke (1987), unlike the introductory psychology students, had no difficulty with the sorting task and therefore did not require pretesting. This difference might be due to the fact that the programmers had spent considerable time thinking about the programming terms or because they were better educated (they were mostly graduate students). At any rate, this difference suggests that little practice should be required to perform these tasks in a typical knowledge elicitation setting.

Link classification and labeling are important components of link interpretation, but in order to fully interpret links, it is necessary to know what inferences are associated with each one. For example, the relation *is-a* is typically associated with the inference that the subordinate concept can inherit properties of the superordinate concept, but what inferences can be made from the *controls* or *does* relations? The problem could be even greater when concepts and relations are specific to a particular domain. The research presented here does not address this issue, but there are several possible approaches to the elicitation of this information from domain experts. Experts could be asked to associate each obscure labeled relation with an analogous relation for natural language concepts—for example, "this relation is like the relation between *robin* and *bird*." Then it could be assumed that the inferences that hold for the common relations are similar to the domain-specific relations. However, this strategy still requires knowledge of inferences associated with the common relations so future research is required.

In summary, the use of scaling techniques by themselves enables the researcher to address structural issues regarding memory organization. Without concept elicitation and interpretation methodologies, the information content of the structure is impoverished. Consequently, such enhancements combined with Pathfinder and other scaling techniques comprise a knowledge elicitation methodology that is better suited for this application than is scaling alone. Empirically driven knowledge elicitation is particularly appealing in comparison to the highly subjective and time-consuming protocol analysis and interview techniques. Ongoing research is directed toward expanding this methodology to include other techniques as well as toward validating various aspects of the methodology. In addition, there are plans to automate the methodology in the form of a knowledge acquisition and representation tool kit.

Chapter 17

A Structural Assessment of Classroom Learning

Timothy E. Goldsmith and Peder J. Johnson

A most basic and long standing concern of philosophers, psychologists, and educators is the problem of knowledge elicitation and representation. How do we assess and represent an individual's knowledge? Philosophers, when asking these questions, have usually expressed an interest in general or world knowledge. Psychologists and educators, on the other hand, have often been more interested in the problem of assessing and representing a person's knowledge of some particular topic or area. It is this problem, as it arises in the assessment of classroom types of knowledge, that is the concern of the present chapter. Knowledge assessment and representation, as carried out in the classroom, appears as a relatively straightforward matter. Knowledge is assessed by simply asking factual questions and is represented by presenting the individual's relative standing in terms of a percentile. We begin with a critique of this conventional approach to assessing and representing classroom knowledge.

The two processes, assessment and representation, are obviously related. In our view the approach to representation is more fundamental in that assumptions regarding the organization of knowledge have implications for how we assess knowledge. The representation of classroom knowledge is usually in terms of percentage correct, or in the case of standardized tests, performance may be converted to a percentile ranking within some designated population. In some instances performance may be analyzed into subscales, but for the most part classroom learning is represented in terms of a unidimensional scale reflecting the student's relative standing in her class. Percentage correct may be perfectly adequate for representing certain types of knowledge (e.g., a student's knowledge of the capitals of the 50 states) where the conceptual relationships among the knowledge elements are not particularly relevant. In this case it may be safe to assume that the "facts" comprising the domain are independent and additive. An important property of more interesting domains of knowledge involves the relationships or organization of the elements which compose the domain. We contend that for domains of this type it is this *configural* property of knowledge that must be assessed and represented. A percentile ranking may be very convenient in assigning grades to students, but it tells us very little regarding what the student knows or does not know. The fundamental problem with conventional educational assessment procedures is that they are in principle, incapable of explicitly representing the abstract nature of conceptual domains. A basic assumption in cognitive psychology is that knowledge entails an understanding of the interrelationships among concepts and that this organizational property of knowledge can best be captured with structural representations (e.g., Bower, 1972; Collins & Quillian, 1969). It is our aim to develop and evaluate a set of procedures that capture this configural property of knowledge.

Turning to the problem of assessing classroom knowledge, historically, evaluation has occurred by presenting various recognition (e.g., true-false or multiple choice) or recall (e.g., essay) types of questions that are directly related to the relevant content area. It is generally recognized that this method of assessment has several potential problems. First, performance on recall and even recognition tests may be impaired by retrieval problems. Because conventional test procedures depend very heavily on episodic memory, they are subject to the influence of context specific cues on retrieval processes. As a consequence we often hear students complain that they knew the answer but could not recall it, or they did not know what was being asked or that they misunderstood the question. All of these complaints have sufficient validity that should concern us.

A second problem is that much of what we expect students to learn is implicit knowledge that is difficult, if not impossible, to state in an explicit manner. To the extent that some relevant domain-specific knowledge is implicit, it may be extremely difficult to assess that knowledge with direct questions. This problem is most obvious in the case of procedural knowledge. Third, there is the problem of developing tests that are objective and easily scored, while at the same time assessing the more abstract or conceptual aspects of knowledge. This is the obvious tradeoff between multiple choice and essay questions. Multiple-choice exams, while objective and easy to score, require considerable expertise in their development if they are to assess conceptual knowledge. On the other hand, essay questions may assess conceptual knowledge more readily, but the scoring is likely to require a considerable degree of expertise. For this reason essay exams are often considered too time-consuming and resource demanding to be used on a large scale.

To summarize, we are suggesting that the ideal approach for assessing and representing classroom knowledge would be objective, reliable, require minimal retrieval demands, and most important reflect the student's conceptual organization of the domain. We now turn to some earlier research that takes a structural approach to assessment and representation of classroom learning.

Structural Assessment Approaches to Learning

The limitations of conventional testing methods point to the need for procedures that assess and represent the conceptual properties of classroom learning. This need was clearly recognized as early as 1972 by Shavelson and his colleagues (Geeslin & Shavelson, 1975; Shavelson, 1972, 1974; Shavelson & Geeslin, 1973; Shavelson & Stanton, 1975) who took the view that classroom instruction is most properly seen as the communication of a specific structure that is implicit within the curriculum of a subject matter. It was Shavelson's assumption that "a structure of a subject matter, ultimately, rests in the minds of the 'great scientists.' This structure is communicated through the scientists' writings in journals and advanced textbooks as well as through informal communication channels" (Shavelson, 1974, p. 232).

Shavelson's goal was to assess the effects of classroom instruction by determining whether the student's organization of the material became more similar to an expert's over the course of instruction. In his research Shavelson attempted to assess the structure of a classroom topic (e.g., high school physics) using a variety of techniques, such as word associations, card sorting, and a graph construction method. Distance measures were calculated on the basis of these data, which in turn were subject to a hierarchical cluster analysis to derive an underlying structure.

Shavelson's results were generally encouraging, showing that a student's structure did become more congruent with an expert's structure. Whereas this program of research has not had an obvious impact on educational assessment, it has provided the impetus for a slowly developing literature investigating the relationship between derived cognitive structures and domain performance (Champagne, Klopfer, Desena, & Squires, 1981; Diekhoff, 1983; McKeithen, Reitman, Reuter, & Hirtle, 1981; Naveh-Benjamin, McKeachie, Lin, & Tucker, 1986; Thro, 1978). It is not our aim to provide a comprehensive review of this literature, however, some sense of the direction taken in the more recent work can be gained by describing two of these studies.

Champagne et al. (1981) assessed students' cognitive structures of physical geology with a procedure they called the *Concept Structure Analysis Technique*. This technique involved having the students arrange a set of core concepts spatially on a large piece of paper. While the students arranged the concepts, the researcher, guided by the student, labeled subgroupings of concepts. In a related study, Naveh-Benjamin et al. (1986) used a modification of Reitman and Reuter's (1980) ordered tree technique to assess changes in student's cognitive structure of the material presented in a course on the psychology of aging. The results from both of these studies were generally positive in that they found structures to become more similar to experts as a consequence of training, and the structure of students who earned the higher grades were more similar to the experts' structure.

In summary, a number of studies have been conducted showing that individual differences in levels of domain performance are related to differences in derived cognitive structures. However, all of these studies suffered from one or more of the following problems. First, the data on which the cognitive structures were derived were averaged across subjects and therefore failed to assess individual students. If this approach is eventually to be used in the classroom it must be applicable to individual students. Second, the assessment procedures often required the students to report directly the organization of their structure. There are reasonable concerns (e.g., Nisbett & Wilson, 1977) as to whether we have direct conscious access to cognitive structure. Third, the derived structures were often assumed to be hierarchical. Hierarchical structures may be appropriate for some domains, but not all. It would be preferable not to constrain the solution to a hierarchical representation. Finally, the basis for comparing the similarity of structures was often subjective. Ideally we would have an objective quantitative means of assessing the similarity of representations. In the next section we define more precisely the nature of the knowledge representations that we employ and then go on to describe the methods to derive and use these representations to assess classroom learning.

Empirically Derived Knowledge Representations

In his paper on the fundamental aspects of cognitive representations, Palmer (1978) noted that the field is "obtuse, poorly defined, and embarrassingly disorganized" (p. 259). Although more than a decade has passed since Palmer made these observations, there is ample evidence to suggest that his observations remain valid. Heeding Palmer's criticism we attempt, in this section, to describe how we conceptualize the representations we use in the present work.

We begin, as Palmer (1978) did, by distinguishing between what we may think of as an individual's actual knowledge of some domain and some inferred representational model of this knowledge. We assume that the actual knowledge comprises a set of relevant data structures and processes that we shall refer to as the cognitive system. As an individual becomes more expert in a domain the cognitive system is assumed to be modified in some manner. Although the precise changes in the data structures and/or processes may be indeterminate (Anderson, 1978), they are assumed to result in certain changes in domain-relevant performance. Of the many behavioral changes that may occur with the acquisition of expertise, we are particularly interested in judgments of relatedness among central concepts within the relevant domain. This set of relatedness judgments are, in this sense, a reflection or a representation of the state of the cognitive system. Although there are a variety of transformations, such as multidimensional scaling (MDS), that can be performed on these proximity data, these different transformations are simply alternative representations of the state of the cognitive system. Hence, we view both a set of proximity values and any transformation of those values as simply ways of characterizing a functioning cognitive system rather than as representations of the cognitive system's actual data structure.

In our approach, the preference of one transformation over another is determined first by its validity as a predictor of domain performance and second by its representational simplicity relative to the complete data. In the present work we are specifically interested in determining whether Pathfinder (Schvaneveldt, Durso, & Dearholt, 1987, 1989) representations have greater predictive validity than raw proximity data or MDS-derived representations.

A second distinction in our interpretation of knowledge representations concerns the relationship among the elements represented. Earlier we made the argument that an important aspect of knowledge involves the configural relationships among the concepts of the domain. That is, to be knowledgeable of a domain implies that the important concepts are interrelated and organized in some particular configuration or class of configurations. In order to discuss this issue of concept interrelatedness, we will employ network representations; however, our view of configurality also extends to other types of transformations of proximity data.

Our interpretation of the relationships among concepts as revealed in a network representation differs in two important ways from other representational approaches with which the reader may be more familiar. First, in contrast to network models of language comprehension where it was necessary to explicitly label the type of link that connected various nodes (Collins & Quillian, 1969), the link in the present networks are unlabeled. This raises the question of whether there can be any semantics in a network with unlabeled links (Woods, 1975). We assume that to the extent that a derived representation has predictive validity it also contains a degree of semantic relevance. Second, links between specific nodes in the network do not necessarily have any direct causal implications regarding domain performance. Instead, we assume that it is the pattern of links that is meaningful. This assumption is in contrast to rule-based representational approaches, where relations often do have direct performance implications. What then is implied by network representations as we are interpreting them? We hope to show that structural properties of networks reflect general associative information regarding the state of a cognitive system.

Given our assumption regarding the configural character of knowledge, it follows that we would want to assess the configural properties of network representations. In particular, in comparing the knowledge representations of two individuals we would want to assess the degree of their configural similarity. We assume that the configural properties of a network are not directly obtainable but rather must somehow be interpreted. For our purposes this interpretation process is described by a metric that assesses the similarity of two networks. Goldsmith and Davenport (Chapter 5, this volume) report a set-theoretic method for defining structural similarity between graphs, and we employ this method later to assess the configural similarity of two networks.

To summarize, a central thesis of this chapter is the idea that configural properties of representations reflect important characteristics of an individual's cognitive system. We further assume that these configural characteristics can be compared in network representations by employing a method for assessing structural similarity between graphs. Below we test the hypothesis that representations of domain knowledge have functional utility by attempting to predict individual differences in performance within a domain from the structural similarity of an individual's representation to some idealized referent representation.

Methodological Issues

First, we turn to some general issues that arise with the method that we are proposing to use to assess domain-specific knowledge. We discuss here the choice of a procedure for collecting proximity data on a set of domain concepts, the particular type of transformations performed on these data, and the methods by which different representations are compared.

A variety of techniques for obtaining proximity data have been previously used, ranging from sorting to memory recall tasks. We have chosen to use direct judgments of concept relatedness as the basis for obtaining conceptual representations. Our choice of relatedness ratings is based in part on the past successes of this method, much of which comes from rating the similarities of directly perceived physical objects. There exists extensive literature demonstrating that such direct similarity ratings are useful for studying perception, memory, and learning (e.g., Shepard, 1974). The validity of similarity judgments as they apply to semantic concepts is more indirect, but here too there has been some success especially in discriminating experts from novices. For example, different levels of expertise have been discriminated among fighter pilots (Schvaneveldt, Durso, Goldsmith, Breen, Cooke, Tucker, & DeMaio, 1985) and computer programmers (Cooke, 1983; Cooke & Schvaneveldt, 1988) on the basis of representations that were in turn derived from pairwise similarity ratings. This suggests that the ratings were sensitive to properties of the cognitive system related to domain performance.

In addition, judgments of relatedness or similarity occur naturally in the course of learning and performing in a domain. The journey from novice to expert may be viewed as a continuous sequence of analysis and synthesis, with each successive cycle providing a more differentiated and integrated cognitive system. In this regard the basic processes of generalization and discrimination play a fundamental role in the acquisition of knowledge. Judgments about what is alike and what is different would appear capable of reflecting fundamental properties of the developing cognitive system. Perhaps William James (1890/1981) had something like this in mind when he said, "the sense of sameness is the very keel and backbone of our thinking" (p. 434).

The validity of relatedness judgments ultimately rests on their ability to provide meaningful results, and this issue raises the question of transforming proximity data. Are transformations of proximities more useful for understanding psychological phenomena than the raw proximities themselves? The history of scaling suggests that the answer to this question is yes, and so a major effort of our work in classroom assessment has been to determine which transformations of concept ratings have greatest predictive validity. In particular we compare raw proximity ratings, MDS spatial representations of the ratings, and Pathfinder representations of the ratings.

A third issue that arises is how to compare different representations. To assess the effects of classroom learning on cognitive structure requires that we have some means of objectively comparing a student's conceptual representation with a referent representation that is assumed to reflect a desired organization of the domain's concepts. In our work a student's structural representation of course concepts is compared to the instructor's representation of the same concepts.

In the case of raw proximity values, we simply calculate the Pearson correlation coefficient between corresponding values in the two sets of proximities. For MDS spatial representations, we first calculate the euclidean distance between all pairs of points in the n-dimensional space and then calculate the Pearson correlation coefficient between corresponding distances in different representations. For Pathfinder networks we employ two measures of similarity. The first is similar to the one just described with MDS representations, but instead of euclidean distances in space we use graph-theoretic distances between nodes in the derived networks. The second similarity measure for networks is the technique mentioned previously and described by Goldsmith and Davenport (Chapter 5, this volume). Briefly, this technique employs a set-theoretic method to compare corresponding neighborhood regions of two networks. The method computes for any two networks a single quantitative index of closeness called C. The values of C range from zero to one.

The issues of how to represent concept ratings of relatedness and then how to compare these representations have important implications for our work. As stated previously, we believe that an important property of conceptual representations is their configural nature. If true, then it is the pattern of interrelationships among a set of concepts that should prove useful for differentiating among individuals with differing levels of knowledge. Hence, we want to employ transformations of concept ratings that preserve or uncover configural relationships in the data and then to use methods for comparing these representations that are sensitive to configural information. We believe that C offers such a method for assessing structural (i.e., configural) similarity, and therefore hypothesize that C applied to Pathfinder networks will indeed result in higher validity for predicting levels of knowledge in students than other representation/comparison methods.

Assessing Classroom Learning

We turn next to an empirical study that investigated the feasibility of assessing classroom learning with empirically derived knowledge structures. The basic purpose of the study was to measure student's knowledge structures over the course of learning and to assess whether the degree of agreement between the students' structure and the instructor's structure was indicative of classroom performance as measured by conventional testing techniques. We hypothesized that students whose structures more closely match the instructor's will indeed be more knowledgeable and hence perform better on standard examinations. We further hypothesized that a measure of representational similarity that assesses configural relationships of knowledge elements will be more predictive of performance than one not based on configural information.

Method

Domain. The knowledge domain was a 16-week sophomore/junior-level college course on psychological research techniques with a primary focus on the analysis and design of experiments. Prior to taking the course, each student had completed an introductory course in probability and statistics. An initial set of concepts considered to be central to experimental design was selected by the instructor, and then suggestions from other faculty members who taught courses in statistics and design were obtained resulting in a revised set of 30 concepts. The final set of concepts is provided in the Appendix. Student performance in the course was measured by three exams and two papers totaling 480 points.

17 Structural Assessment of Learning

Subjects. A total of 40 students participated in the study with 20 students coming from each of two separate courses taught in different semesters by the same instructor. The students were primarily college juniors and seniors.

Procedure. The purpose of the concept rating project was explained to students at the beginning of the semester. They were told they would be rating the relatedness of 435 pairs ($n(n-1)/2$) of concepts and that these ratings would be used to assess their knowledge of the course material. Students were told that they could earn up to 20 extra course points by performing well on the tasks. At the end of the semester extra points were assigned on the basis of the degree of agreement between their structures and the instructor's.

Students were asked to judge the relatedness of each pair of concepts using a 7-point scale where 1 corresponded to less related and 7 to more related. At the beginning of each rating session, students were shown the complete set of concepts and were encouraged to pick out some pairs that were highly related and some they thought were quite unrelated to serve as anchors. They were also told to use the full range of the scale in making their ratings. Because students would be unfamiliar with some of the concepts at the beginning of the semester, they were asked to consider their confidence in their knowledge of the concepts while making relatedness judgments. Specifically, they were told that ratings from the ends of the scale (e.g., 1, 2, 6, and 7) implied that they were more certain of the meaning of those concepts, whereas ratings from the middle part of the scale (e.g., 3, 4, and 5) could reflect both medium relatedness or uncertainty about the meaning of the concepts.

Students were instructed to give quick intuitive judgments of relatedness rather than performing a lengthy and deliberate analysis of the concept pairs. On average, students took about one hour to complete the set of 435 ratings.

Each pair of concepts appeared left-right centered below the rating scale. A bar marker appeared initially at rating 4 for each concept pair. The bar marker could then be moved with the left and right directional keys on the computer keyboard until it was above the desired rating. Pressing the space bar accepted the rating for the current pair and presented the next pair of concepts. The presentation order of the concept pairs was randomized individually for each subject. Additionally, the left-right order of the concepts was randomized for each pair.

Students rated the concepts on approximately the 1st, 8th, and 15th weeks of the semester. Each student performed the task individually and at their own convenience on microcomputers located around campus. The course instructor also rated the concepts to provide a referent structure against which to compare students.

Results

Data from both classes were combined and analyzed together. The concept ratings yielded proximities by subtracting each rating from eight. These proximities were then analyzed by Kruskal's (1964) nonmetric MDS procedure and Pathfinder. In the case of MDS, an elbow criterion test yielded four dimensions as optimal and so all subsequent MDS analyses are based on four dimensions. PFNETs(∞, $n-1$) were derived on the same datasets.

Once the MDS and Pathfinder representations were obtained, the similarity between each student's representation and the instructor's was determined using the methods described previously. Comparisons were made from each student's data using each of four different knowledge indices: correlations on raw proximities, correlations on MDS distances, correlations on Pathfinder graph-theoretic distances, and Pathfinder networks assessed by C. To simplify reporting of the results, we abbreviate these as PRX, MDS, PFR, and PFC, respectively.

The first question we turn to is how frequently students used the seven values along the rating scale. Table 1 shows the frequency distribution of relatedness ratings for both the instructor and students. Particularly striking about these data are the stability of the three distributions over the 15 weeks of the semester. Also, the distribution of the students' ratings was quite similar to the instructor's, with the surprising exception that the instructor tended to use extreme ratings (1 and 7) less frequently than students. Both students and the instructor were more inclined to rate concepts as related (5, 6, or 7) than as unrelated (1, 2, or 3). This may be a function of the domain or the particular set of concepts selected.

We turn next to how well students agreed with themselves and with the instructor as a function of time in the semester. Table 2 shows mean agreement measures assessed both within students and between students and instructor on each knowledge index. Keep in mind that the agreement between networks as measured by C units is not directly comparable to the correlation coefficients. Notice first that some agreement exists among students and with the instructor even at the beginning of the semester. This is not too surprising because all of the students had taken a previous course in probability and statistics and many of the concepts were already known.

Table 1. Frequency distribution of relatedness ratings for students at the 1st, 8th, and 15th week of the semester and for the course instructor.

Dataset	Relatedness Rating						
	1	2	3	4	5	6	7
Week 1	.06	.13	.16	.14	.20	.18	.13
Week 8	.06	.13	.16	.12	.22	.17	.14
Week 15	.08	.15	.16	.11	.19	.16	.15
Instructor	.02	.14	.20	.14	.25	.21	.04

Table 2. Mean agreement of representations[a] within students and between students and instructor at the 1st, 8th, and 15th week of the semester.

Knowledge Index	Student-Student			Student-Instructor		
	Week 1	Week 8	Week 15	Week 1	Week 8	Week 15
PRX	.24	.35	.30	.26	.32	.34
MDS	.28	.23	.43	.39	.49	.54
PFR	.19	.23	.25	.24	.29	.32
PFC	.30	.34	.41	.43	.45	.50

[a]PRX - correlation on raw proximities
MDS - correlation on MDS distances
PFR - correlations on Pathfinder graph-theoretic distances
PFC - Pathfinder similarity assessed by C

Since students in the same course learn a common knowledge base across the semester, we would expect the degree of agreement to increase among students over the semester. This trend clearly existed for all of the knowledge indices from the beginning to end of the semester. However, the mid-semester correlations on proximities and MDS distances fall outside of their beginning-to-end ranges. One explanation for this finding is that students in the middle part of the semester may still view the material quite differently as a result of differing past orientations to the domain. Students may also differ both in their rate of assimilation of the material and in their development of strategies for organizing the material. However, by the end of the semester a sufficient number of shared conceptual experiences has occurred to ensure that a fairly homogeneous view of the domain emerges.

More important, perhaps, is the change in agreement across the semester between students and instructor. We assume that the common knowledge base learned by students is, at least to some extent, that of the instructor's and learning occurs when students agree with the instructor, not necessarily when they agree with one another. Here we find a consistent trend of increasing agreement over time for all of the indices. Notice also that the magnitude of increase over time is greater between students and instructor than within students. The change was most dramatic with MDS representations which might lead one to speculate that if MDS better reflects changes in representations across learning, the degree of agreement between a student's and instructor's MDS representations would be a better index of the student's level of knowledge. We turn to this question next.

Agreement as assessed by the various knowledge indices between each student and instructor was computed based on the student's end-of-the-semester ratings. The last set of ratings was analyzed because they should best reflect the student's overall knowledge of the domain. Pearson product-moment correlations were then computed between each knowledge index and the student's earned course points at the end of the semester. Table 3 shows the resulting correlations. The correlations were all significant ($p < .01$).

Table 3. Correlations (and squared correlations) of instructor-student agreement and final course points for students.

Knowledge Index	Correlation
PRX	.61 (.37)
MDS	.54 (.29)
PFR	.66 (.43)
PFC	.74 (.55)

The correlation coefficient between a student's and instructor's concept ratings (PRX) accounts for 37% of the variance associated with the student's final course grade. Therefore, concept ratings themselves appear to be an indicator of a student's knowledge. Of more interest, however, is whether scaling algorithms, such as MDS and Pathfinder, are able to extract from these ratings information that would allow even better performance predictions. The answer appears to vary. Distances from MDS were slightly poorer than proximities in predicting performance, whereas Pathfinder distances were better than the raw proximities. Comparison of Pathfinder networks with C (PFC) provided even better predictions than with correlations (PFR).

One way of looking closer at the relative contribution of each knowledge index to predicting final course points is to examine partial correlations. Table 4 gives the correlation between each index and final points with the variance contributed by each other index partialed out. Consider first Pathfinder networks that have been compared using C. PFC correlates significantly with final points even when each of the other indices is held constant. However, none of the other indices correlate significantly with course grades when the variance contributed by PFC is held constant. This pattern of findings strongly suggests that Pathfinder networks, as assessed by C, are uniquely capturing important predictive variance in the concept ratings. Consider next Pathfinder networks as assessed by correlations. PFR is a significant predictor when proximities and MDS is each held constant, but not PFC.

Therefore, taken together, these results imply that Pathfinder networks do indeed contain unique predictive variance over the proximity ratings and MDS, and that a configural assessment of networks is a better index for assessing network similarity than correlations. Apparently, C better reflects commonalties between structures that happen to be important in assessing knowledge. We assume that the characteristics common to a student's structure and instructor's structure that are predictive of knowledge attainment exist at a global or configural level within those representations. This, of course, is exactly the type of information that C is assumed to be good at assessing.

Table 4. Partial correlations between Knowledge Index 1 and final course points with Knowledge Index 2 partialed out.

Knowledge Index 1	Knowledge Index 2			
	PRX	MDS	PFR	PFC
PRX	-	.34*	.30	.15
MDS	.03	-	.29	.12
PFR	.43**	.52**	-	.17
PFC	.54**	.61**	.46**	-

$^*p < .05$
$^{**}p < .01$

Consider next the results from MDS. Spatial structures did not significantly predict course performance when the other variables were partialed out. MDS has been successful in previous application for representing physical or continuous relations. Our results may indicate a specific limitation of MDS for assessing conceptual-level relations. Some corroboration for this conclusion is found in work by Cooke, Durso, and Schvaneveldt (1986) who compared MDS and Pathfinder in predicting recall data.

In summary, we conclude that knowledge representations based on college students' concept ratings do indeed offer a valid assessment of their classroom learning. Students' conceptual representations appear to reflect changes in their learning over the course of instruction and become increasingly similar to the instructor's conceptual representation. Further, the extent to which a student's representation matches that of the instructor at the end of the semester is a good index of how much knowledge the student has learned about the domain of study. As hypothesized, predictive ability depends on how the structure is represented and how structural similarity is assessed. Pathfinder networks assessed by C appear to offer the best indication of student performance.

General Discussion and Conclusions

The primary aim of this undertaking was to investigate the possibility of using knowledge representations derived from relatedness ratings of domain concepts as a means of assessing classroom knowledge. The primary assumption guiding the work was that expertise in abstract domains, such as statistics and experimental design, requires an understanding of the interrelationships among the concepts, which we have referred to as the configural property of knowledge. The results that bear most directly on this thesis were the partial correlations which showed that all significant variance in course performance was captured by PFC. When PFC was partialed out, none of the other predictor variables accounted for a significant proportion of variance, but with these other predictors partialed out, PFC continued to significantly account for the variance in course performance. Of particular importance is the finding that PFC accounted for variance in classroom performance that was not captured by PFR. This points up the importance of the metric by which network similarity is measured. The superiority of PFC over PFR is seen as support for the idea that configural properties of representations do indeed capture important aspects of a cognitive system.

The success of the C measure is, of course, dependent on the validity of the relatedness ratings and the Pathfinder representation of these relations. The significant relationship between raw proximity data and classroom performance corroborates previous results indicating that relatedness ratings are a valid measure of domain knowledge (e.g., Diekhoff, 1983; Schvaneveldt, Durso, Goldsmith, et al., 1985; Thro, 1978). The superior predictive power of the PFC measure over the raw proximity data and the MDS representations points up the value of the Pathfinder-derived network representations. Previous studies (Schvaneveldt, Durso, Goldsmith, et al., 1985; Stephens, 1987) have also found network representations to be better predictors of domain performance than MDS spatial representations. As suggested by Reitman and Reuter (1980) it may be the case that network structures are superior to spatial structures in representing conceptual domains.

Comparison to Expert-Novice Research

The cognitive structural approach to classroom assessment, as exemplified by the work beginning with Shavelson, has an obvious connection with much of the expert-novice research. Both approaches take a structural representational view of domain-specific knowledge. There is, however, an interesting difference between these two areas of research. The expert-novice work, as this descriptor denotes, usually compares groups that differ widely in skill levels. Perhaps as a consequence of the extreme differences between experts and novices in training and experience, the inferred cognitive differences have often been discussed in qualitative terms. For example, computer programmers have been shown to organize programming concepts either semantically or syntactically depending on skill level (Adelson, 1981).

Our results show that it is possible to discriminate among students within a level of expertise and describe these finer-grain differences of performance along a quantitative continuum (i.e., similarity to the instructor's representation). Although this is not to be interpreted as suggesting that qualitative distinctions are unnecessary, it does introduce the possibility that at some levels of analysis, expert-novice distinctions can be seen as a continuous transition. More important, the present findings suggest that relatively small differences in expertise can be discriminated with a cognitive representational approach.

Application of a Structural Assessment Approach

Our approach to assessment is similar in many ways to that of traditional psychometric tests. In both cases the representation is based on a sample of performance that serves as an index of knowledge. In the case of an achievement test, the performance sample is often a direct assessment of domain knowledge (e.g., do you know this fact?), whereas in the present case the performance is a sample of a person's judgments of the general associations that exist among some subset of the domain's relevant concepts.

Assuming that future work continues to support the validity of the present approach, it may be appropriate to consider some advantages of the structural assessment approach as an assessment tool. As noted earlier the present technique avoids a number of problems that are often associated with standard examination procedures. The structural assessment procedure is less dependent on: (a) recall of episodic information; (b) idiosyncratic misinterpretations of questions; and (c) students' ability to articulate the relevant knowledge. The technique also has the potential of being completely automated. This would allow for large-scale application with rapid, objective scoring. Automation would also allow students to perform self-evaluations at any juncture in training. Finally, the approach has a wide range of potential application in that it can be applied to any domain that involves understanding the interrelations of some specifiable set of concepts.

Future Directions

Over the time of conducting and reporting this research we have had numerous thoughts on improving and extending the approach used in the present work. In closing we shall mention briefly what we believe may be the more important of these ideas and also respond to some common criticisms of our work.

One of the most frequently voiced criticisms of this approach relates to the use of the instructor as the standard of comparison. In defense, one can contend that this only makes explicit what happens implicitly in the design and scoring of exams. It could also be argued that the instructor is as good a model as any for relatively novice students taking a sophomore-level class. However, a more positive reply to this criticism is that it is not necessary to use the instructor. A number of possibilities can be explored, ranging from other individual experts, top students, or even families of desired representations. In an analysis not reported here, we found that PFC comparisons among the students themselves allowed us to discriminate good from poor students. More specifically, when we selected a subgroup of students having the highest between-student PFC scores, these students turned out also to be the better students based on final course points.

A second matter concerns the reliability of our assessments. People performing the ratings task often comment on how uncertain and subjective these ratings appear to them. We have looked at the correlations between repeated ratings of the same concept pairs and found them to average around .60. This may appear somewhat low for a reliability coefficient relative to what is usually reported as test reliability. However, there are several important differences between standard test reliability and reliability of relatedness ratings. First, in the present case the correlation of .60 reflects item stability for an individual's ratings, whereas test reliability reflects subject stability within a sample of subjects. In this regard these correlations are not directly comparable measures of reliability. Second, the actual mean difference in ratings across repeated items tends to be quite small (e.g., 1.03 for subjects using a 7-point rating scale) and only 9% of the absolute differences were greater than two ratings apart. Therefore, it appears that students are more consistent in their ratings than they may think they are. Finally, if PFC is used to predict classroom performance, then the reliability in question is of the final predictions, and this can be directly assessed only by determining the stability of Pathfinder representations as measured by C. In this regard, rating consistency is only indirectly related to the reliability of the predictor variable.

Another issue is the selection of the set of concepts to be rated. How many concepts should be sampled and on what basis should they be selected? In the work reported here, only 30 concepts were used for the simple reason that 435 ratings seemed to be close to the upper limits of time and fatigue for a single hour session of ratings. The basis for selecting concepts was that they were important and representative of the material covered in the course. We are currently exploring the effects of sample size on predictive validity. Not surprisingly, we find that predictive validity is much more variable with smaller samples. However, on the basis of preliminary results it appears possible to attain relatively high predictive validity with as few as 10 concepts. We are investigating whether there might be some principled basis for selecting concepts that maximize predictive validity.

Finally, we have speculated about using empirically derived knowledge representations as a guide to teaching. If a domain of knowledge, as a field of study, has a structure that is more or less shared implicitly or explicitly by experts in the field, then it should be the goal of instruction and training to communicate this structure as effectively and efficiently as possible. Is it possible that an individual's personal representation of concepts may help identify particular deficiencies in her knowledge of the domain? Perhaps so, but this type of intervention would seem to require an analysis of individual concepts in a representation. If the important property of concept representations really is of a configural nature, then such an analysis may in fact not be meaningful.

Appendix

Set of 30 Core Concepts from Experimental Design

analysis	main effect
assignment	matching
between subjects	model
block	order effect
confound	orthogonal
control	random
counterbalance	replication
covariance	significance
data	subject
design	theory
distribution	treatment
error	validity
experiment	variable
hypothesis	variance
interaction	within subjects

Chapter 18

Representation of Problem Schemata*

Lisa A. Onorato

Early "brute force" theories in problem solving (Newell & Simon, 1972) concentrated on developing powerful heuristics to reduce a problem's search space. The more recent "knowledge-based" trend, however, recognizes the importance of the potential solver's problem representation (e.g., Greeno, 1977; Hayes & Simon, 1974). In this respect, problem difficulty is not attributed to the size of the problem search space, but rather to the problem solver's interpretation of the problem. Simply, people have difficulties in solving a problem because they have not formed an optimal internal representation of it.

The purpose of the present study is to examine what is included in a problem solver's representation from the original problem statement and to determine whether Pathfinder networks are useful for capturing a "snap-shot" glimpse of this problem interpretation.

Schemata and Interpreting Problems

The notion of a "schema" is at the core of this research. A schema is an abstract high-level knowledge structure containing "slots" that are filled by specific elements. The schema is said to be structured in that it shows the relationships among these component slots. The importance of this cognitive structure has been emphasized throughout the literature on expert-novice differences in problem solving (e.g., Chi, Feltovich, & Glaser, 1981; Larkin, McDermott, Simon, & Simon, 1980; Schoenfeld & Hermann, 1982).

It is important to identify both the problem structures and the processes by which information from the problem statement is transformed into the problem schema (cf., Dellarosa, 1985). It is unlikely that a problem solver will duplicate a problem statement exactly in memory. Rather, the solver will probably delete some irrelevant information from the stated problem in addition to adding some pre-existing knowledge or inferences in interpreting the problem.

Simon and Hayes (1976) developed a computer program, Understand, that attempts to simulate one aspect of this representation-formulation process. Understand makes judgments on the importance and relevance of concepts in the problem statement and creates a problem representation from selected portions of the original problem text. Although the internal representation corresponds closely to the givens and the structure of the presented problem statement, not all aspects from the text are included in the resulting schema.

One question at hand concerns exactly what is taken from the problem text. In this respect, problem-solving research shares objectives with research in text comprehension. Problem statements, whether written or verbal, are necessarily a form of prose. Research on memory for prose has shown that when an appropriate schema is activated, subjects

---

*I thank Tom Dayton and Roger Schvaneveldt for their thoughtful and extensive comments on an earlier draft. I also thank Barbara Lawrence for her advice on data analysis.

have a more unified knowledge base with which to understand text (Bransford & Johnson, 1972; Dooling & Lachman, 1971; Owens, Bower, & Black, 1979; Thorndyke, 1977).

It has been suggested that one particular aspect that serves as a schema frame is the particular goal or motive of a character within a text. Owens, Bower, and Black (1979) have shown that the reader will scan the text primarily for the goal-relevant facts. Other details receive less attention and are established only weakly in memory. Like readers of text, problem comprehenders also organize their schemata around their goal by extracting goal-oriented material from the text. Because their particular goal is problem solution, it is hypothesized that their schemata should be organized around a framework reflecting the steps necessary to achieve that goal.

One problem-solving model that includes text comprehension has been developed by Kintsch and Greeno (1985). These authors present a processing model that simulates the construction of cognitive representations for word arithmetic problems. Building on van Dijk and Kintsch's work (1983), the model contains two main components. The text base consists of a microstructure pertaining to the text's concepts and a macrostructure corresponding to the essential ideas these concepts convey. The second component, the problem model, contains representations of the relations and entities expressed in the text base. This problem model "reflects knowledge of the information needed to solve the problem" (p. 111). In forming the problem model, the reader will delete irrelevant concepts from the text base, in addition to adding inferences not included. This problem model becomes especially important for complex story problems where it is necessary to distinguish between the real problem and the presented cover story.

In the present research, one such complex story problem was studied. The reported experiment demonstrates that solvers with different problem solution goals not only abstract different portions from the problem text, but also organize this information in a manner unlike solvers with a separate solution goal. In addition, this research demonstrates that Pathfinder networks are useful for capturing these different problem schemata.

Hypotheses and Rationale for Experiment

The particular problem under study in the present work (the "horse race problem") was one that could lend itself to the formation of dual representations. Specifically, the nature of the problem was such that two very different questions could be answered from the information presented in one cover story. Because the questions were quite different (one presented a mathematical speed/distance problem, the other, a more abstract puzzle based on sorting out incorrect inferences), it was predicted that the cognitive structures would reflect these differences.

The actual cover story used for both questions was stated as follows:

A stranger approaches two men sitting on the side of the road with their horses. One horse is a mustang, the other an appaloosa. The men discuss their long and unusual horse race. The first hour of the race, they traveled at a constant rate of six m.p.h. But since they had agreed that the owner of the horse that crossed the finish line first would lose, they slowed down. For the next half hour they traveled at half the original speed. Eventually, when they could go no slower, the horses halted and the men got off. The sun begins to set in the distance. The stranger remarks on the race. Then each of the two men jumps on one of the horses. The men speed the last half mile at a rate four times as fast as the original speed. The sun sets behind a hilltop as the stranger continues on his way. (Adapted from Poser #53 in Kaplan, 1963)

18 Problem Schemata

This same problem cover story yielded two different problem questions. The Stranger question was stated as follows: "What might the stranger have suggested, consistent with the men's original agreement, to make the men speed toward the finish line?" (answer: The stranger suggested the men switch horses). In comparison, the Equation question asked, "How many total miles did the men race?" (answer: $6 + 1.5 + .5 = 8$ miles).

The present experiment employed three experimental conditions. Subjects in the two Problem conditions were required to produce a solution to one of the two questions. The Problem question was presented prior to the cover story in order to activate the appropriate schema in a top-down fashion during the initial reading of the problem story. The Stranger problem subjects ($n = 6$) read the Stranger question, followed by the problem cover story, and then were required to produce a solution to the stated problem. The Equation problem subjects ($n = 6$) read the Equation question, followed by the problem cover story, and then attempted to answer this stated problem. After solutions were attempted, both of these Problem groups were required to recall as much of the initial problem story as possible. The control or Read subjects ($n = 6$), on the other hand, were only required to read the original cover story without either of the two problem questions. They were not required to solve a problem, but rather, were only required to recall as much of the original problem cover story as possible.

It was hypothesized that the resulting cognitive structures would depend upon the experimental condition. Specifically, Problem subjects should develop a closely knit schema of ideas central to their solution, with irrelevant concepts attached loosely to this core. Predictions for Read subjects are less simple to make. It might be the case that even without a guiding problem question they can recognize one of the inherent problem statements and their schemata would be organized accordingly. This possibility seems plausible, considering the active nature of text comprehenders. However, it might also be the case that no higher order framework will be activated for Read subjects. In this case, Read subjects will have stored only disconnected fragments in memory, with no efficient way to access all of them in free recall. Perhaps a little of each problem story will be reflected in the resulting schema.

Characteristics of the Cover Story

In a separate study, 33 judges were presented with the problem text and were asked to answer both problem questions. Each problem was appropriately difficult for the group under study (9 out of 33 subjects found the correct solution to the Stranger problem; 12 out of 33 solved the Equation problem correctly).

After attempting both problem solutions, subjects then provided a rating for each of the 12 statements in the problem text. Specifically, these judges indicated whether a given sentence was related to understanding only the Stranger problem (S), only the Equation problem (E), Both problems (B), or Neither problem (N). The results of this analysis are shown in Table 1. Conveniently, the ratings resulted in three statements for each of the four categories, presented in an ungrouped fashion in the original text. These ratings imply that only half of the statements are relevant to the Equation problem [the three Equation only (six mph, half speed, half mile) plus three Both (stranger approaches, men discuss, horses halt) statements]. Similarly, only six statements are relevant to the Stranger problem [the three Stranger only (agreement, stranger remarks, men jump) plus three Both (stranger approaches, men discuss, horses halt) statements]. Consequently, exactly half of the statements are not relevant to the solving of the problem. For the Equation problem, the Stranger only (agreement, stranger remarks, men jump) and Neither (mustang, sun sets, stranger continues) statements do not apply. For the Stranger problem, the Equation only (six mph, half speed, half mile) and Neither (mustang, sun sets, stranger continues) statements are not relevant.

Table 1. Classification and abbreviations of story sentences.

Classification and Abbreviation	Sentence from Story
B - stranger approaches	A stranger approaches two men sitting on the side of the road with their horses.
N - mustang	One horse is a mustang, the other an appaloosa.
B - men discuss	The men discuss their long and unusual race.
E - six mph	The first hour of the race, they traveled at a constant rate of six mph.
S - agreement	But since they had agreed that the owner of the horse that crossed the finish line first would lose, they slowed down.
E - half speed	For the next half hour they traveled at half the original speed.
B - horses halt	Eventually, when they could go no slower, the horses halt and the men got off.
N - sun sets	The sun begins to set in the distance.
S - stranger remarks	The stranger remarks on the race.
S - men jump	Then each of the two men jumps on one of the horses.
E - half mile	The men speed the last half mile at a rate four times the original speed.
N - stranger continues	The sun sets behind a hilltop as the stranger continues on his way.

Recall Study

Subjects in the Stranger and Equation conditions read their respective problem questions followed by the problem story and then attempted a problem solution, showing as much work as possible on the paper provided. Read subjects were asked to read the problem story "until they were sure they fully understood it." Then, all subjects were asked to "recall (into a tape recorder) as much of the original problem story as possible." It was predicted that the recall performance would reflect the particular assigned experimental condition.

Specifically, subjects who had a schema for the Stranger problem should recall more stranger relevant concepts than equation concepts, because when they read the story they extracted information from the text by trying to fit information into their extant schema, which had slots only for stranger information. Slots should not be available for irrelevant information, so should not be stored in a form organized for easy recall. In contrast, subjects who had a prior schema for the Equation problem should recall more equation concepts than stranger concepts because only equation information is in an easily accessible form in memory. Finally, Read subjects who were lacking an organized prior schema should show no recall preference for concept type because they had no bias in memory.

18 Problem Schemata

Between group recall predictions also follow the same rationale. Stranger subjects should recall more stranger concepts than Equation and Read subjects, and Equation subjects should recall more equation concepts than Stranger and Read subjects. Based on this prediction, it also follows that Read subjects should recall fewer concepts overall.

Recall order was assessed by two judges who were blind to experimental condition. Coding of verbal protocols is a particularly difficult task, but based on the following set of coding rules, judges reached 100% agreement. Specifically, the main concepts from each of the above statements were listed. These main ideas were specified by the two judges as follows:

stranger approaches	stranger approaches (meets, sees, etc.) two men two men are sitting on the side of the road two men have two horses
mustang	two types of horses OR mustang OR appaloosa
men discuss	men discuss a race
six mph	one hour OR six mph
agreement	the men make an agreement (essentially that the winner loses)
half speed	next half hour OR half speed OR three mph
horses halt	men go slower and slower OR horses halt OR men jump off
sun sets	sun begins to set
stranger remarks	stranger remarks (comments, says something, etc.)
men jump	men jump back on the horses
half mile	four times as fast OR last half mile
stranger continues	sun sets OR stranger continues

Subjects were given credit for the particular statement whenever the gist of its main concept was recalled. In this manner, a given statement could be represented at various times throughout the protocol. The coding was fairly strict, however. According to this coding scheme, for example, in order for subjects to receive credit for recalling the fourth statement (six mph), they were required to mention the actual speed (six mph); they were not given credit if they vaguely mentioned just "some speed."

As the reader will note, this analysis is quite different from typical propositional analyses of semantic relations (e.g., Kintsch, 1974). Specifically, not every proposition receives credit. This limited propositional analysis was required for several reasons: to keep the number of main ideas relevant to both problems equal; to highlight the important aspects of the text; and to keep the number of concepts from becoming too large and unwieldy. For example, if a subject vaguely remembered that the horses were traveling some speed but could not remember the exact number, then it can be assumed that this subject is lacking important information from the problem statement relevant to producing the correct solution. If this subject was given the same credit as a subject who had actually recalled that the exact speed was six mph, then the transcription would not reflect these differences. Rather than assigning degree of accuracy ratings for each proposition, for simplicity, in the present analysis, it was decided that either the subject recalled the main point of the statement or did not recall it at all.

Recall Analysis

The specific items recalled by each group were examined. Table 2 presents the mean number of statements recalled by the experimental group for the four statement categories.

Table 2. Mean number of items recalled in each concept category by each condition.

	Concept Category				
	Stranger Related		Stranger Unrelated		
Instruction Condition	Equation Related (Both)	Equation Unrelated (Stranger)	Equation Related (Equation)	Equation Unrelated (Neither)	Average
Stranger	2.06	2.50	1.75	1.25	1.89
Equation	1.37	1.69	2.69	1.69	1.86
Read	2.25	2.06	1.94	1.69	1.99
Average	1.89	2.08	2.13	1.54	

An analysis of variance was performed on a 2 × 2 × 3 two-within, one-between subjects design. The two-within subjects variables pertained to statement category: Stranger (Related or Unrelated) and Equation (Related or Unrelated). The between subject variable was experimental Condition (Stranger, Equation, or Read).

An ANOVA performed on the above design yielded only two significant effects. A significant interaction for the Stranger Category × Condition was found ($F(2,45) = 8.37$, $p < .001$). This interaction is presented in Table 3.

Table 3. Mean number of items recalled by experimental condition for *Stranger Related* and *Stranger Unrelated* concepts.

	Concept Category		
Condition	Stranger Related (Both/Stranger)	Stranger Unrelated (Equation/Neither)	Average
Stranger	2.28	1.50	1.89
Equation	1.53	2.19	1.86
Read	2.16	1.82	1.99
Average	1.99	1.84	

Simple effects analyzed for Stranger Related items were significant ($F(2,45) = 6.09$, $p < .01$). A Newman-Keulls analysis indicated that although the Stranger condition did not differ from the Read condition, both groups recalled significantly more Stranger Related items than the Equation condition. No simple effect was found for the Stranger Unrelated items, although this comparison did approach significance ($p < .07$), hinting that the Equation condition recalled more Stranger Unrelated concepts than the Read or Stranger groups.

In addition to the above interaction, the Stranger Category × Equation Category interaction was also significant ($F(1,45) = 10.08$, $p < .01$). This interaction is presented in Table 4.

Table 4. Mean number of items recalled by concept category.[a]

	Equation Related	Equation Unrelated	Average
Stranger Related	(Both) 1.89	(Stranger) 2.08	1.99
Stranger Unrelated	(Equation) 2.13	(Neither) 1.54	1.83
Average	2.01	1.81	

[a]Category abbreviation in parentheses

Paired t-tests indicated significant differences for the Stranger Unrelated simple effect ($t(47) = 2.96$, $p < .01$). Specifically, Equation concepts were more likely to be recalled than Neither concepts. No simple effect was found for the Stranger Related items: Stranger Only concepts were just as likely to be recalled as Both concepts.

No main effects for Condition, Stranger category, or Equation category were found. In addition, although the Stranger × Equation × Condition interaction was insignificant, the Equation × Condition interaction approached significance ($p < .10$). This Equation × Condition interaction showed that, although not statistically significant, the Equation subjects had a tendency to recall more Equation Related and less Equation Unrelated concepts than the Stranger and Read groups.

Recall Conclusion and Discussion

As the above analysis indicates, essentially all groups are recalling the same number of items. Interestingly, there were significant differences in the specific concepts recalled: Although groups recalled about the same number of concepts, the particular concepts recalled depended on condition. Specifically, both the Stranger and Read conditions recalled significantly more Stranger related items than the Equation subjects, and Equation subjects tended to recall more Equation related items than the Stranger and Read subjects.

These findings met the predictions made by the literature on memory for prose. When given a specific theme for a passage, items relevant to that theme are more likely to be recalled. In problem solving, the particular "theme" is represented by the problem solution required of the potential solver. As the above recall analysis indicates, subjects do recall more concepts that are relevant to their problem goal—the concepts required to produce the solution. This explanation also helps us understand why the concepts labeled Neither were the least likeliest to be recalled, as these concepts are not required for either solution. At this point in the analysis, Read subjects are more like Stranger subjects than they are like Equation subjects. Implications for this result will be discussed as further analyses unfold.

Pathfinder Analysis

The cognitive structures of the three experimental conditions were assessed through Pathfinder Graphs. In these graphs, nodes represent specific concepts. The particular concepts in the present research represent the main ideas derived from the problem cover story, as previously summarized. Links between nodes represent relationships among these concepts. Because the distance between two concepts is the same regardless of direction, the term "edge" will be used to refer to this specific undirected link. The graph distances used in the following analyses were the minimum number of edges connecting two nodes.

Not all subjects recalled all of the concepts. To provide a recall distance for each concept pair, it was necessary to decide how to handle distances between pairs in which at least one member of the pair was not recalled. For this research, there were 12 concepts. Without repeats, the largest distance between any two concepts is 11 items. An example would be the distance between concept one and concept 12 in a protocol in which every item was recalled in the original order. Examination of individual protocols indicated that the largest distance obtained was, in fact, 11. Therefore, missing distance estimates were assigned a value of 12—one greater than the largest actual distance obtained.

The 66 (i.e., $12 \times 11/2$) pairwise recall distances obtained for each subject were then combined with the data from the other subjects in the same group, to yield an average distance matrix for each of the three groups of subjects. These averaged matrices were submitted to Pathfinder. The simplest networks were derived using $q = n-1$ and $r = \infty$. The resulting graphs for the three experimental conditions are shown in Figures 1-3.

Node placement is identical across all three figures. The figures do differ, however, in the particular edges that connect the nodes. The number of edges obtained for the average Stranger, Equation, and Read graphs are 12, 12, and 11, respectively.

By visually inspecting the graphs, one can see that the three differ with respect to the "highest-degree" nodes, or nodes with the most edges attached. Specifically, the highest-degree nodes in the Equation graph are all relevant only to the Equation problem: *six mph, half speed,* and *half mile*. The highest-degree nodes in both the Read and Stranger graphs are *agreement* (related to Stranger only) and *horses halt* (related to both problems).

These highest-degree nodes might be indicative of higher order schema slots as described in Anderson and Pichert (1978). They provide the framework to which less important elements are loosely attached. Perhaps these nodes are indicative of specific subgoals that solvers need to achieve. Or, it might be that these are the important concepts or the "attention getters" in the recall paradigm. The previously discussed recall analyses supplement this interpretation: The category of these highest-degree nodes also resulted in high-degree recall for that experimental condition.

Looking at the directly relevant problem nodes in the three graphs, one can see that the number of edges connecting the three Equation nodes (*six mph, half speed, half mile*) is smaller for the Equation graph (one, one, and two edges apart) than for the Stranger (one, three, and four edges apart) and Read (two, three, and three edges apart) graphs. Similarly, the Stranger graph illustrates that Stranger concepts (*agreement, stranger remarks, men jump*) were more closely linked (one, two, and three edges apart) than they were in the Equation graph (two, three, and three edges apart), although they did not differ in distance with the Read group.

18 Problem Schemata

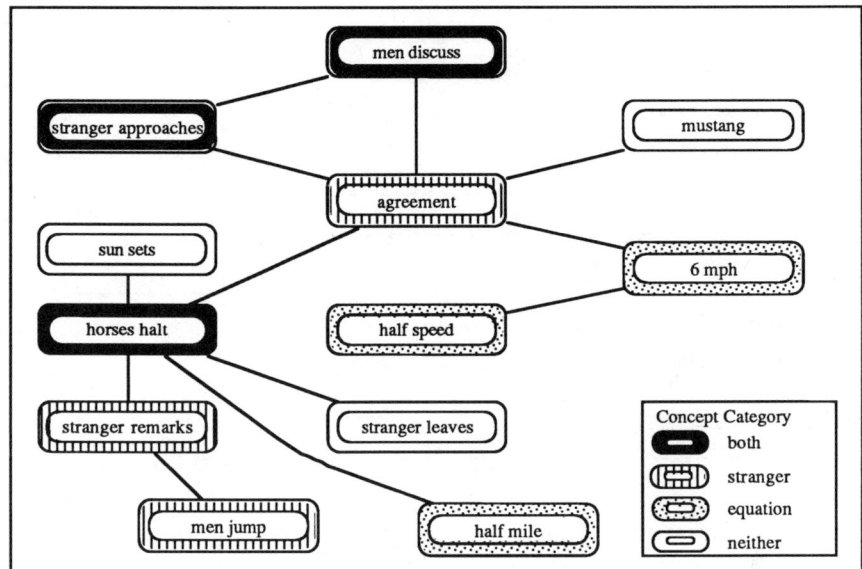

Figure 1. Pathfinder graph from Stranger Group average ratings.

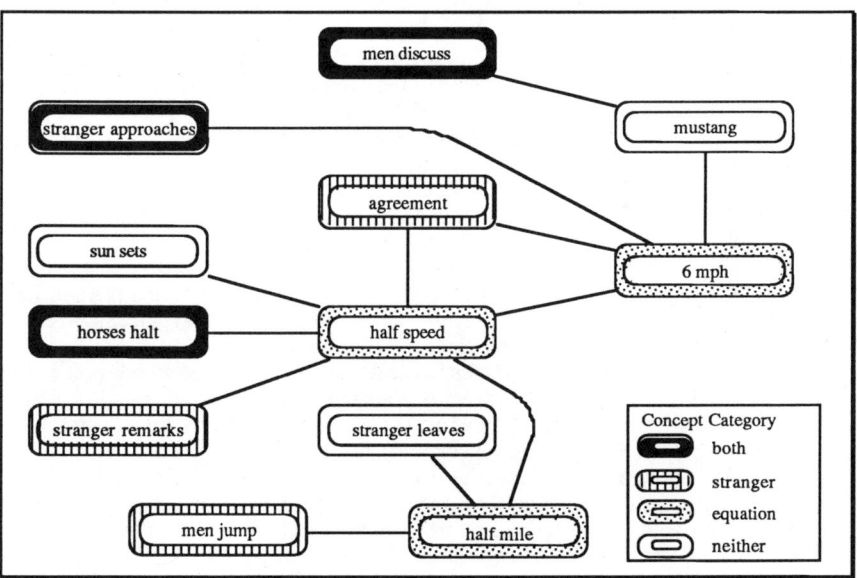

Figure 2. Pathfinder graph from Equation Group average ratings.

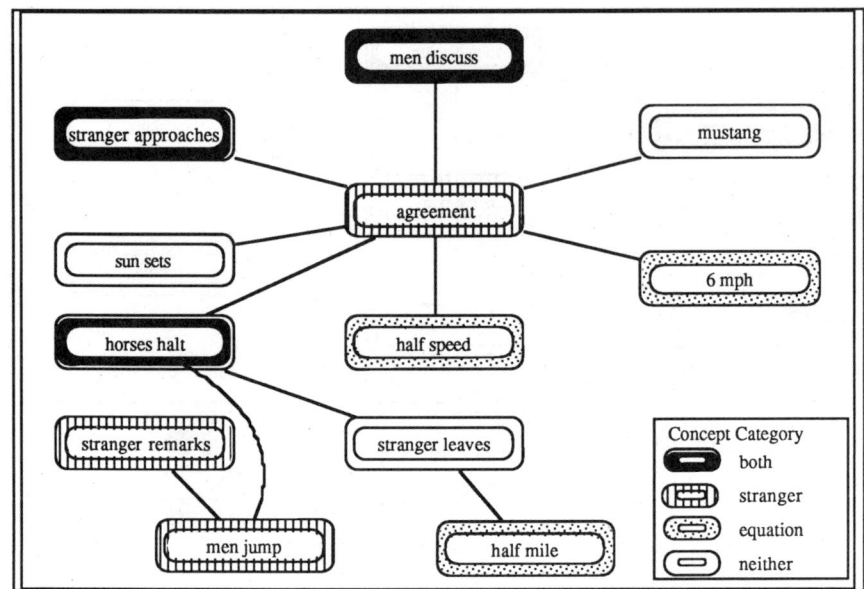

Figure 3. Pathfinder graph from Read Group average ratings.

An ANOVA performed on the original recall data supplements these findings. Specifically, one can pair each of the three Equation concepts with each other and examine the actual recall distances obtained for each of these three pairs across groups. An ANOVA indicated that the groups did differ in how far apart the two items in the pair were recalled from each other: *(six mph-half speed)*: $(F(2,45) = 4.44, p < .01)$; *(six mph-half mile)*: $(F(2,45) = 10.33, p < .001)$; *(half mile-half speed)*: $(F(2,45) = 7.61, p < .001)$. Analytical comparisons indicated that Equation subjects had significantly shorter pairwise distances for all three combinations than did Read or Stranger subjects.

Similarly, an analysis of the three Stranger pairwise distances showed differences in recall performance, but only for two of the three pairs: *(agreement-stranger remarks)*: $(F(2,45) = 4.94, p < .01)$; *(agreement-men jump)*: $(F(2,45) = 5.61, p < .01)$. Analytical comparisons indicate that for the first pair, Stranger subjects recalled the two items closer in time than did the Equation and Read subjects. For the second pair, Stranger subjects only differed significantly from the Equation subjects.

Returning to Figures 1-3, as the graphs illustrate, the concepts labeled relevant to Both problems are attached to different nodes across the three graphs. Specifically, in the Equation graph, Both nodes are closer to Equation nodes than to any of the three other types of nodes. Similarly, in the Stranger and Read graphs, the Both nodes are more likely to be directly attached to Stranger nodes than to any other node type. This finding makes sense in terms of schema theory described above. Once text comprehenders have an idea about the meaning of the prose, it is possible to interpret incoming text around that idea. Ambiguous information, or information that is appropriate to two competing problems, in this case, migrates to other information relevant to the problem at hand.

The above qualitative description was confirmed quantitatively through ANOVA's performed on the original recall data. Each of the three concepts relevant to Both problems were first paired with each of the three Stranger relevant concepts. Of these nine combinations, for five of them, Stranger subjects had shorter recall distances between the pair then did Equation subjects, although only one shorter than Read subjects. Similarly, the three Both concepts were then paired with the three Equation concepts. Of these nine pairs, Equation subjects had significantly shorter recall distances between three of them than did Stranger subjects, although only one shorter than Read subjects.

The above findings illustrate that subjects within the two Problem groups are performing quite differently from each other. Specifically, subjects appear to be chunking problem relevant information together. These closely knit subschemata form the core of the problem representation. Subjects are hence quite capable of disentangling problem relevant information from irrelevant distractors in the cover story. Memory organization is quite efficient in that problem solvers need only pay attention to the core of the representation, and can successfully ignore the less central slots.

A final analysis examined the similarity of the graphs for the three groups (see Goldsmith & Davenport, Chapter 5, this volume). The intersection/union ratio which ranges from 0 (no commonality) to 1 (identical) was used. The average intersection/union ratios for the three comparisons were as follows: Stranger-Equation = .29; Stranger-Read = .57; Equation-Read = .35. An ANOVA indicated that these differences were significant ($F(2,33) = 9.38$, $p < .001$). A paired t-test showed that the comparison Stranger-Equation versus Stranger-Read was significant ($t(11) = 4.23$, $p < .01$), and that Stranger-Read versus Equation-Read was significant ($t(11) = 2.94$, $p < .05$). These results indicated that there were differences in the way the three groups represented the relationships among the 12 concepts. Specifically, the Stranger-Read representations were significantly more alike than any of the other pairs.

Conclusion and Discussion

In this research, problem representation was captured by empirically derived networks for three experimental groups. Although the two problem groups read the same problem cover story, each group read the story with a different goal in mind. The resulting PFNETs were organized around and reflected these separate goals. Specifically, goal relevant information was highly interwoven, forming the center of the networks with irrelevant information loosely attached to this schematic core.

Previous research on memory for prose has indicated that readers of text are more likely to recall thematically relevant portions of the text in comparison to unrelated aspects of the prose. It has been suggested that the use of a schema is responsible for such biases. Specifically, upon being given a theme for the passage, a high-level knowledge structure or schema becomes activated. This mechanism contains slots readily filled with thematically relevant concepts. Unrelated concepts have no slots in this overall schematic representation, and are thus stored in unconnected fragments in memory. Because they are not available in a highly accessible form, these unrelated concepts result in hindered recall performance. In contrast, those concepts that have been organized into a meaningful high-order schema are more readily available during recall. Although this schema explanation makes sense at an intuitive level, confirmation of its existence has been difficult due to an inability to directly examine these knowledge structures.

Just as the schema has been considered the main factor in determining recall for prose, it has also been suggested that this knowledge structure is the most important determinant of problem-solving performance. In fact, the knowledge-based analysis in problem solving recognizes the importance of the problem solver's representation. As in memory for prose, however, techniques for capturing one's problem representation have not been as direct as desired.

The present research has combined the above two areas of research by examining the extraction of problem relevant information from text. The obtained results support the notion of a schema-like mechanism for influencing problem perception and the way incoming information is being interpreted. For example, this research has shown that readers of problems are more likely to recall aspects of the problem story that are relevant to the overall theme. In problem solving, the goal state or the solution required serves as this schematic organizer. This research also demonstrates that Pathfinder is useful for capturing these knowledge representations.

Pathfinder networks demonstrated the major differences between problem solvers with different goals. Although the Stranger group and the Equation group both read the same cover story, this story resulted in different memory organizations for the two groups, as captured by the empirically derived networks. Specifically, the problem representation was organized around the specific goal or solution required of its potential solver. Oddly, few differences were found between the subjects who had the Stranger theme for the problem story and those who had no theme at all. The literature on memory for prose would predict that these latter Read subjects would have no overall schema activated with which to understand the incoming text. Thus, their recall performance should have been hindered. This prediction was not met in the current experiment, however. In fact, the resulting Read network was highly similar to the Stranger network, although differed considerably from the Equation network. One explanation for this finding is that even without being given a problem question, Read subjects could recognize an underlying problem statement without being cued for it directly. One reason why the Stranger problem might have jumped out at its reader is because this problem was so unusual and was actually the original problem to which the Equation problem was added. This explanation seems plausible, as many of the Read subjects actually asked the experimenter what the stranger suggested after the experiment was over. Further research on different problem cover stories would enhance our understanding of this result. It might be especially interesting to capture the resulting schemata for problem groups that read the same problem story made up of two story problems rather than one story problem and one mathematical problem.

Further research might also examine the ability to retrieve irrelevant concepts. For example, in this research, problem subjects were more likely to recall relevant concepts than irrelevant concepts. This experiment did not determine whether or not irrelevant information was completely unavailable, however. It might be the case, for example, that separate schema slots exist for unimportant concepts and these ideas are simply stored as such. It might be interesting to see if more accurate recall could be obtained by allowing some time to elapse and then prompting subjects with the problem question opposite to the problem they actually worked on. In this way, one could also obtain information that would help disentangle whether or not the problem question aided schema formation during encoding versus during retrieval (i.e., by providing an appropriate framework with which to access or retrieve the slots). Once again, in directly comparing the differences between the schema formed during encoding versus the schema formed during retrieval, Pathfinder networks would be useful.

Chapter 19

A Measure of the Knowledge Reorganization Underlying Insight*
Tom Dayton, Francis T. Durso, and Jack D. Shepard

No phenomenon has remained as inscrutable under the probing eye of cognitive methodologies as has insight. The sudden rush to solution, the click of comprehension, continues to fascinate us. This is not merely because insight has been credited for saving Hiero II the price of an alloyed crown, for structuring the benzene molecule, and for leading to velcro, but because it is a threat to the analytic, incremental Zeitgeist of information-processing theory. Intuitively, the solution to an insight puzzle does not seem to be obtained in the same way as other puzzles or problems. To share in the intuition, consider a problem used by Bowden (1985):

> A woman walked for twenty minutes on the surface of a lake without sinking into the water. She was not using any form of flotation device, such as a boat or a raft. How did she manage to do it? (p. 285)

The Phenomenology

The most apparent characteristic of insight is the "aha!" or "eureka!" phenomenology (Ellen, 1982, p. 324; Metcalfe & Wiebe, 1987). Although the "aha!" has implicitly been a defining feature of insight in all research discussions, no one studied it explicitly until Metcalfe (1986a, 1986b; Metcalfe & Wiebe, 1987). The feature of insight problems that makes them appear special to subjects and researchers alike is the phenomenology, and the goal of formal work on insight has been to explain (or explain away) that phenomenology. The Gestaltists regarded insight not as a process, but as the experience that accompanies problem solving when knowledge is dramatically reorganized (Dominowski, 1981; Ellen, 1982, p. 324). Despite the primacy and centrality of the phenomenology in any definition of insight, its quantification waited until Metcalfe (1986a) measured subjects' initial confidence in their ability to solve insight problems versus confidence in their ability to solve memory questions. Subjects were poor at predicting whether they would eventually solve the insight problems, but were good at predicting whether they could answer the memory questions: "Insightful solutions could not be predicted in advance, which would be expected if insight problems were solved by a sudden 'flash of illumination'" (Metcalfe, 1986b, p. 239).

Metcalfe (1986b) refined her earlier (1986a) dependent measure by looking at changes in subject confidence during insight problem-solving sessions. Her subjects repeatedly estimated how close they were to the correct solution (how "warm" they were). Subjects who eventually submitted the correct solution gave consistently low warmth-ratings

---

*We thank Kimberly Stanford-Way for help in data collection, and Roger Schvaneveldt, Tim Goldsmith, and Lisa Onorato for comments on the manuscript.

throughout the session, then gave high warmth-ratings in the few seconds just before they proposed the correct solution. In contrast, subjects who were destined to submit incorrect solutions gradually became more confident through the session. Metcalfe and Wiebe (1987) went on to show that solution of insight problems was accompanied by a sudden increase in rated warmth at the time of solution, but noninsight problems had rated warmth increasing gradually up to the solution. Thus, correct solution of insight problems—but not of failed attempts, noninsight problems, or problems requiring only memory retrieval—was accompanied by a dramatic change in metacognition that presumably corresponded to the "aha!" phenomenology. "These findings indicate in a straightforward manner that insight problems are, at least subjectively, solved by a sudden flash of illumination" (Metcalfe & Wiebe, 1987, p. 243).

Fixedness

What produces this phenomenology? Fixedness is often blamed for preventing such an "aha!" (e.g., Scheerer, 1963), and Weisberg and Alba (1981a, 1981b) have concluded that some Gestaltists went even further to define insight as a breaking of the fixation. We think that fixedness is somewhat independent of insight, though insight problems often have some aspect that creates fixedness. Subjects are fixated when they are incapable of abandoning a substantial but incorrect knowledge structure, and that is what prevents them from adopting the correct structure (Scheerer, 1963). In the lake problem, subjects are fixated on the idea that the lake is liquid; they do not solve the problem until they question that assumption. The story's explicit denial of any form of flotation device entices subjects to assume that the lake surface was liquid by getting them to assume that a flotation device would be appropriate.

There is evidence that fixation breaking is not sufficient for insight. Weisberg and Alba (1981a, 1981b, 1982) showed that subjects often did not solve the nine-dot problem even when the experimenters broke the subjects' fixation by telling them to draw lines outside of the square. Problems like the nine-dot problem and match-triangle problem have a nontrivial verification component after the illumination stage (Wallas, 1926). The problems logically require additional steps after the fixation is broken. Because we do not necessarily equate insight with the breaking of fixation, it is useful to minimize the verification stage to focus on illumination. The lake problem is from that class of problems not requiring additional steps: *The lake was frozen* is both the key assumption that reorganizes knowledge, and the answer to the problem's question, "How did she manage to do it?" Solution is rapid and direct because no further steps are necessary after the insight; the insight *is* the solution.

If insight is not equivalent to overcoming fixation, then what does underlie insight? "When familiar objects acquire a place in a new perceptual organization, they become new things and have new meanings. It is this change in meaning that gives us the jolt and to which the term *insight* is applied" (Ellen, 1982, p. 324). We think that this new organization need not arise by changing a previous strong organization; it is not necessary for a fixation to exist and to be broken. A dramatic change in organization can also occur from a previous state of disorganization. In other words, subjects may fail to solve a problem without being fixated if they simply have no strong knowledge structure to begin with.

Reorganization of Knowledge

Realizing that the lake was frozen is more than the addition of one more piece to the puzzle; it is more than beginning to think that the lake is not liquid. It is a reorganization of knowledge, a new view of the material, a new gestalten. This reorganization is the underlying mechanism we are exploring as the explanation of the insight phenomenology.

19 Insight

Metcalfe (1986a, 1986b; Metcalfe & Wiebe, 1987) provided the empirical support necessary to justify inclusion of the phenomenology in a definition of insight, but no one has filled the corresponding gap for reorganization of knowledge. This absence of relevant data—pro or con—led Weisberg and Alba to conclude that the terms *insight* and *fixation* are no better than other descriptors of insight problem solving, because "the only way we know that a subject was fixated was that the subject did not solve the problem. However, the reason that the problem was not solved was that the subject was fixated, which puts us in a circle. A parallel circularity concerns the use of insight" (1981a, p. 188).

Dominowski noted that the obstacle to utilization of the insight and fixedness constructs was not merely absence of data, but inability to gather relevant data because of the limited tools available for knowledge measurement. Although he wrote of fixedness, a similar argument could be made for insight: "If some means could be found to distinguish varying degrees of adherence to ideas (separate from whether a person solves the problem), then fixation could serve some useful theoretical purpose. Otherwise, we are probably better off without it, as Weisberg and Alba suggest" (Dominowski, 1981, p. 197). In this chapter we describe our use of Pathfinder to capture subjects' knowledge of our insight problem. We then apply the method to show a dramatic difference in knowledge organization between solvers and nonsolvers of the problem, thereby providing some empirical support for reorganization as the cause of the insight phenomenology.

An Experiment

The Puzzle
We chose the following puzzle as the insight problem because (a) When we first heard it, we experienced an "aha!" that we attributed to crystallization of knowledge in a novel way; (b) it is the same type of problem used by authors such as Bowden (1985); and (c) rapidity and directness of solution are not at issue, because the insight *is* the solution. Like Metcalfe and Wiebe (1987), we acknowledge that our criteria for labeling this an insight problem are ill defined. Subjects were told,

> A man walks into a bar and asks for a glass of water. The bartender pulls a shotgun on the man. The man says "thank you" and walks out. What missing piece of information would cause the puzzle to make sense?

Rating Task
Regardless of the experimental condition or whether the puzzle was ultimately solved, each of our subjects made their knowledge available by providing pairwise relatedness judgments among 14 terms that were relevant to the puzzle. To prevent confounding of asymmetries of association, we randomized the left-right presentation order of pair members. In addition, half of the subjects made similarity judgments (high numbers for high similarity) and half of the subjects made dissimilarity judgments. Within each subject's list of 91 pairs, the order of pairs was randomized. All subjects were told to use a scale of 1 through 10.

Some of the 14 terms to be judged came directly from the puzzle: *man, bar, bartender, glass of water, shotgun, thank you*. Other terms were absent from the puzzle but were relevant to the solution: *surprise* (the man was surprised by the shotgun), *remedy* (the shotgun was the remedy of the man's ailment), *relieved* (the man was relieved to be cured), *friendly* (the bartender was friendly, not hostile, and that was the reason he pulled the shotgun), *loaded* (the shotgun need not have been loaded in order to effect the surprise, but needed to be loaded if the bartender intended to shoot the man), *paper bag* (a paper bag can be used as a hiccough cure). The final two terms, *pretzels* and *TV*, had nothing to do with the insight that the man had the hiccoughs, but represented items typically present in a bar.

We used Pathfinder (Schvaneveldt, Durso, & Dearholt, 1989) to turn the pairwise relatedness judgments into graphs. If subjects' insight was accompanied by dramatic reorganization of their knowledge structures, the graphs of the solvers should produce poor correlations with those of the nonsolvers. We selected the 14 terms carefully so that there were several plausible organizations. For example, *remedy* and *relieved* should be remote from the rest of the graph of a subject who does not know the solution, but should be central in the graph of a subject who knows that the *shotgun, water,* and *paper bag* can be hiccough *remedies* that lead to *relief*. Microanalysis of the graphs' specific links could then explain low correlations between solver and nonsolver graphs; we could make specific predictions because of our judicial choice of the 14 terms. For example, we expected that a subject who did not know the solution might think of *paper bag* only as a container for *pretzels*, whereas a subject who knew the hiccough solution might think of *paper bag* as another means to *remedy* hiccoughs, and thereby to acquire *relief*.

Subject Groups

There were four groups of subjects:

1. Story Only: These subjects were told the puzzle story and immediately judged relatedness of the 14 words, without being asked for the solution and without being allowed to ask questions. They were not even told it was a puzzle. Afterward they were told it was a puzzle and asked if they knew the solution; none did.

2. Active Nonsolver: This group did the rating task after failing to solve the puzzle despite asking yes-no questions for two hours.

3. Passive Nonsolver: Subjects in this condition were each yoked to a counterpart in the solver condition: Each listened to an audio tape of one solver being told the puzzle and asking yes-no questions. They listened to the entire session until immediately before the solver asked the critical question that directly preceded the solution; thus, passive nonsolvers did not hear the solution. We did not allow them to ask any questions, and at the end of their sessions none could provide the solution when asked.

4. Solver: This group comprised subjects who had successfully solved the puzzle after asking yes-no questions for up to two hours. Examples of the questions and answers are: "Does the man drive to the bar?"—"No." "Is the man bigger than the bartender?"—"No." "Does the man get what he wants?"—"Yes." "Does the man have any animals with him?"—"No." "Does the bartender intend to shoot the man?"—"No." "Does the bartender understand what the man wants?"—"Yes." The subjects in this group rated the stimuli after they achieved the solution.

Each of the four groups ultimately contained six subjects. One of the subjects intended for the passive nonsolver group solved the puzzle and was therefore replaced by a subject who did not solve it. All subjects were students in Introductory Psychology at the University of Oklahoma, who participated in partial fulfillment of a research requirement for the class.

Conversion of Ratings to Pathfinder Graphs

We converted all the ratings to dissimilarities, with 10 for maximum dissimilarity between pair members and 1 for minimum dissimilarity. The form of the resulting data was a 14 × 14 symmetric matrix for each subject, with zeros on the diagonal. The individual subject matrices were averaged to yield four mean dissimilarity matrices, one for each of the groups of subjects.

For each subject group, we submitted the mean dissimilarity matrix to the Pathfinder algorithm (Schvaneveldt et al., 1989) as implemented in Pascal on a Zenith PC-compatible microcomputer with an 8087 math coprocessor. We used the Pathfinder r parameter equal to infinity so as to make only ordinal assumptions about the subjects' ratings. We set the q parameter at $n-1 = 13$ to achieve the simplest graph for our chosen value of r.

Results and Discussion

Macroanalysis

Pictures. The Pathfinder solutions (PFNETs) for the four subject groups are represented as graphs in Figures 1 through 4. Each of the 14 terms relevant to the bartender problem is a node, with links between nodes showing that the concepts represented by those terms are connected in the subjects' gestalten. The four knowledge structures are apparently different. The story-only graph had 15 links (Figure 1), the active nonsolver had 14 (Figure 2), the passive nonsolver 15 (Figure 3), and the solver had 13 (Figure 4). The solver PFNET is a tree, but all the others contain cycles.

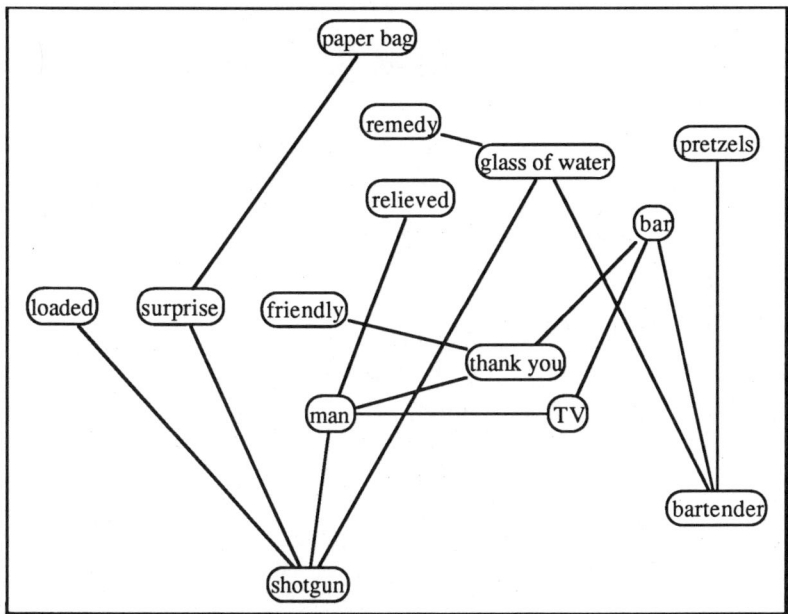

Figure 1. PFNET($r = \infty$, $q = n - 1 = 13$) for the Story-Only group.

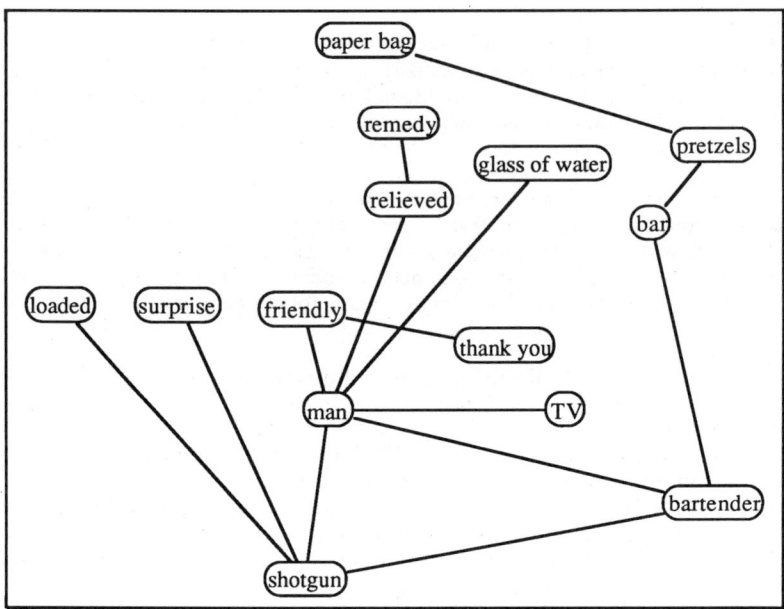

Figure 2. PFNET($r = \infty$, $q = n-1 = 13$) for the Active Nonsolver group.

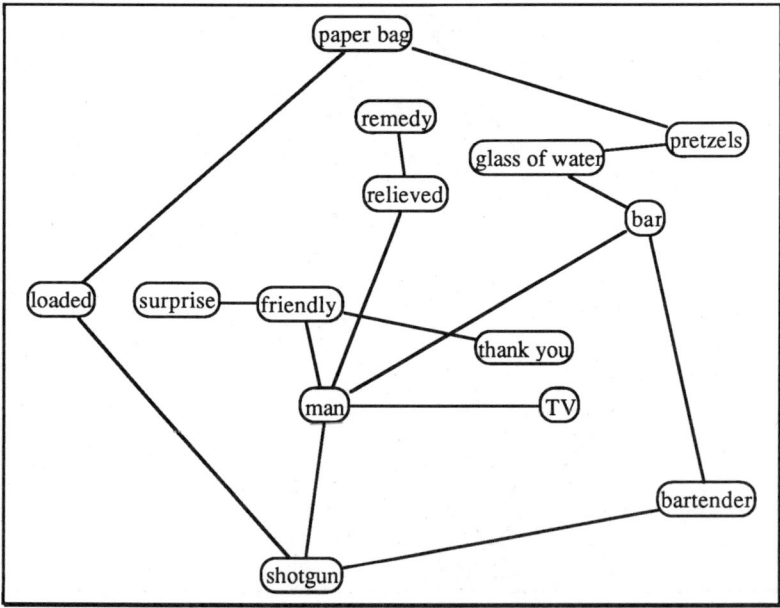

Figure 3. PFNET($r = \infty$, $q = n-1 = 13$) for the Passive Nonsolver group.

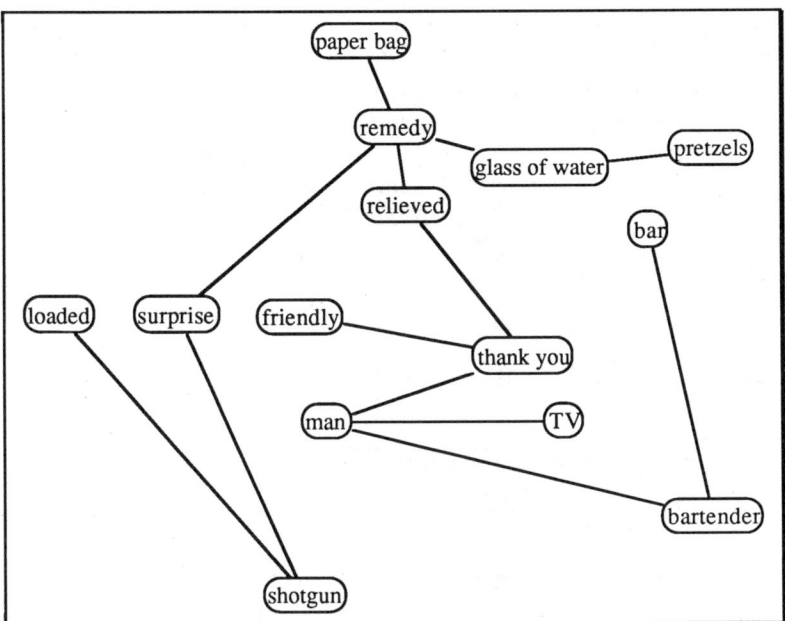

Figure 4. PFNET($r = \infty$, $q = n - 1 = 13$) for the Solver group.

Correlations among graph links. Gross differences among the graphs of the four subject groups were quantified by correlating the upper triangular halves of the adjacency matrices with each other. (An adjacency matrix contains ones and zeros to show presence and absence of links.) Table 1 displays the proportions of variance in each graph that are accounted for by the other graphs. There were several uninteresting links common to all the graphs, such as *man* to *TV*, and *friendly* to *thank you*, which resulted in all the correlations being significantly greater than 0.

The correlations reveal that the similarities between groups' knowledge structures met our expectations. Despite the passive nonsolvers' exposure to the same information that the solvers heard, passive nonsolvers were less correlated with solvers than with active nonsolvers. This supports insight's sudden and dramatic restructuring of knowledge instead of gradual change, because if information were slowly accumulated, passive nonsolvers should have been very similar to solvers. Indeed, there is more similarity between the two nonsolver groups than between any other groups. All other correlations were similar and equally low. The difference between nonsolvers and solvers cannot be due simply to the quality of the information they received, because the active and passive nonsolvers had similar organizations despite their potentially different information, whereas the passive nonsolvers and solvers had less similar organizations despite their virtually identical information. In fact, passive nonsolvers were less similar to solvers than were any other subjects, though these differences were not significant. Just as solvers were unlike any other group, the story-only group had an organization different from any group that attempted to solve the puzzle. Thus, hearing a story created one organization, attempting to solve the puzzle created another, and solving the puzzle created a third. This reorganization implied that going from the attempt to the solution was accompanied by the "aha!" phenomenology, but that the change in going from the story to the mere attempt was not.

Table 1. Proportion of shared variance among Pathfinder graphs (squared correlations[a] on adjacency matrices).

	Story Only	Active Nonsolver	Passive Nonsolver	Solver
Story Only	1.00	.15	.08	.17
Active Nonsolver	.15	1.00	.40	.19
Passive Nonsolver	.08	.40	1.00	.11
Solver	.17	.19	.11	1.00

[a]All correlations are significantly greater than zero ($p < .01$). The correlation between Active and Passive Nonsolvers ($r^2 = .40$) is significantly higher than all other correlations except Active with Solver, which is marginal ($p = .066$). Comparisons of correlations used $p \leq .05$, via Fisher's z-transform.

Graph-theoretic indices of the focal node. We used the graph theory constructs of highest degree, median, and center to identify each graph's central node (see Table 2). The degree of a node is simply the number of links incident to the node. The median and center nodes are the least distant from all other nodes, but in different senses. The median is the node with the shortest average distance to all other nodes, whereas the center is the one with the shortest maximum (i.e., minimax) distance to any other node. An analogy will clarify this distinction. If one wanted to place a hospital in a city's graph of roads so that travel time to the hospital was minimized for the most people—so the *average* travel time was minimized—then the hospital should be at the graph's median. But suppose instead that one was trying to place a cardiac arrest ambulance station so that all patients would be reached in less than the critical first four minutes. It might matter little that many people would require close to the four-minute limit, and that few people would be reached in less than two minutes. Because the four-minute limit is a catastrophic threshold, the average time is not important. The ambulance station should be at the road graph's center instead of its median.

The median and center for each group of subjects in Table 2 were computed from that group's matrix of path distances among all the nodes. Each node-to-node path was composed only of links in that group's Pathfinder solution. For instance, the Story-Only group's path distance from *pretzels* to *man* was computed by finding all possible paths through Figure 1, between those two nodes. The length of each of those possible paths was defined as its number of links. The shortest of all those possible paths was selected as *the* path between *pretzels* and *man*, and its length was entered at the path-distance matrix (14 nodes by 14 nodes) *pretzels* column intersection with the *man* row. From this column we found the largest distance from *pretzels* to any other node. From all 14 such column maximums we selected the smallest, and used its column's node as the center—the node whose maximum distance to any other node is smaller than all other nodes' maximum distances. To find the median node, we averaged the distances within each column, found the smallest of those 14 column means, and chose that column's node as the one with the smallest mean distance to all other nodes.

The most relevant feature of Table 2 is that *remedy* and *relieved* are the focus of the solver graph, but not of any other graph. This is consistent with our belief that *remedy* and *relieved* are of little relevance to someone who does not know the solution to the bartender problem. The other groups oriented their graphs around *man, shotgun, bar,* or *bartender*. The solver group's insight did not just strengthen the solution-relevant nodes but considerably restructured the graph around them. Table 2 is consistent with the correlations in its implications that the solvers were different from either nonsolver group, and that the two nonsolver groups were similar. Unlike the correlations, Table 2 suggests that the story-only people were more like the nonsolvers than they were like the solvers.

Table 2. Three graph-theoretic versions of the focal nodes.

Subject Group	Highest Degree Node[a]	Median Node[b]	Center Node[c]
Story Only	Man & Shotgun	Shotgun	Shotgun
	4	1.9	3
Active Nonsolver	Man	Man	Bartender
	6	1.8	3
Passive Nonsolver	Man	Man	Man, Bar, & Shotgun
	5	1.8	3
Solver	Remedy	Remedy & Relieved	Relieved
	4	2.5	4

[a] The node(s) with the greatest number of links. The highest degree is given below the node names.
[b] The node(s) with the smallest average distance to all other nodes. The smallest average distance is given below the node names.
[c] The node(s) with the smallest maximum-distance-to-any-other-node. The smallest maximum distance is given below the node names.

Microanalysis of Solvers and Nonsolvers

The intergroup correlations (Table 1) revealed that the graphs of passive and active nonsolvers were similar to each other but different from the solvers' graphs. The graph-theoretic indices of centrality (Table 2) showed that both active and passive nonsolver graphs were built around the same nodes despite the differences in information the two groups received. The nonsolver graphs also had different central nodes than did the solver graphs. Now we will pinpoint the specific links that distinguish solvers from nonsolvers, with the aid of Figure 2. Figure 2 shows only links that differ between solvers and nonsolvers, where "nonsolvers" include both the passive and the active. It is most useful to interpret the differences by focusing on the links added by the solvers, represented in the figure by thick lines.

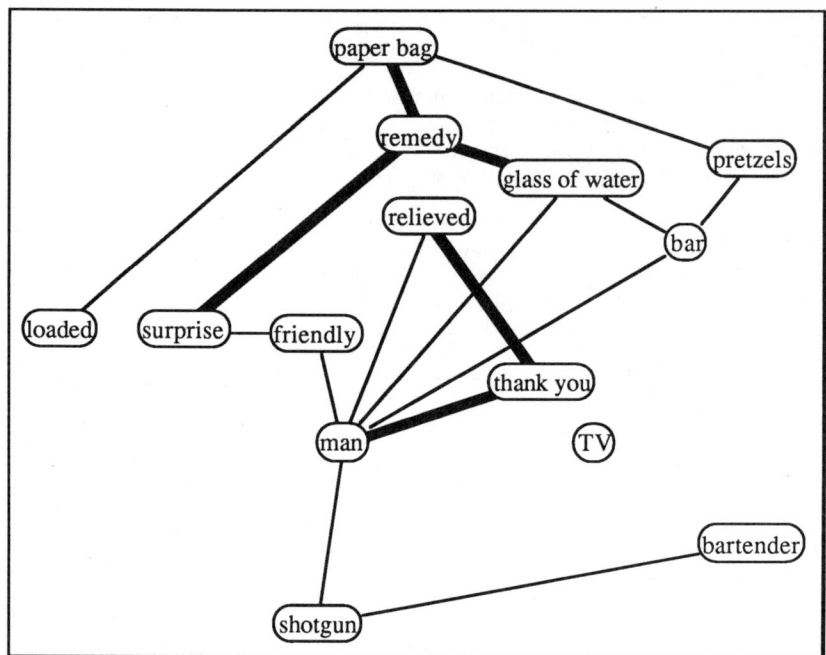

Figure 5. Differences between solver and nonsolver graphs. Active and passive nonsolvers are combined. Thick lines are links present in solvers but not in nonsolvers. Thin lines are links present in nonsolvers but not in solvers.

The concrete reason for the abstract graph-theoretical conclusion (Table 2) that *remedy* and *relieved* were central to the solver graph and not to the nonsolver graphs is that for the nonsolvers, *remedy* was connected only to *relieved*, and *relieved* was attached to the rest of the graph only by its connection to *man* (see Figure 2). We had predicted that nonsolvers would not be thinking of the paper bag or glass of water as instruments for a remedy; we put *paper bag* in the list of terms to be judged for just this reason, because it could easily be considered a container for pretzels. Another important deviation of nonsolvers from solvers is that *shotgun* was not connected to *surprise* for passive nonsolvers, and though active nonsolvers tied *shotgun* to *surprise*, they did not join *surprise* to *remedy* or to *relieved*. We think nonsolvers might have viewed the shotgun primarily as a possession of the bartender, and associated *remedy* and *relieved* to the graph only loosely, as attributes of *man*. In contrast, solvers might have seen the paper bag and glass of water as remedies, and surprise as the vehicle for the shotgun remedy. By these three additional links (*remedy* to *paper bag*, *glass of water*, and *surprise*), the solvers made *remedy* a central concept.

It is also apparent that *relieved* and *remedy* switched from mere attributes of *man* for nonsolvers, to reasons for the man to say "thank you" for solvers: The nonsolver links from *relieved* to *man* disappeared for the solvers, who instead directly connected *relieved* to *thank you*. Also, nonsolvers connected *thank you* to the rest of the graph through its mundane association to *friendly*, which was perhaps seen as an attribute of *man*. Solvers, in contrast, might have seen *thank you* as an action of *man* (*man* linked directly to *thank you*) and *thank you* as a consequence of *relieved*. Solvers also dropped links from *friendly* to *man* and to *surprise*, changing *friendly* from a hub into an isolate, connected to the graph only through its link to *friendly*.

These solution-coincident, sudden changes in knowledge structure undoubtedly required preparation of the ground by subtler shifts resulting from the long question and answer period. Yaniv and Meyer (1987, p. 189), among others, suggested that incubation effects in problem solving are due to the problem solver encountering new stimuli that trigger chains of associations. This spreading activation could improve access to memories critical to problem solution—memories that before were only partially accessible. There is evidence that when people are trying to solve insight problems, they fail to use relevant memories because they cannot retrieve them (Lockhart, Lamon, & Gick, 1988; Perfetto, Bransford, & Franks, 1983). Perhaps an insightfully sudden change of organization occurs when the set of necessary memories is finally available so that a novel stimulus can quickly and easily cause activation propagation through them to form or to reach the answer. An analogy is a seed crystal's ability to crystallize a solution explosively only when the solution is supersaturated.

Conclusions

In this chapter we showed the practicality of our Pathfinder-based method for revealing detailed knowledge patterns. Our approach is the practical support for an operationalization of knowledge reorganization that we think should be added to Metcalfe and Wiebe's (1987) quantitative assessment of the phenomenology to form a two-part, measurable, definition of insight.

We have shown that people who solve an insight problem have a much different knowledge organization than do people who do not solve it or who are unaware of the problem. We have also shown that the correct organization is not achieved merely by exposure to the relevant information, and we believe this is evidence that the difference in structure between problem solvers and nonsolvers is due to a sudden and substantial shift in connections as contended by the Gestaltists (e.g., Dominowski, 1981; Ellen, 1982), rather than to an incremental response to accumulated information (e.g., Weisberg & Alba, 1982). Reorganization of knowledge obviously *can* occur incrementally—the nonsolver groups were more similar to each other than to the story-only group—so the rarity of insight implies that incremental reorganization is not sufficient for insight.

Reorganization alone was never thought to produce the "aha!" experience. The key is the jolt, the click, the suddenness of the reorganization. In current models of information processing, a massive reorganization can take place by the importation of a schema into the comprehension situation. Together with the existing goals of the situation, such importation can have dramatic effects on cognitive processing (e.g., Bransford & Johnson, 1972; Smith, Adams, & Schorr, 1978). We speculate that in problem solving it is the importation of the schema that causes the breathtaking reorganization. With our more sophisticated understanding of cognitive representations this can be considered incremental, but it is certainly not incremental in the same sense as increments in other problem-solving situations. In this view it may be a step, but it is a very large step. What is one small step technically, can be one giant leap phenomenologically.

References

Adelson, B. (1981). Problem solving and the development of abstract categories in programming languages. *Memory & Cognition, 9*, 422-433.

Aho, A. V., Hopcroft, J. E., & Ullman, J. D. (1974). *The design and analysis of computer algorithms.* Reading, MA: Addison Wesley.

Anderson, J. R. (1978). Arguments concerning the representations for mental imagery. *Psychological Review, 85*, 249-277.

Anderson, J. R. (1982). Acquisition of cognitive skill. *Psychological Review, 89*, 369-406.

Anderson, J. R. (1983). *The architecture of cognition.* Cambridge, MA: Harvard University Press.

Anderson, J. R., & Bower, G. H. (1973). *Human associative memory.* Washington, DC: V. H. Winston & Sons.

Anderson, R. C., & Pichert, J. W. (1978). Recall of previously unrecallable information following a shift in perspective. *Journal of Verbal Learning and Verbal Behavior, 17*, 1-12.

Anisfeld, M., & Knapp, M. (1968). Association, synonymity, and directionality in false recognition. *Journal of Experimental Psychology, 77*, 171-179.

Asher, S. R., & Dodge, K. A. (1986). Indentifying children who are rejected by their peers. *Developmental Psychology, 22*, 444-449.

Asher, S. R., Singleton, L. L., Tinsley, B. R., & Hymel, S. (1979). A reliable sociometric measure for preschool children. *Developmental Psychology, 15*, 443-444.

Baird, J. C., & Noma, E. (1978). *Fundamentals of scaling and psychophysics.* New York: John Wiley & Sons.

Ball, G. H., & Hall, D. J. (1970). Some implications of interactive graphic computer systems for data analysis and statistics. *Technometrics, 12*, 17-31.

Barsalou, L. W. (1982). Context-independent and context-dependent information in concepts. *Memory & Cognition, 10*, 82-93.

Barsalou, L. W. (1983). Ad hoc categories. *Memory & Cognition, 11*, 211-227.

Barsalou, L. W. (1985). Ideals, central tendency and frequency of instantiation. *Journal of Experimental Psychology: Learning, Memory and Cognition, 11*, 629-654.

Barsalou, L. W. (1987). The instability of graded structure: Implications for the nature of concepts. In U. Neisser (Ed.), *Concepts and conceptual development: Ecological and intellectual factors in categorization* (pp. 101-140). New York: Cambridge Univ. Press.

Bates, M. J. (1986). Subject access in on line catalogs: A design model. *Journal of the American Society for Information Science, 37*, 357-386.

Belkin, N. J., & Croft, W. B. (1987). Retrieval techniques. *Annual Review of Information Science and Technology, 22*, 109-146.

Belkin, N. J., & Kwasnik, B. H. (1986). Using structural representations of anomalous states of knowledge for choosing document retrieval strategies. *ACM SIGIR Conference Proceedings on Research and Development in Information Retrieval,* 11-22.

Belkin, N. J., Oddy, R. N., & Brooks, H. M. (1982). ASK for information retrieval: Part I. Background and theory. *Journal of Documentation, 38,* 61-71.

Bertrand-Gastaldy, S., & Davidson, C. H. (1986). Improved design of graphic displays in thesauri – through technology and ergonomics. *Journal of Documentation, 42,* 225-251.

Billingsley, P. A. (1982). Navigation through hierarchical menu structures: Does it help to have a map? *Proceedings of the 26th Annual Meeting of the Human Factors Society,* 103-107.

Bobrow, R. J., & Webber, B. L. (1980). Knowledge representation for syntactic/semantic processing. *Proceedings of the First Annual National Conference on Artificial Intelligence,* 316-323. Stanford University.

Bocker, H., Fischer, G., & Nieper, H. (1986). The enhancement of understanding through visual representations. *CHI '86 Proceedings,* 44-50.

Bousfield, W. A., & Sedgewick, C. H. W. (1944). An analysis of sequences of restricted associative responses. *Journal of General Psychology, 30,* 149-165.

Bowden, E. M. (1985). Accessing relevant information during problem solving: Time constraints on search in the problem space. *Memory & Cognition, 13,* 280-286.

Bower, G. H. (1972). A selective review of organizational factors in memory. In E. Tulving & W. Donaldson (Eds.), *Organization of memory.* New York: Academic Press.

Brachman, R. J. (1977). What's in a concept: Structural foundations for semantic networks. *International Journal of Man-Machine Studies, 9,* 127-152.

Brachman, R. J. (1979). On the epistemological status of semantic networks. In N. Findler (Ed.), *Associative networks: Representation and use of knowledge by computers.* New York: Academic Press.

Bransford, J. D., & Johnson, M. K. (1972). Contextual prerequisites for understanding: Some investigations of comprehension and recall. *Journal of Verbal Learning and Verbal Behavior, 11,* 717-726.

Brooks, H. M. (1987). Expert systems and intelligent information retrieval. *Information Processing and Management, 23,* 367-382.

Brooks, R. A., & Binford, T. O. (1980). Interpretative vision and restriction graphs. *Proceedings of the First Annual National Conference on Artificial Intelligence,* 21-27. Stanford University.

Brown, S. W. (1987). *The representation of space by rats: Evidence for a distinction between places and paths.* Unpublished doctoral dissertation, University of Oklahoma.

Burt, R. S., & Minor, M. J. (1983). *Applied network analysis: A methodological introduction.* Beverly Hills: Sage Publication.

Bush, V. (1945). As we may think. *Atlantic Monthly, 176,* 101-108.

Butler, K. A., & Corter, J. E. (1986). The use of psychometric tools for knowledge acquisition: A case study. In W. Gale (Ed.), *Artificial intelligence and statistics* (pp. 295-319). Reading, MA: Addison-Wesley.

Caramazza, A., Hersh, H., & Torgerson, W. (1976). Subjective structures and operations in semantic memory. *Journal of Verbal Learning and Verbal Behavior, 15,* 103-107.

References

Card, S. K., Moran, T. P., & Newell, A. (1983). *The psychology of human-computer interaction.* Hillsdale, NJ: Erlbaum.

Carre, B. (1979). *Graphs and networks.* Oxford: Clarendon Press.

Carroll, J. M., Mack, R. L., & Kellogg, W. A. (1988). Interface metaphors and user interface design. In M. Helander (Ed.), *Handbook of human-computer interaction* (pp. 67-85). Amsterdam, the Netherlands: North-Holland.

Carroll, J. M., & Mazur, S. (1986). Lisa Learning. *IEEE Computer, 19,* 35-49.

Carroll, J. M., & Olson, J. R. (1988). Mental models in human-computer interaction: Research issues about what the user of software knows. In M. Helander (Ed.), *The handbook of human-computer interaction.* Amsterdam, the Netherlands: North-Holland.

Chaffin, R., & Herrmann, D. J. (1984). The similarity and diversity of semantic relations. *Memory & Cognition, 12,* 134-141.

Champagne, A. B., Klopfer, L. E., Desena, A. T., & Squires, D. A. (1981). Structural representations of students' knowledge before and after science instruction. *Journal of Research in Science Technology, 18,* 97-111.

Chappell, S. L., & Sexton, G. A. (1985). Advanced concepts flight simulation facility. *Proceedings of the Third Symposium on Aviation Psychology,* 22-25. Colombus, OH.

Charney, D. (1987). *Comprehending non-linear text: The role of discourse cues and reading strategies.* Paper presented at Hypertext '87, Chapel Hill, NC.

Chase, W. G., & Simon, H. A. (1973). Perception in chess. *Cognitive Psychology, 4,* 55-81.

Chi, M. T. H., Feltovich, P. J., & Glaser, R. (1981). Categorization and representation of physics problems by experts and novices. *Cognitive Science, 5,* 121-152.

Chi, M. T. H., & Koeske, R. D. (1983). Network representation of a child's dinosaur knowledge. *Developmental Psychology, 19,* 29-39.

Christofides, N. (1975). *Graph theory: An algorithmic approach.* New York: Academic Press.

CODASYL (Programming Language Committee, Conference on Data Systems Languages). (1971). *CODASYL data base task group report.*

Cohen, B. H., Bousfield, W. A., & Whitmarsh, G. A. (1957). *Cultural norms for items in 43 categories* (Tech. Rep. No. 22. ONR Contract Nonr-631(00)). Stamford, CT: University of Connecticut.

Cohen, P. R., & Kjeldsen, R. (1987). Information retrieval by constrained spreading activation in semantic networks. *Information Processing and Management, 23,* 255-268.

Coie, J. D., Dodge, K. A., & Coppotelli, H. (1982). Dimensions and types of social status: A cross-age perspective. *Developmental Psychology, 18,* 557-570.

Collins, A. M., & Loftus, E. F. (1975). A spreading activation theory of semantic processing. *Psychological Review, 82,* 407-428.

Collins, A. M., & Quillian, M. R. (1969). Retrieval time from semantic memory. *Journal of Verbal Learning and Verbal Behavior, 8,* 240-247.

Conklin, J. C. (1987). Hypertext: An introduction and survey. *IEEE Computer, 7,* 17-41.

Constantine, A. G., & Gower, J. C. (1978). Graphical representation of asymmetric matrices. *Applied Statistics, 27,* 297-304.

Cooke, N. M. (1983). *Memory structures of expert and novice computer programmers: Recall order vs. similarity ratings.* Unpublished M. A. thesis, New Mexico State University.

Cooke, N. M. (1987). *The elicitation of units of knowledge and relations: Enhancing empirically-derived semantic networks.* Unpublished doctoral dissertation, New Mexico State University.

Cooke, N. M., & McDonald, J. E. (1986). A formal methodology for acquiring and representing expert knowledge. *IEEE Special Issue on Knowledge Representation, 74,* 1422-1430.

Cooke, N. M., & McDonald, J. E. (1987). The application of psychological scaling techniques to knowledge elicitation for knowledge-based systems. *International Journal of Man-Machine Studies, 26,* 533-550.

Cooke, N. M., Durso, F. T., & Schvaneveldt, R. W. (1985). *Measures of memory organization and recall.* Memorandum in Computer and Cognitive Science, MCCS-11 Computing Research Laboratory, New Mexico State University.

Cooke, N. M., Durso, F. T., & Schvaneveldt, R. W. (1986). Recall and measures of memory organization. *Journal of Experimental Psychology: Learning, Memory, and Cognition, 12,* 538-549.

Cooke, N. M., & Schvaneveldt, R. W. (1988). Effects of computer programming experience on network representations of abstract programming concepts. *International Journal of Man-Machine Studies, 29,* 407-427.

Cooper, W. S. (1969). Is interindexer consistency a hobgoblin? *American Documentation, 20,* 268-278.

Craven, T. C. (1984). Thesaural relations in a concept-network management system for customizing of permuted index displays. *Information Processing and Management, 20,* 603-610.

Crouch, D. (1986). The visual display of information in an information retrieval environment. *ACM SIGIR Conference Proceedings on Research and Development in Information Retrieval,* 58-67.

Date, C. J. (1981). *An introduction to database systems* (3rd ed.). Reading, MA: Addison-Wesley.

Dearholt, D., & Gonzales, N. (1987). A database/knowledge structure for robotics vision systems. *Proceedings of the NASA Workshop on Space Telerobotics, 2,* 135-143. Pasadena, CA.

Dearholt, D. W., Schvaneveldt, R. W., & Durso, F. T. (1985). *Properties of networks based on proximities.* Memorandum in Computer and Cognitive Science, MCCS-85-14, Computing Research Laboratory, New Mexico State University.

Deese, J. (1965). *The structure of associations in language and thought.* Baltimore, MD: Johns Hopkins Press.

De Groot, A. M. B. (1983). The range of automatic spreading activation in word priming. *Journal of Verbal Learning and Verbal Behavior, 22,* 417-436.

Dellarosa, D. (1983). *The role of comprehension processes and analogical reasoning in the development of problem-solving expertise.* Unpublished doctoral dissertation, University of Colorado.

References

Dellarosa, D. (1985). *Abstraction of problem-type schemata through problem comparison* (Tech. Rep. No. 146). Boulder, CO: Institute of Cognitive Science, University of Colorado.

Diekhoff, G. M. (1983). Testing through relationship judgments. *Journal of Educational Psychology, 75*, 227-233.

Dodge, K. A., & Somberg, D. R. (1987). Hostile attributional biases among aggressive boys are exacerbated under conditions of threats to self. *Child Development, 58*, 213-224.

Dominowski, R. L. (1981). Comment on "An examination of the alleged role of 'fixation' in the solution of several 'insight' problems" by Weisberg and Alba. *Journal of Experimental Psychology: General, 110*, 193-198.

Dooling, D. J., & Lachman, R. (1971). Effects of comprehension on retention of prose. *Journal of Experimental Psychology, 88*, 216-222.

Doyle, L. (1961). Semantic roadmaps for literature searchers. *Journal of the Association for Computing Machinery, 8*, 367-391.

Dubes, R., & Jain, A. K. (1977). *Models and methods in cluster validity* (Tech. Rep. No. TR-77-05). East Lansing: Computer Science, Michigan State University.

Dumais, S. T. (1988). Textual information retrieval. In M. Helander (Ed.), *Handbook of human-computer interaction* (pp. 673-700). North-Holland: Elsevier.

Dumais, S. T., Furnas, G. W., Landauer, T. K. Deerwester, S., & Harshman, R. (1988). Using latent semantic analysis to improve access to textual information. *Proceedings of CHI '88*, 281-285.

Ekman, G. (1954). Dimensions of color vision. *Journal of Psychology, 38*, 467-474.

Ekstrand, B. R. (1966). Backward associations. *Psychological Bulletin, 65*, 50-64.

Ellen, P. (1982). Direction, past experience, and hints in creative problem solving: Reply to Weisberg and Alba. *Journal of Experimental Psychology: General, 111*, 316-325.

Englebart, D. C., & English, W. K. (1968). A research center for augmenting human intellect. *AFIPS Conference Proceedings, 33*, 395-410.

Ericsson, K. A., & Simon, H. A. (1984). *Protocol analysis: Verbal reports as data.* Cambridge, MA: MIT Press.

Esposito, C. (1988). *Time, space, and statistical improvements on the Pathfinder network generation algorithm.* Unpublished doctoral dissertation, New Mexico State University.

Even, S. (1979). *Graph algorithms.* Potomac, MD: Computer Science Press.

Fahlman, S. E. (1979). *NETL: A system for representing and using real world knowledge.* Cambridge, MA: MIT Press.

Fairchild, K. M., Poltrock, S. E., & Furnas, G. W. (1988). SemNet: Three-dimensional graphic representations of large knowledge bases. In R. Guindon (Ed.), *Cognitive science and its applications to human-computer interaction* (pp. 201-233). Hillsdale, NJ: Erlbaum.

Festinger, L. (1954). A theory of social comparison processes. *Human Relations, 7*, 117-140.

Fidel, R. (1986). Towards expert systems for the selection of search keys. *Journal of the American Society for Information Science, 37*, 37-44.

Fienberg, S. (1977). *The analysis of cross-classified categorical data.* Cambridge, MA: MIT Press.

Fikes, R., & Hendrix, G. (1977). A network-based knowledge representation and its natural deduction system. *Proceedings of the Fifth International Joint Conference on Artificial Intelligence.* Cambridge, MA.

Fillenbaum, S., & Rapoport, A. (1971). *Structures in the subjective lexicon.* New York: Academic Press.

Fischer, G. (1987). *Intelligent support systems for Hyperknowledge.* Boulder, CO: Department of Computer Science, University of Colorado.

Fletcher, C. R. (1984). *Surface forms, textbases, and situation models: Recognition memory for three types of textual information.* Unpublished manuscript, University of Colorado.

Ford, D. L., Jr., Nemiroff, P. M., & Pasmore, W. A. (1977). Group decision-making performance as influenced by group tradition. *Small Group Behavior, 8,* 223-228.

Fowler, R. H., & Dearholt, D. W. (1989). *Pathfinder networks in information retrieval.* Memorandum in Computer and Cognitive Science, MCCS-89-167, Computing Research Laboratory, New Mexico State University.

Frei, H. P., & Jauslin, J. F. (1983). Graphical presentation of information and services: A user-oriented interface. *Information Technology: Research and Development, 2,* 23-42.

Freund, J. E. (1971). *Mathematical statistics.* Englewood Cliffs, NJ: Prentice-Hall.

Friedman, H. P., & Rubin, J. (1967). On some invariant criteria for grouping data. *Journal of the American Statistical Associaton, 62,* 1159-1178.

Friendly, M. L. (1977). In search of the M-gram: The structure of organization in free recall. *Cognitive Psychology, 9,* 188-249.

Friendly, M. L. (1979). Methods for finding graphic representations of associative memory structures. In C. R. Puff (Ed.), *Memory organization and structure* (pp. 85-129). New York: Academic Press.

Fromkin, V., & Rodman, R. (1983). *An introduction to language* (3rd ed.). New York: CBS College Publishing.

Furnas, G. W. (1986). Generalized fisheye views. *CHI '86 Proceedings,* 16-23.

Furnas, G. W., Landauer, T. K., Gomez, L. M., & Dumais, S. T. (1983). Statistical semantics: Analysis of the potential performance of keyword information systems. *The Bell System Technical Journal, 62,* 1753-1806.

Furnas, G. W., Landauer, T. K., Gomez, L. M., & Dumais, S. T. (1987). The vocabulary problem in human-system communication. *Communications of the ACM, 30,* 964-971.

Galil, Z. (1986). Efficient algorithms for finding maximum matching in graphs. *Computing Surveys, 18,* 128-143.

Gammack, J. G. (1987a). Different techniques and different aspects on declarative knowledge. In A. L. Kidd (Ed.), *Knowledge acquisition for expert systems: A practical handbook.* New York: Plenum Press.

Gammack, J. G. (1987b). Modeling expert knowledge using cognitively compatible structures. *Proceedings of the Third International Conference on Expert Systems.* Oxford: Learned Information.

References

Gammack, J. G. (1988). *Eliciting expert conceptual structure using converging techniques.* Unpublished doctoral thesis, University of Cambridge.

Gammack, J. G., & Anderson, A. (submitted). Constructive interaction in knowledge engineering.

Gammack, J. G., & Young, R. M. (1985). Psychological techniques for eliciting expert knowledge. In M. A. Bramer (Ed.), *Research and development in expert systems.* London: Cambridge University Press.

Geeslin, W. E., & Shavelson, R. J. (1975). Comparison of content structure and cognitive structure in high school students learning of probability. *Journal of Research in Mathematics Education, 6,* 109-120.

Gibson, E. J. (1969). *Principles of perceptual learning and development.* New York: Appleton-Century-Crofts.

Glass, A. L., & Holyoak, K. J. (1975). Alternative conceptions of semantic memory. *Cognition, 3,* 313-339.

Gluck, M. A., & Corter, J. E. (1985). Information, uncertainty and the utility of categories. *Proceeding of the Seventh Annual Cognitive Science Society,* 283-287.

Graham, R. L. (1987). A similarity measure for graphs – reflections on a theme of Ulam. *Los Alamos Science,* Special Issue, 114-212.

Greeno, J. G. (1977). Process of understanding in problem solving. In N. J. Castellan, D. B. Pisoni, & G. R. Potts (Eds.), *Cognitive theory* (Vol. 2). Hillsdale, NJ: Erlbaum.

Gregory, D. (1986). Delimiting expert systems. *IEEE Transactions on Systems, Man and Cybernetics, 16,* 834-843.

Griffith, R. L. (1982). Three principles of representation for semantic networks. *ACM Transactions on Database Systems, 7,* 417-442.

Gruenewald, P. J., & Lockhead, G. R. (1980). The free recall of category examples. *Journal of Experimental Psychology: Human Learning and Memory, 6,* 225-240.

Hakimi, S. L., & Yau, S. S. (1964). Distance of a matrix graph and its realizability. *Quarterly Journal of Applied Mathematics, 22,* 305-317.

Halasz, F., & Moran, T. (1983). Mental models and problem solving in using a calculator. *CHI '83 Proceedings,* 212-216.

Hall, J., & Williams, M. S. (1966). A comparison of decision-making performances in established and ad hoc groups. *Journal of Personality and Social Psychology, 3,* 214-222.

Hall, J., & Williams, M. S. (1970). Group dynamics training and improved decision making. *Journal of Applied Behavioral Science, 6,* 27-32.

Hallinan, M. T. (1982). Classroom racial composition and children's friendship. *Social Forces, 61,* 56-72.

Hammond, N., Morton, J., MacLean, A., & Barnard, P. (1983). Fragments and signposts: Users' models of the system. *Proceedings of the 10th International Symposium on Human Factors in Telecommunications,* 81-88. Helsinki, Finland.

Hansell, S., & Karweit, N. (1983). Curricular placement, friendship networks, and status attainment. In J. L. Epstein & N. Karweit (Eds.), *Friends in school* (pp. 141-161). New York: Academic Press.

Harary, F. (1969). *Graph theory.* Reading, MA: Addison-Wesley.

Harary, F., Norman, R. Z., & Cartwright, D. (1965). *Structural models: An introduction to the theory of directed graphs.* New York: Wiley.

Harshman, R. A., Green, P. E., Wind, Y., & Lundy, M. E. (1982). A model for the analysis of asymmetric data in marketing research. *Marketing Science, 1,* 205-242.

Hayes, J. R., & Simon, H. A. (1974). Understanding problem instructions. In L. W. Gregg (Ed.), *Knowledge and cognition.* Hillsdale, NJ: Erlbaum.

Hayes-Roth, F., Waterman, D. A., & Lenat, D. B. (1983). *Building expert systems.* London: Addison-Wesley.

Hays, W. L. (1973). *Statistics for the social sciences* (2nd ed.). New York: Holt, Rinehart & Winston.

Herndon, W. C. (1988). Graph codes and a definition of structural similarity. *Computers & Mathematics with Applications, 15,* 303-309.

Herrmann, D. J., Shoben, E. J., Klun, J. R., & Smith, E. E. (1975). Cross-category structure in semantic memory. *Memory & Cognition, 3,* 591-594.

Humphreys, M., & Greeno, J. G. (1970). Interpretation of the two-stage analysis of paired-associate memorizing. *Journal of Mathematical Psychology, 7,* 275-292.

Hutchins, E. L., Hollan, J. D., & Norman, D. A. (1986). Direct manipulation interfaces. In D. A. Norman & S. W. Draper (Eds.), *User centered system design* (pp. 87-124). Hillsdale, NJ: Erlbaum.

Hutchinson, J. W. (1981). *Network representations of psychological relations.* Unpublished doctoral dissertation, Stanford University.

Hutchinson, J. W. (1989). NETSCAL: A network scaling algorithm for nonsymmetric proximity data. *Psychometrika, 54,* 25-51.

Hutchinson, J. W., & Lockhead, G. R. (1977). Similarity as distance: A structural principle for semantic memory. *Journal of Experimental Psychology: Human Learning and Memory, 3,* 660-678.

Hyman, I. E., Jr., & Rubin, D. C. (1988). *Memorabeatlia: Quantitative and qualitative analyses of memory for Beatles' lyrics.* Paper presented at the meeting of the American Psychological Association, Atlanta, GA.

Isaac, P. D., & Poor, D. D. S. (1974). On the determination of appropriate dimensionality in data with error. *Psychometrika, 39,* 91-109.

Jain, A. K., & Dubes, R. C. (1988). *Algorithms for clustering data.* Englewood Cliffs, NJ: Prentice-Hall.

James, W. (1890/1981). *The principles of psychology.* Cambridge, MA: Harvard University Press.

Jenkins, J. J., & Russell, W. A. (1952). Associative clustering during recall. *Journal of Abnormal and Social Psychology, 47,* 818-821.

Jenkins, J. J., Mink, W. D., & Russell, W. A. (1958). Associative clustering as a function of verbal association strength. *Psychological Reports, 4,* 127-136.

Johnson, S. C. (1967). Hierarchical clustering schemes. *Psychometrika, 32,* 241-254.

Johnson-Laird, P. N., Herrmann, D. J., & Chaffin, R. (1984). Only connections: A critique of semantic networks. *Psychological Bulletin, 96,* 292-315.

Jones, W. P. (1986). On the applied use of human memory models: The Memory Extender personal filing system. *International Journal of Man-Machine Studies, 25,* 191-228.

Jones, W. P., & Furnas, G. W. (1987). Pictures of relevance: A geometric analysis of similarity measures. *Information Processing and Management, 38,* 420-442.

Kahneman, D., & Miller, D. T. (1986). Norm theory: Comparing reality to its alternatives. *Psychological Review, 93,* 136-153.

Kaplan, P. (1963). *Posers.* New York: Harper & Row.

Kay, D. S., & Black, J. B. (1984). Changes in knowledge representations of computer systems with experience. *Proceedings of the 28th Annual Human Factors Society Meeting,* 963-967. San Antonio, TX.

Kay, D. S., & Black, J. B. (1985). The evolution of knowledge representations with increasing expertise in using systems. *Proceedings of the Seventh Annual Conference of the Cognitive Science Society,* Irvine, CA.

Kellogg, W. A., & Breen, T. J. (1987). Evaluating user and system models: Applying scaling techniques to problems in human-computer interaction. *Human Factors in Computing Systems and Graphics Interface Conference Proceedings,* 303-308

Kennard, R. W., & Stone, L. A. (1969). Computer aided design of experiments. *Technometrics, 11,* 137-148.

Keppel, G. (1982). *Design and analysis: A researchers handbook* (2nd ed.). Englewood Cliffs, NJ: Prentice-Hall.

Kieras, D. E., & Bovair, S. (1984). The role of a mental model in learning to operate a device. *Cognitive Science, 8,* 255-273.

Kieras, D. E., & Polson, P. (1985). An approach to the formal analysis of user complexity. *International Journal of Man-Machine Studies, 22,* 365-394.

Kintsch, E. H. (1988). *The role of mental representations of text in the development of summarization strategies.* Unpublished doctoral dissertation, University of Denver.

Kintsch, W. (1974). *The representation of meaning in memory.* Hillsdale, NJ: Erlbaum.

Kintsch, W., & Greeno, J. (1985). Understanding and solving word arithmetic problems. *Psychological Review, 92,* 109-129.

Kirkpatrick, S., Gelatt, C. D., Jr., & Vecchi, M. P. (1983). Optimization by simulated annealing. *Science, 220,* 671-680.

Klovdahl, A. S., & Brindle, J. (1987). TRANS_NET. *Connections, 10,* 176.

Knoebel, A., Dearholt, D., & Schvaneveldt, R. (1988). *Empty nexus graphs.* Paper presented at the American Mathematical Society meetings, Las Cruces, NM.

Knoke, D., & Kuklinski, J. H. (1982). *Network analysis.* Beverly Hills, CA: Sage Publications.

Knoke, D., & Wood, J. R. (1981). *Organized for action.* New Brunswick, NJ: Rutgers University Press.

Krumhansl, C. L. (1978). Concerning the applicability of geometric models to similarity data: The interrelationship between similarity and spatial density. *Psychological Review, 85,* 445-463.

Kruskal, J. B. (1956). On the shortest spanning subtree of a graph and the traveling salesman problem. *Proceedings of the American Mathematical Society, 7,* 48-50.

Kruskal, J. B. (1964). Nonmetric multidimensional scaling: A numerical method. *Psychometrika, 29*, 115-129.

Kruskal, J. B. (1977). Multidimensional scaling and other methods for discovering structure. In K. Enslein, A. Ralston, & H. Wilf (Eds.), *Statistical methods for digital computers*. New York: Wiley.

Kruskal, J. B., & Wish, M. (1978). *Multidimensional Scaling*. Beverly Hills, CA: Sage University Press.

Landauer, T. K. (1975). Memory without organization: Properties of a model with random storage and undirected retrieval. *Cognitive Psychology, 7*, 495-531.

Larkin, J., McDermott, J., Simon, D., & Simon, H. (1980). Expert and novice performance in solving physics problems. *Science, 208*, 1335-1342.

Lawson, E. D. (1964). Reinforced and non-reinforced four-man communication nets. *Psychological Reports, 14*, 287-296.

Leavitt, H. J. (1951). Some effects of certain communication patterns on group performance. *Journal of Abnormal and Social Psychology, 46*, 38-50.

Lesk, M. (1986). *Automatic sense disambiguation using machine readable dictionaries: How to tell a pine cone from an ice cream cone.* Paper presented at the SIGDOC Conference (Association of Computing Machinery), Toronto.

Lesk, M. (1987). *Can machine-readable dictionaries replace a thesaurus for searches in on-line catalogs?* Paper presented at the OED Conference, Waterloo.

Lieblich, I., & Arbib, M. A. (1982). Multiple representations of space underlying behavior. *The Behavioral and Brain Sciences, 5*, 627-640.

Linde, L. (1986). The information-seeker's mental model of the content of a data base. *Proceedings of the Third European Conference on Cognitive Ergonomics*. Cite Universitaire de Paris, France.

Ling, R. F. (1972). On the theory and construction of k-clusters. *Computer Journal, 15*, 326-332.

Lingoes, J. C. (1973). *The Guttman-Lingoes nonmetric program series*. Ann Arbor, MI: Mathesis Press.

Lockhart, R. S., Lamon, M., & Gick, M. L. (1988). Conceptual transfer in simple insight problems. *Memory & Cognition, 16*, 36-44.

Lorch, R. F. (1981). Effects of relation strength and semantic overlap on retrieval and comparison processes during sentence verification. *Journal of Verbal Learning and Verbal Behavior, 20*, 593-610.

MacQueen, J. (1967). Some methods for classification and analysis of multivariate observations. *Proceedings of the Fifth Berkeley Symposium, 1*, 281-297.

Mannes, S. M., & Kintsch, W. (1987). Knowledge organization and text organization. *Cognition and Instruction, 4*, 91-115.

Marriot, F. H. (1971). Practical problems in a method of cluster analysis. *Biometrics, 27*, 501-514.

Marshall, G. R., & Cofer, C. N. (1970). Single-word free association norms for 328 responses from the Connecticut cultural norms for verbal items in categories. In L. Postman & G. Keppel (Eds.), *Norms of word association* (pp. 321-360). New York: Academic Press.

References

Martin, J. (1977). *Computer data-base organization* (2nd ed.). Englewood Cliffs, NJ: Prentice-Hall.

Masson, M. E. J., Hill, W. C., Conner, J., & Guindon, R. (1988). Misconceived misconceptions? *CHI '88 Proceedings*, 151-156.

Maxwell, K. J. (1983). *A scripts analysis of fact retrieval from memory*. Unpublished doctoral dissertation, New Mexico State University.

Mayer, R. E. (1987). Cognitive aspects of learning and using a programming language. In J.M. Carroll (Ed.), *Interfacing thought: Cognitive aspects of human-computer interaction* (pp. 61-79). Cambridge, MA: MIT Press.

McClelland, J. L., & Rumelhart, D. E. (1986). *Parallel distributed processing* (Vol. 2). Cambridge, MA: MIT Press.

McCloskey, M., & Glucksberg, S. (1979). Decision processes in verifying category membership statements: Implications for models of semantic memory. *Cognitive Psychology, 11*, 1-37.

McDonald, D. R. (1987). *Drawing inferences from expository texts*. Unpublished doctoral dissertation, New Mexico State University.

McDonald, J. E., Dayton, T., & McDonald, D. R. (1988). Adapting menu layout to tasks. *International Journal of Man-Machine Studies, 28*, 417-436.

McDonald, J. E., Dearholt, D. W., Paap, K. R., & Schvaneveldt, R. W. (1986). A formal interface design methodology based on user knowledge. *CHI '86 Proceedings*, 285-290.

McDonald, J. E., Molander, M. E., & Noel, R. W. (1988). Color-coding categories in menus. *CHI '88 Proceedings*, 101-106.

McDonald, J. E., & Schvaneveldt, R. W. (1988). The application of user knowledge to interface design. In R. Guindon (Ed.), *Cognitive science and its applications for human-computer interaction* (pp. 289-338). Hillsdale, NJ: Erlbaum.

McDonald, J. E., Stone, J. D., Liebelt, L. S., & Karat, J. (1982). Evaluating a method for structuring the user-system interface. *Proceedings of the 26th Annual Meeting of the Human Factors Society*, 551-555.

McKeithen, K. B., Reitman, J. S., Rueter, H. H., & Hirtle, S. C. (1981). Knowledge organization and skill differences in computer programmers. *Cognitive Psychology, 13*, 307-325.

McRae, D. J. (1971). MICKA, a FORTRAN IV iterative K-means cluster analysis program. *Behavioral Science, 16*, 423-424.

Mendenhall, W., McClave, J., & Ramey, M. (1977). *Statistics for psychology*. North Scituate: Duxbury Press.

Metcalfe, J. (1986a). Feeling of knowing in memory and problem solving. *Journal of Experimental Psychology: Learning, Memory, and Cognition, 12*, 288-294.

Metcalfe, J. (1986b). Premonitions of insight predict impending error. *Journal of Experimental Psychology: Learning, Memory, and Cognition, 12*, 623-634.

Metcalfe, J., & Wiebe, D. (1987). Intuition in insight and noninsight problem solving. *Memory & Cognition, 15*, 238-246.

Mettrey, J. (1987). An assessment of tools for building large KB systems. *AI Magazine, 8*, 81-89.

Meyer, D. E., & Schvaneveldt, R. W. (1971). Facilitation in recognizing pairs of words: Evidence of a dependence between retrieval operations. *Journal of Experimental Psychology, 90*, 227-234.

Meyer, D. E., & Schvaneveldt, R. W. (1976). Meaning, memory structure and mental processes. In C. N. Cofer (Ed.), *The structure of human memory* (pp. 54-89). San Francisco: W. H. Freeman.

Miller, G. A., & Nicely, P. A. (1955). An analysis of perceptual confusions among some English consonants. *Journal of the Acoustical Society of America, 27*, 338-352.

Miyamoto, S., Oi, K., Abe, O., Katsuya, A., & Nakayama, K. (1986). Directed graph representations of association structures: A systematic approach. *IEEE Transactions on Systems, Man, and Cybernetics, 16*, 53-61.

Murphy, G. L., & Medin, D. L. (1985). The role of theories in conceptual coherence. *Psychological Review, 92*, 289-316.

Murphy, G. L., & Wright, J. C. (1984). Changes in conceptual structure with expertise: Differences between real-world experts and novices. *Journal of Experimental Psychology: Learning, Memory, and Cognition, 10*, 144-155.

Navathe, S. B., & Fry, J. P. (1976). Restructuring of large databases: Three levels of abstraction. *ACM Transactions on Database Systems, 1*, 138-158.

Naveh-Benjamin, M., McKeachie, W. J., Lin, Y., & Tucker, D. G. (1986). Inferring students' cognitive structures and their development using the "ordered tree technique." *Journal of Educational Psychology, 78*, 130-140.

Nelson, D. L. (1981). Many are called but few are chosen: The influence of context on the effects of category size. In G. H. Bower (Ed.), *The psychology of learning and motivation: Advances in research and theory* (Vol. 15, pp. 129-162). New York: Academic Press.

Nelson, D. L., McEvoy, C. L., & Friedrich, M. A. (1982). Extralist cuing and retrieval inhibition. *Journal of Experimental Psychology: Learning, Memory, and Cognition, 8*, 89-105.

Newell, A., & Simon, H. A. (1972). *Human problem solving*. Englewood Cliffs, NJ: Prentice-Hall.

Nilsson, N. (1965). *Learning machines: Foundations of trainable pattern-classification systems*. New York: McGraw-Hill.

Nisbett, R. E., & Wilson, T. D. (1977). Telling more than we can know: Verbal reports on mental processes. *Psychological Review, 84*, 231-259.

Norman, D. A. (1983). Some observations on mental models. In D. Gentner & A. Stevens (Eds.), *Mental models* (pp. 7-14). Hillsdale, NJ: Erlbaum.

Novak, G. S. (1977). Representations of knowledge in a program for solving physics problems. *Proceedings of the Fifth International Joint Conference on Artificial Intelligence*. Cambridge, MA.

Ortony, A. (1979). Beyond literal similarity. *Psychological Review, 86*, 161-180.

Osgood, C., Suci, G. J., & Tannenbaum, P. H. (1957). *The measurement of meaning*. Urbana, IL: University of Illinois Press.

Owens, J., Bower, G. H., & Black, J. B. (1979). The "soap opera" effect in story recall. *Memory & Cognition, 7*, 185-191.

References

Paap, K. R., & Roske-Hofstrand, R. J. (1988). Design of menus. In M. Helander (Ed.), *Handbook of human-computer interaction* (pp. 203-235). Amsterdam, the Netherlands: North-Holland.

Palmer, S. E. (1978). Fundamental aspects of cognitive representation. In E. Rosch & B. B. Lloyd (Eds.), *Cognition and categorization* (pp. 259-303). Hillsdale, NJ: Erlbaum.

Peery, J. C. (1979). Popular, amiable, isolated, rejected: A reconceptualization of sociometric status in preschool children. *Child Development, 50*, 1231-1234.

Perfetto, G. A., Bransford, J. D., & Franks, J. J. (1983). Constraints on access in a problem solving context. *Memory & Cognition, 11*, 24-31.

Perrig, W., & Kintsch, W. (1985). Propositional and situational representations of text. *Journal of Memory and Language, 24*, 503-518.

Perry, T. B. (1987). *The relation of adolescent self-perceptions to their social relationships.* Unpublished doctoral dissertation, University of Oklahoma.

Quillian, M. R. (1967). Word concepts: A theory and simulation of some basic semantic capabilities. *Behavioral Science, 12*, 410-430.

Quillian, M. R. (1969). The teachable language comprehender. *Communications of the ACM, 12*, 459-476.

Ramsey, H. R., & Grimes, J. D. (1983). Human factors in interactive computer dialog. *Annual Review of Information Science and Technology, 18*, 30-50.

Reitman, J. S., & Reuter, H. H. (1980). Organization revealed by recall orders and confirmed by pauses. *Cognitive Psychology, 12*, 554-581.

Rips, L. J. (1975). Inductive judgments about natural categories. *Journal of Verbal Learning and Verbal Behavior, 14*, 665-681.

Rips, L. J., Shoben, E. J., & Smith, E. E. (1973). Semantic distance and the verification of semantic relations. *Journal of Verbal Learning and Verbal Behavior, 12*, 1-20.

Robertson, S. E., & Sparck Jones, K. (1976). Relevance weighting of search terms. *Journal of Documentation, 33*, 1-14.

Roistacher, R. C. (1974). A microeconomic model of sociometric choice. *Sociometry, 37*, 219-238.

Rosch, E. (1973). On the internal structure of perceptual and semantic categories. In T. E. Moore (Ed.), *Cognitive development and acquisition of language*. New York: Academic Press.

Rosch, E. (1975). Cognitive representations of semantic categories. *Journal of Experimental Psychology: General, 104*, 192-233.

Rosch, E., Mervis, C. B., Gray, W. D., Johnson, D. M., & Boyes-Braem, P. (1976). Basic objects in natural categories. *Cognitive Psychology, 8*, 382-439.

Roske-Hofstrand, R. J., & Paap, K. R. (1986a). Cognitive networks as a guide to menu organization: An application in the automated cockpit. *Ergonomics, 29*, 1301-1312.

Roske-Hofstrand, R. J., & Paap, K. R. (1986b). *Cognitive JNDs: Implications for scaling techniques using semantic judgments.* Paper presented at the 27th Annual Psychonomic Society Meeting, New Orleans, LA.

Rouse, W. B., & Morris, N. M. (1985). *On looking into the black box: Prospects and limits in the search for mental models* (Tech. Rep. No. 85-2). Atlanta: School of Industrial and Systems Engineering, Georgia Institute of Technology.

Rubin, D. C. (1980). Fifty-one properties of 125 words: A unit analysis of verbal behavior. *Journal of Verbal Learning and Verbal Behavior, 19*, 736-755.

Rubin, D. C. (1983). Associative asymmetry, availability, and retrieval. *Memory & Cognition, 11*, 83-92.

Rubin, D. C., & Friendly, M. (1986). Predicting which words get recalled: Measures of free recall, availability, goodness, emotionality, and pronunciability for 925 nouns. *Memory & Cognition, 14*, 79-94.

Rubin, D. C., & Olson, M. J. (1980). Recall of semantic domains. *Memory & Cognition, 8*, 354-360.

Rumelhart, D. E., & Abrahamson, A. A. (1973). A model for analogical reasoning. *Cognitive Psychology, 5,* 1-28.

Rumelhart, D. E., & McClelland, J. L. (1986). *Parallel distributed processing* (Vol. 1). Cambridge, MA: MIT Press.

Rumelhart, D. E., Smolensky, P., McClelland, J. L., & Hinton, G. E. (1986). Schemata and sequential thought processes in PDP models. In J. McClelland, D. Rumelhart, & the PDP Research Group (Eds.), *Parallel distributed processing: Explorations in the microstructure of cognition* (Vol. 2, pp. 7-57). Cambridge, MA: MIT Press.

Salton, G. (1968). *Automatic information organization and retrieval.* New York: McGraw-Hill.

Salton, G. (1986). Another look at automatic text-retrieval systems. *Communications of the ACM, 29*, 648-656.

Salton, G., & McGill, M. J. (1983). *Introduction to modern information retrieval.* New York: McGraw-Hill.

Sampson, G. (1986). A stochastic approach to parsing. *Proceedings of COLING, The 11th International Conference on Computational Linguistics,* 151-155. Bonn, Germany.

Schaeffer, B., & Wallace, R. (1969). The comparison of word meanings. *Journal of Experimental Psychology, 86*, 144-152.

Schank, R. (1972). Conceptual dependency: A theory of natural language understanding. *Cognitive Psychology, 3*, 552-631.

Schank, R. C., Collins, G. C., & Hunter, L. E. (1986). Transcending inductive category formation in learning. *The Behavioral and Brain Sciences, 9*, 639-686.

Scheerer, M. (1963). Problem-solving. *Scientific American, 208*, 118-128.

Schkolnick, M. (1977). A clustering algorithm for hierarchical structures. *ACM Transactions on Database Systems 2*, 27-44.

Schmalhofer, F., & Glavanov, D. (1986). Three components of understanding a programmer's manual: Verbatim, propositional, and situational representations. *Journal of Memory and Language, 25*, 279-294.

Schmolze, J. G., & Lipkis, T. A. (1983). Classification in the KL-ONE knowledge representation system. *Proceedings of the Eighth International Joint Conference on Artificial Intelligence.*

References

Schoenfeld, A., & Herrmann, D. (1982). Problem perception and knowledge structures in expert and novice mathematical problem solvers. *Journal of Experimental Psychology: Learning, Memory and Cognition, 8*, 484-494.

Schvaneveldt, R. W., Dearholt, D. W., & Durso, F. T. (1988). Graph theoretic foundations of Pathfinder networks. *Computers & Mathematics with Applications, 15*, 337-345.

Schvaneveldt, R. W., & Durso, F. T. (1981). *General semantic networks*. Paper presented at the annual meeting of the Psychonomic Society, Philadelphia, PA.

Schvaneveldt, R. W., Durso, F. T., & Dearholt, D. W. (1985). *Pathfinder: Scaling with network structures*. Memorandum in Computer and Cognitive Science, MCCS-85-9, Computing Research Laboratory, New Mexico State University.

Schvaneveldt, R. W., Durso, F. T., & Dearholt, D. W. (1987). *Pathfinder: Networks from proximity data*. Memorandum in Computer and Cognitive Science, MCCS-87-9, Computing Research Laboratory, New Mexico State University.

Schvaneveldt, R. W., Durso, F. T., & Dearholt, D. W. (1989). Network structures in proximity data. In G. H. Bower (Ed.), *The psychology of learning and motivation: Advances in research and theory* (Vol. 24, pp. 249-284). New York: Academic Press.

Schvaneveldt, R. W., Durso, F. T., Goldsmith, T. E., Breen, T. J., Cooke, N. M., Tucker, R. G., & DeMaio, J. C. (1985). Measuring the structure of expertise. *International Journal of Man-Machine Studies, 23*, 699-728.

Schvaneveldt, R. W., Durso, F. T., & Mukherji, B. R. (1982). Semantic distance effects in categorization tasks. *Journal of Experimental Psychology: Learning, Memory, and Cognition, 8*, 1-15.

Schvaneveldt, R. W., & Goldsmith, T. (1985). *ACES: Air combat expert simulation*. Memorandum in Computer and Cognitive Science, MCCS-85-34, Computing Research Laboratory, New Mexico State University.

Schvaneveldt, R. W., & Meyer, D. E. (1973). Retrieval and comparison processes in semantic memory. In S. Kornblum (Ed.), *Attention and performance IV* (pp. 395-409). New York: Academic Press.

Schwartz, R. M., & Humphreys, M. S. (1973). Similarity judgments and free recall of unrelated words. *Journal of Experimental Psychology, 101*, 10-15.

Shavelson, R. J. (1972). Some aspects of the correspondence between content structure and cognitive structure in physics instruction. *Journal of Educational Psychology, 63*, 225-234.

Shavelson, R. J. (1974). Methods for examining representations of subject-matter structure in a student's memory. *Journal of Research for Science Teaching, 11*, 231-249.

Shavelson, R. J., & Geeslin, W. E. (1973). A method for examining subject-matter structure in written material. *Journal of Structural Learning, 4*, 101-109.

Shavelson, R. J., & Stanton, G. C. (1975). Concept validation: Methodology and application to three measures of cognitive structure. *Journal of Educational Measurement, 12*, 67-85.

Shaw, M. E. (1954a). Some effects of problem complexity upon problem solution efficiency in different communication nets. *Journal of Experimental Psychology, 48*, 211-217.

Shaw, M. E. (1954b). Some effects of unequal distribution of information upon group performance in various communication nets. *Journal of Abnormal and Social Psychology, 49,* 547-553.

Shaw, M. E. (1964). Communication networks. In L. Berkowitz (Ed.), *Advances in experimental social psychology* (Vol. 1, pp. 111-147). New York: Academic Press

Shaw, M. E. (1981). *Group dynamics: The psychology of small group behavior* (3rd ed.). New York: McGraw-Hill.

Shepard, R. N. (1962). The analysis of proximities: Multidimensional scaling with an unknown distance function (Parts 1 & 2). *Psychometrika, 27,* 125-140, 219-246.

Shepard, R. N. (1974). Representation of structure in similarity data: Problems and prospects. *Psychometrika, 39,* 373-421.

Shepard, R. N., & Arabie, P. (1979). Additive clustering: Representation of similarities as combinations of discrete overlapping properties. *Psychological Review, 86,* 87-123.

Shiffrin, R. M., & Schneider, W. (1977). Controlled and automatic human information processing: II. Perceptual learning, automatic attending, and a general theory. *Psychological Review, 84,* 127-190.

Shneiderman, B. (1982). The future of interactive systems and the emergence of direct manipulation. *Behavior and Information Technology, 1,* 237-256.

Shoben, E. J. (1976). A verification of semantic relations in a same-different paradigm: An asymmetry in semantic memory. *Journal of Verbal Learning and Verbal Behavior, 15,* 365-379.

Simon, H., & Hayes, J. (1976). The understanding process: Problem isomorphs. *Cognitive Psychology, 8,* 165-190.

Slavin, R. E., & Hansell, S. (1983). Cooperative learning and intergroup relations: Contact theory in the classroom. In J. L. Epstein & N. Karweit (Eds.), *Friends in school* (pp. 93-114). New York: Academic Press.

Sleeman, D., & Brown, J. S. (1982). *Intelligent tutoring systems.* London: Academic Press.

Smith, E. E., Adams, N., & Schorr, D. (1978). Fact retrieval and the paradox of interference. *Cognitive Psychology, 10,* 438-464.

Smith, E. E., Shoben, E. J., & Rips, L. J. (1974). Structure and process in semantic decisions. *Psychological Review, 81,* 214-241.

Sokal, R. R., & Sneath, P. H. A. (1963). *Principles of numerical taxonomy.* San Francisco: W. H. Freeman.

Sowa, J. F. (1984). *Conceptual structures: Information processing in mind and machine.* Reading, MA: Addison-Wesley.

Sparck Jones, K. (1986). *Synonymy and semantic classification.* Edinburgh, Scotland: Edinburgh University Press.

Stephens, D. L. (1987). *Use of cognitive structure in predicting test achievement and ideational creativity in biology students.* Unpublished master's thesis, Univ. of Georgia.

Stevens, S. S. (1951). Mathematics, measurement, and psychophysics. In S. S. Stevens (Ed.), *Handbook of experimental psychology.* New York: Wiley.

Stevens, S. S. (1975). *Psychophysics.* Somerset, NJ: John Wiley & Sons.

References

Thiel, U., & Hammwohner, R. (1987). Informational zooming: An interactive model for the graphical access to text knowledge bases. *ACM SIGIR Conference Proceedings on Research and Development in Information Retrieval*, 45-56.

Thorndike, E. L. (1932). *The fundamentals of learning*. New York: Teachers College, Bureau of Publications.

Thorndike, R. L. (1973). Who belongs in a family? *Psychometrika, 18*, 267-276.

Thorndyke, P. W. (1977). Cognitive structures in comprehension and memory of narrative discourse. *Cognitive Psychology, 9*, 77-110.

Thro, M. P. (1978). Relationship between associative and content structure of physics concepts. *Journal of Educational Psychology, 70*, 971-978.

Tolman, E. C. (1932). *Purposive behavior in animals and men*. New York: Appleton-Century-Crofts.

Townsend, J. T. (1971). Theoretical analysis of an alphabetic confusion matrix. *Perception & Psychophysics, 9*, 40-50.

Tullis, T. S. (1985). Designing a menu-based interface to an operating system. *CHI '85 Proceedings*, 73-78.

Tulving, E. (1962). Subjective organization in free recall of "unrelated" words. *Psychological Review, 69*, 344-354.

Tversky, A. (1977). Features of similarity. *Psychological Review, 84*, 327-352.

Tversky, A., & Gati, I. (1982). Similarity, separability, and the triangle inequality. *Psychological Review, 89*, 123-154.

Tversky, A., & Hutchinson, J. W. (1986). Nearest neighbor analysis of psychological spaces. *Psychological Review, 91*, 3-22.

Ullman, J. D. (1982). *Principles of database systems* (2nd ed.). Rockville, MD: Computer Science Press.

van Dam, A. (1988). Hypertext '87 keynote address. *Communications of the ACM, 31*, 887-895.

van Dijk, T. A., & Kintsch, W. (1983). *Strategies of discourse comprehension*. New York: Academic Press.

Waern, Y. (1987). Mental models in learning computerized tasks. In M. Frese, E. Ulich, & W. Dzida (Eds.), *Psychological issues of human-computer interaction in the work place*. Amsterdam, the Netherlands: Elsevier Science Publishers.

Wallace, W. T., & Rubin, D. C. (1988a). Memory of a ballad singer. In M. M. Gruenberg, P. E. Morris, & R. N. Sykes (Eds.), *Practical aspects of memory: Current research and issues: Vol. 1. Memory in everyday life* (pp. 257-262). New York: Wiley.

Wallace, W. T., & Rubin, D. C. (1988b). "The wreck of the old 97": A real event remembered in song. In U. Neisser & E. Winograd (Eds.), *Remembering reconsidered: Ecological and traditional approaches to the study of memory* (pp. 283-310). Cambridge: Cambridge University Press.

Wallas, G. (1926). *The art of thought*. New York: Harcourt Brace.

Waltz, D. (1972). *Generating semantic descriptions from drawings of scenes with shadows*. Unpublished doctoral thesis, MIT.

Waterman, D. A. (1986). *A guide to expert systems*. Reading, MA: Addison-Wesley.

Weaver, C. A., III, & Kintsch, W. (1985). *Textual and non-textual factors in situation model formation* (Tech. Rep. No. 150). Boulder, CO: Institute of Cognitive Science, University of Colorado.

Weber, E. H. (1846). Der Tastsinn und das Gemeinfuhl. In R. Wagner (Ed.), *Handworterbuch der Physiologie* (Vol. 3, pp. 481-588). Braunschweig: Vieweg.

Weisberg, R. W., & Alba, J. W. (1981a). An examination of the alleged role of "fixation" in the solution of several "insight" problems. *Journal of Experimental Psychology: General, 110*, 169-192.

Weisberg, R. W., & Alba, J. W. (1981b). Gestalt theory, insight, and past experience: Reply to Dominowski. *Journal of Experimental Psychology: General, 110*, 199-203.

Weisberg, R. W., & Alba, J. W. (1982). Problem solving is not like perception: More on Gestalt theory. *Journal of Experimental Psychology: General, 111*, 326-330.

White, H. C., Boorman, A., & Breiger, R. L. (1976). Social structures from multiple networks. I. Blockmodels of roles and positions. *American Journal of Sociology, 81*, 730-780.

Whorf, B. L. (1956). Science and linguistics. In J. B. Carroll (Ed.), *Language, thought, and reality: Selected writings of Benjamin Lee Whorf.* Cambridge, MA: MIT Press.

Winograd, T., & Flores, F. (1986). *Understanding computers and cognition: A new foundation for design.* Norwood, NJ: Ablex.

Winston, P. H. (1970). *Learning structural descriptions from examples.* Unpublished doctoral thesis, MIT.

Woods, W. A. (1975). What's in a link: Foundations for semantic networks. In D. G. Bobrow & A. M. Collins (Eds.), *Representation and understanding: Studies in cognitive science* (pp. 35-82). New York: Academic Press.

Woods, W. A., Kaplan, R. M., & Nash-Webber, B. (1972). *The lunar sciences natural language information system: Final report* [prepared for The Language Research Foundation]. Cambridge, MA: Bolt Beranek & Newman, Inc.

Wright, J. C., Giammarino, M., & Parad, H. W. (1986). Social status in small groups: Individual-group similarity and the social "misfit." *Journal of Personality and Social Psychology, 50*, 523-536.

Yaniv, I., & Meyer, D. E. (1987). Activation and metacognition of inaccessible stored information: Potential bases for incubation effects in problem solving. *Journal of Experimental Psychology: Learning, Memory, and Cognition, 13*, 187-205.

Young, F. W., Takane, Y., & Lewyckyj, R. (1978). ALSCAL: A non-metric multidimensional scaling program with several individual differences options. *Behavioral Research Methods and Instrumentation, 10*, 451-453.

Young, R. M. (1983). Surrogates and mappings: Two kinds of conceptual models of interactive devices. In D. Gentner & A. Stevens (Eds.), *Mental models* (pp. 35-52). Hillsdale, NJ: Erlbaum.

Graph Theory and Pathfinder Primer

Graph theory is the mathematical study of structures consisting of *nodes* with *links* connecting some pairs of nodes (Carre, 1979; Christofides, 1975; Harary, 1969). Terminology in graph theory varies somewhat from one source to another. Our terms represent a distillation of various sources with adaptations to our purposes.

A *graph G* consists of *nodes* and *links*. The nodes are a finite set, e.g., $\{1, 2, ..., n\}$, and the links are a subset of the set of all node pairs. For example, the node pairs (1, 2), (4, 3), (7, 1), designate links between the first and the second node in each pair. The nodes connected by a link are known as *endpoints* of the link. A link is *incident* to a node if the node is an endpoint of the link. The *degree of a node* is the number of links incident to the node. A graph can be displayed by a diagram in which nodes are shown as points, and links are indicated by lines or arrows connecting appropriate pairs of points.

A graph may be either directed or undirected. A *directed graph* (sometimes referred to as a *digraph*) has directed links (or *arcs*). The order of the nodes in a pair designating an arc specifies a direction for the arc which is regarded as going *from* the first (or *initial*) node *to* the second (or *terminal*) node. In diagrams of directed graphs, arcs are represented as arrows extending from the initial node to the terminal node. An *undirected graph* has undirected links (or *edges*). The nodes in a pair designating an edge are regarded as unordered. In diagrams of undirected graphs, edges are represented as lines connecting appropriate nodes. In our usage, the terms *graph* and *link* refer to the general case which includes both directed and undirected graphs.

A *walk* is an alternating sequence of nodes and links such that each link in the sequence connects the nodes that precede and follow it in the sequence. For example, given nodes $\{1, 2, 3, 4\}$, the sequence, 3, (3,2), 2, (2,1), 1, (1,4), 4, specifies a walk, while the sequence, 3, (3,2), 2, (1,4), 4, (2,1), 1, does not. A walk can be specified by the sequence of nodes which it visits in which case the existence of the appropriate links is assumed. For the example walk specified above, the node sequence is 3,2,1,4. The *length of a walk* corresponds to the number of links in the walk. A walk is a *path* if all the nodes in the walk are distinct. A link is a path of length 1. A *cycle* is a closed path with all nodes distinct except the first and last nodes, which are identical.

A *connected graph* contains a path between any two nodes. A *tree* is a connected graph with no cycles. An undirected tree with n nodes has exactly $n-1$ edges, and it contains exactly one path between any two nodes. A *complete graph* has all possible links.

Links may have positive real numbers (weights, distances, or costs) associated with them in which case the graph is known as a *network*. The graph corresponding to a network is obtained by deleting the weights. The graph represents the structure of a network, and the weights associated with links in a network, provide quantitative information to accompany that structure. The *weight* of link (i,j) is designated by w_{ij}. A graph may be regarded as a network with all link weights equal to one (1). In a network, the *weight of a path* is the sum of the weights associated with the links in the path. A *geodesic* is a minimum weight path connecting two nodes. The *distance* between two nodes is the weight of a geodesic connecting the nodes. The *minimal spanning tree* (Kruskal, 1956) of an undirected network consists of a subset of the edges in the network such that the subgraph is a tree and the sum of the link weights is minimal over the set of all possible trees.

Various characteristics of graphs are conveniently represented by matrices. A graph G can be represented by the *adjacency matrix A*, the $n \times n$ matrix with $a_{ij} = 1$ if G contains the link (i,j) and $a_{ij} = 0$ otherwise. A network is similarly represented by the *network adjacency matrix* A with $a_{ii} = 0$, $a_{ij} = w_{ij}$, $i \neq j$ if the network contains the link (i, j), otherwise $a_{ij} = \infty$. The *reachability matrix* of G is the $n \times n$ matrix in which the ij^{th} entry is 1 if there is a path in G from node i to node j and is 0 otherwise. The *distance matrix* D of a network is the $n \times n$ matrix in which d_{ij} is the (minimum) distance from node i to node j in a network. If there is no path from node i to node j (a disconnected network), $d_{ij} = \infty$. The distance matrix of a graph contains the (minimum) number of links between pairs of nodes. The distance matrix is not necessarily symmetric, but it will be symmetric if the network consists of undirected links. A link in a network is *redundant* if the network obtained by removing the link yields the same distance matrix as the original network.

Pathfinder Networks

Pathfinder networks are derived from proximity data.[1] In defining Pathfinder networks (PFNETs), it is helpful to conceptualize proximity data as a complete network[2] with the weight on each link equal to the proximity between the entities connected by the link. Call this network the DATANET. The DATANET is a direct representation of the proximities, but because of the density of links in the network, it is not very informative. The essential idea underlying Pathfinder networks is that a link in a DATANET is a link (with the same weight) in a PFNET if and only if the link is a minimum weight path in the DATANET. Equivalently, we can say that the PFNET has the same distance matrix as the DATANET, but the PFNET has the minimum number of links needed to yield that distance matrix.

A variety of different PFNETs can be derived from a given set of proximity data. A particular PFNET is determined by the values of two parameters, r and q. These two parameters represent generalizations of the usual definition of distances in networks. The r parameter determines how the weight of a path is computed from the weights on links in the path. The q parameter limits the number of links allowed in paths.

The r Parameter

Usually, in graph theory, the weight of a path is the sum of the weights of the links in the path. When link weights are obtained from empirical data, it may not be justifiable to compute path weight in this way because that computation assumes ratio-scale measurement (cf. Stevens, 1951). For computing distances in DATANETs, we need a distance function that will permit computations of distances in networks with different assumptions about the level of measurement associated with the proximities. From the perspective of deriving networks from proximities, such a distance function should preserve ordinal relationships between link weights and path weights for all permissible transformations of

---

[1] Similarity, relatedness, and psychological distance are closely related concepts indicating the degree to which things "belong together" psychologically. Proximity is a general term which represents these concepts as well as other measurements, both subjective and objective, of the relationship between pairs of entities. In this chapter, we use the term, proximity, to refer to such measurements. In the techniques we propose, the measurements have the direction of distances (or distance estimates) so that small values represent similarity, relatedness, or nearness, and large values represent dissimilarity, lack of relatedness, or distance.

[2] The proximity estimates will define a complete network when the set of proximities is complete. Missing data can be handled by using infinity for missing values. Pairs of entities with infinite proximities will never be linked in any PFNET. This fact can also be used to prevent the linking of any two nodes simply by using infinite proximities for the appropriate pairs. PFNETs are not necessarily connected when some of the proximities are infinite.

Primer

the proximities with different assumptions about the level of measurement associated with the proximities. Then, ordinal comparisons of path weights and link weights could be used to determine link membership in PFNETs.

A distance function with the required qualities can be defined by adapting the Minkowski distance measure to computing distances over paths in networks. It can easily be shown that the Minkowski r distance satisfies the requirements of a path algebra for networks as defined by Carre (1979). The r distance function replaces the normal sum with the r distance so that $x + y$ is replaced by $(x^r + y^r)^{1/r}$, $x \geq 0$, $y \geq 0$, $r \geq 1$. Given a path P consisting of k links with weights $w_1, w_2, ..., w_k$, the weight of path P, $w(P)$ becomes:

$$w(P) = \left(\sum_{i=1}^{k} w_i^r \right)^{1/r} \quad \text{where } r \geq 1, \ w_i \geq 0 \text{ for all } i.$$

Note that with $r = 1$, the function corresponds to simple addition (the usual definition of distances in networks). With $r = \infty$, the function is the maximum function. In fact,

$$\lim_{r \to \infty} \left(w_i^r + w_j^r \right)^{1/r} = maximum \left(w_i, w_j \right).$$

Thus with $r = \infty$, computing network distances with the Minkowski r distance only requires maximum (as above) and minimum (for identifying geodesics or minimum weight paths) operations which are order preserving and, therefore, appropriate for ordinal scale measurement. In particular, the ordinal relationships of path weights will be preserved for any nondecreasing transformation of the link weights (proximities).

In summary, the r parameter for PFNETs is the value of r in the Minkowski r distance computation for the weight of a path as a function of the weights of links in the path.

The q Parameter

The distance matrix of a network is usually determined by finding the minimum weight paths regardless of the number of links in those paths. The q parameter is another generalization of this definition of network distance. This parameter places an upper limit on the number of links in paths used to determine the minimum distance between nodes in the DATANET. There are two reasons for using the q parameter, one psychological, and the other representational. From a psychological perspective, there may be some limit on the number of links that could meaningfully connect nodes in a particular domain. This amounts to a limit in the chain of relations that can be constructed relating any two concepts in the domain. This limit can be incorporated into the network generation procedure with the q parameter. The representational motivation for the q parameter is that it provides a method for systematically controlling the density of links in PFNETs. Users of PFNETs may have various reasons for preferring networks of varying density.

With the two parameters r and q, a particular PFNET is identified as PFNET(r, q). The properties of PFNETs and their relation to one another are discussed in detail in Chapter 1.

Glossary

adjacent - Two nodes are adjacent in a graph if and only if they are connected by a link.

adjacency matrix - The adjacency matrix of a graph with n nodes is the $n \times n$ matrix $\mathbf{A}$ with $\mathbf{A}_{ij} = 1$ if nodes i and j are adjacent in the graph, otherwise $\mathbf{A}_{ij} = 0$. The adjacency matrix of a network is the $n \times n$ matrix $\mathbf{A}$ with $\mathbf{A}_{ii} = 0$ and $\mathbf{A}_{ij} = w_{ij}$ (the weight of the link between nodes i and j) if the nodes i and j are adjacent, otherwise $\mathbf{A}_{ij} = \infty$.

arc - An arc is a directed link with a direction from the initial or originating node to the terminal node. In diagrams, arcs are usually shown as arrows pointing from the initial node to the terminal node.

center - A center of a graph or network is a node with minimum *eccentricity*.

clique - A clique is a maximal subgraph with three or more nodes in which every node in the subgraph is connected to every other node in the subgraph.

complement - The complement G´ of a graph G has the same nodes as G, but two nodes in G´ are connected by a link if and only if they are not connected by a link in G. G´ has the opposite set of links to those in G.

complete graph or complete network - A graph or network with all possible links, that is, a link for every pair of nodes.

connected graph - A graph is connected if there is a path between the nodes in every pair of nodes.

cutnode or cutpoint - If removing a node and its incident links results in a disconnected graph, that node is a cutpoint or a cutnode.

cycle - A cycle is a closed path with the same first and last nodes.

degree - The number of links incident to a node is the degree of that node.

density - The number of links in a graph divided by the number of possible links.

diameter - The diameter of a connected graph or network is the length of any longest geodesic.

distance - The distance between nodes in a network is the length of the geodesic connecting the nodes, or equivalently, the distance between two nodes is the length of the minimum-length path connecting the nodes. Distances in networks do not have the same limitations as distances in space. For example, network distances can be asymmetrical (d_{ij} may not be the same as d_{ji}), but network distances do obey the triangle inequality ($d_{ij} \leq d_{ik} + d_{kj}$). See also graph-theoretic distance.

eccentricity - The eccentricity of a node is the maximum distance between that node and all other nodes in a graph or network.

edge - An edge is an undirected link. In diagrams, edges are usually shown as lines drawn between the nodes connected by the edge.

geodesic - The path of minimum length between the nodes in a pair is the geodesic connecting the nodes.

graph - A graph is a finite set of nodes and a subset of pairs of nodes (the links).

graph-theoretic distance - The distance measured by the minimum number of links connecting two nodes in a graph is the graph-theoretic distance. This definition of distance is a special case of the general notion of distance in a network. For graph distance, consider the weight on each link to be 1 (one) and use $r = 1$ in the definition of the length of a path.

incident - All links connected to a node are incident to that node.

indegree - The number of arcs terminating on a node is the indegree of that node.

isomorphic - Two graphs are isomorphic if there exists a one-to-one correspondence between their nodes that results in the same set of links in the two graphs.

length (of a path) - In a graph, the length of a path is the number of links in the path. In a network, the length of a path consisting of k links with weights, $w_1, w_2, ...w_k$ is computed by: $(w_1^r + w_2^r + ... + w_k^r)^{1/r}$, where r is a parameter, $1 \leq r \leq \infty$. Note that $r = 1$ corresponds to simple addition of the link weights. With $r = \infty$, the length of a path can be computed using the maximum function, $\text{Max}(L_1, L_2, ..., L_k)$, which is the limit of the general function as r approaches infinity.

link - A link is a connection between nodes. A particular link can be identified by the pair of nodes it connects (the endnodes of the link). Links can be directed (arcs) or undirected (edges).

loop - A loop is a link connecting a node to itself.

median - A median of a graph or network is a node with the smallest average distance to all other nodes in the graph.

minimum-cost network (MCN) - The Pathfinder network generated with $r = \infty$ and $q = n-1$, where n is the number of nodes.

minimal dominating node set - The minimal set of nodes from which every node in the graph can be reached over, at most, one link.

minimal spanning tree - A minimal spanning ree of a network consists of a subset of the links in the network such that the subset constitutes a tree, and the sum of the link weights in the subset are minimal over all possible subsets.

minimum cycle - A cycle with minimum distance.

network - A network is a graph with nonnegative real numbers (weights) associated with the links. Each link in a network has a weight.

node - Along with links, nodes are the basic units of graphs and networks. A graph is defined as a finite set of nodes with links connecting some pairs of the nodes.

ordinal level measurement - Measurement on a scale in which only the order of the values is thought to be meaningful. Thus, any values which have the same order as the original values constitute an equivalent scale. Operations on values from ordinal scales should only rely on the order of values (see r parameter).

outdegree - The number of arcs originating from a node is the outdegree of that node.

path - A sequence of distinct nodes and connecting links in a graph or network.

Glossary

Pathfinder graph - The graph obtained by deleting the link weights from a Pathfinder Network.

Pathfinder network - The network obtained by deleting from the complete network corresponding to a proximity matrix every link whose weight is larger than the length of the geodesic connecting the endnodes of the link. The weights on the remaining links are the same as the weights on the corresponding links in the complete network.

PFNET(r, q) - A Pathfinder graph or network computed with particular values of the parameters r and q.

proximity - Proximity is a term used to refer to a measure of relationship between two entities. Measures of similarity, relatedness, dissimilarity, distance, conditional probability, or association are all instances of proximity measures. In the context of networks, proximity measures have the direction of distance with small values representing similarity, closeness, or high relatedness, and large values representing dissimilarity, farness, or low relatedness.

q parameter - The q parameter specifies a limit on the number of links allowed in paths as path lengths are determined in deriving a Pathfinder network from proximity data. Only paths of q or fewer links are considered. For a network with n nodes, meaningful values of q range from 2 to $n-1$. With $q = n-1$, there is essentially no limit on the lengths of paths because the longest possible paths have $n-1$ links.

ratio-level measurement - Measurement on a scale with a true zero point and meaningful differences or intervals on the scale. Physical measurement is usually ratio level. A ratio scale is preserved by multiplying by a positive constant (a change of unit). All other transformations distort the ratio properties.

reachable - A node is reachable from another node in a graph if there is a path from the first node to the second in the graph.

r parameter - The r parameter is the value used in the computation of the lengths of paths in determining a Pathfinder network. Meaningful values of r range from 1 through infinity. When $r = \infty$, link membership in Pathfinder networks is determined solely by the order of the proximity data values. Thus, infinite r is appropriate for ordinal-level measurement. Other values of r require ratio-level measurement.

subgraph - A subgraph is obtained by removing a subset of nodes and their incident links from a graph.

tree - A connected graph with no cycles.

triangle inequality - The triangle inequality is satisfied when a set of meaures, d_{ij}, on pairs of points i and j, $d_{ij} \leq d_{ik} + d_{kj}$ for all i,j,k.

weight - The cost or distance associated with a link. A network is a graph with weighted links.

z parameter - The z parameter determines the width of the interval used to make link membership decisions with the FUZZYPF algorithm (see Chapter 3).

Author Index

Numbers in italic indicate the pages on which the complete references can be found.

A

Abe, O., 232, *290*
Abrahamson, A.A., 102, *292*
Adams, N., 277, *294*
Adelson, B., 228, 251, *279*
Aho, A.V., 98, *279*
Alba, J.W., 268, 269, 277, *296*
Anderson, A., 225, *285*
Anderson, J.R., 1, 31, 101, 135, 228, 243, 262, *279*
Anderson, R.C., 262, *279*
Anisfeld, M., 132, *279*
Arabie, P., 22, 26, 98, *294*
Arbib, M.A., 31, *288*
Asher, S.R., 47, *279*

B

Baird, J.C., 67, *279*
Ball, G.H., 94, *279*
Barnard, P., 179, *285*
Barsalou, L.W., 130, 233, 224, *279*
Bates, M.J., 165, *279*
Belkin, N.J., 151, 167, 169, 175, *279, 280*
Bertrand-Gastaldy, S., 172, *280*
Billingsley, P.A., 167, *280*
Binford, T.O., 1, *280*
Black, J.B., 184, 256, *287, 290*

Bobrow, R.J., 1, *280*
Bocker, H., 167, *280*
Boorman, A., 44, *296*
Bousfield, W.A., 34, 122, *280, 281*
Bovair, S., 179, *287*
Bowden, E.M., 267, 269, *280*
Bower, G.H., 101, 241, *279, 280, 290*
Boyes-Braem, P., 29, 34, 95, *291*
Brachman, R.J., 1, *280*
Bransford, J.D., 256, 277, *280*
Breen, T.J., 39, 40, 41, 78, 103, 135, 179, 187, 245, 251, *287, 293*
Breiger, R.L., 44, *296*
Brindle, J., 172, *287*
Brooks, H.M., 151, 167, *280*
Brooks, R.A., 1, *280*
Brown, J.S. 227, *294*
Brown, S.W., 31, *280*
Burt, R.S., 32, 44, *280*
Bush, V., 197, 199, *280*
Butler, K.A., 228, *280*

C

Caramazza, A., 41, 42, 101, 102, 104, *280*
Card, S.K., 180, *281*
Carre, B., 83, *281, 297, 299*
Carroll, J.M., 179, 189, *281*

Cartwright, D., 31, *286*
Chaffin, R., 223, 229, 235, 238, *281, 286*
Champagne, A.B., 243, *281*
Chappell, S.L., 202, *281*
Charney, D., 200, *281*
Chase, W.G., 38, 227, *281*
Chi, M.T.H., 32, 38, 39, 41, 102, 227, 228, 255, *281*
Christofides, N., *281*, 297
CODASYL, 1, *281*
Cofer, C.N., 34, *288*
Cohen, B.H., 34, *281*
Cohen, P.R., 152, *281*
Coie, J.D., 47, *281*
Collins, A.M., 1, 29, 33, 101, 135, 241, 244, *281*
Collins, G.C., 223, 224, *292*
Conklin, J.C., 198, 200, *281*
Conner, J., 179, *289*
Constantine, A.G., 3, *281*
Cooke, N.M., 39, 40, 41, 43, 78, 101, 103, 108, 109, 111, 113, 114, 115, 120, 135, 184, 187, 228, 238, 239, 245, 250, 251, *282, 293*
Cooper, W.S., 165, *282*
Coppotelli, H., 47, *281*
Corter, J.E., 224, 228, *280, 285*
Craven, T.C., 172, *282*
Croft, W.B., 151, 167, 175, *279*
Crouch, D., 167, *282*

D
Date, C.J., 1, *282*
Davidson, C.H., 172, *280*
Dayton, T., 61, 201, *289*
Dearholt, D.W., ix, 1, 2, 8, 16, 31, 34, 53, 56, 75, 89, 94, 102, 111, 115, 121, 135, 148, 154, 177, 185, 201, 221, 244, 270, 271, *282, 284, 287, 289, 293*
Deerwester, S., 152, *283*

Deese, J., 121, 132, *282*
De Groot, A.M.B., 222, *282*
Dellarosa, D., 199, 255, *282, 283*
DeMaio, J.C., 39, 40, 41, 78, 103, 135, 245, 251, *293*
Desena, A.T., 243, *281*
Diekhoff, G.M., 243, 251, *283*
Dodge, K.A., 47, 48, *279, 281*
Dominowski, R.L., 267, 269, 277, *283*
Dooling, D.J., 256, *283*
Doyle, L., 172, *283*
Dubes, R.C., 9, 22, 24, 94, *283, 286*
Dumais, S.T., 151, 152, 165, *283, 284*
Durso, F.T., ix, 1, 2, 8, 16, 33, 34, 39, 40, 41, 53, 56, 75, 78, 95, 101, 102, 103, 104, 108, 109, 111, 113, 114, 115, 120, 121, 135, 148, 154, 185, 221, 244, 245, 250, 251, 270, 271, *282, 293*

E
Ekman, G., 36, *283*
Ekstrand, B.R., 132, *283*
Ellen, P., 267, 268, 277, *283*
Englebart, D.C., 173, *283*
English, W.K., 173, *283*
Ericsson, K.A., 227, *283*
Esposito, C., 29, *283*
Even, S., 14, *283*

F
Fahlman, S.E., 1, *283*
Fairchild, K.M., 167, *283*
Feltovich, P.J., 39, 227, 228, 255, *281*
Festinger, L., 31, *283*
Fidel, R., 166, *283*
Fienberg, S., 185, *284*
Fikes, R., 1, *284*
Fillenbaum, S., 1, 32, 102, 121, *284*
Fischer, G., 167, 199, *280, 284*

Author Index

Fletcher, C.R., 199, *284*
Flores, F., 225, *296*
Ford, D.L., Jr., 49, *284*
Fowler, R.H., 177, *284*
Franks, J.J., 277, *291*
Frei, H.P., 167, *284*
Freund, J.E., 54, 56, *284*
Friedman, H.P., 94, *284*
Friedrich, M.A., 121, *290*
Friendly, M.L., 1, 41, 102, 121, 132, *284, 292*
Fromkin, V., 124, *284*
Fry, J.P., 1, *290*
Furnas, G.W., 152, 165, 167, 175, *283, 284, 287*

G

Galil, Z., 176, *284*
Gammack, J.G., 213, 214, 224, 225, 228, *284, 285*
Gati, I., 2, 5, *295*
Geeslin, W.E., 242, *285, 293*
Gelatt, C.D., Jr., 81, *287*
Giammarino, M., 47, *296*
Gibson, E.J., 124, *285*
Gick, M.L., 277, *288*
Glaser, R., 39, 227, 228, 255, *281*
Glass, A.L., 101, *285*
Glavanov, D., 199, *292*
Gluck, M.A., 224, *285*
Glucksberg, S., 101, *289*
Goldsmith, T.E., 39, 40, 41, 78, 103, 135, 245, 251, *293*
Gomez, L.M., 152, 165, *284*
Gonzales, N., 89, *282*
Gower, J.C., 3, *281*
Graham, R.L., 77, *285*
Gray, W.D., 29, 34, 95, *291*
Green, P.E., 3, *286*

Greeno, J.G., 120, 255, 256, *285, 286, 287*
Gregory, D., 225, *285*
Griffith, R.L., 1, *285*
Grimes, J.D., 173, *291*
Gruenewald, P.J., 121, 132, *285*
Guindon, R., 179, *289*

H

Hakimi, S.L., 1, 2, 5, *285*
Halasz, F., 179, *285*
Hall, D.J., 94, *279*
Hall, J., 49, *285*
Hallinan, M.T., 47, 49, *285*
Hammond, N., 179, *285*
Hammwohner, R., 167, *295*
Hansell, S., 47, 48, *285, 294*
Harary, F., 31, 95, *286, 297*
Harshman, R., 152, *283*
Harshman, R.A., 3, *286*
Hayes, J.R., 255, *286, 294*
Hayes-Roth, F., 227, *286*
Hays, W.L., 184, 185, *286*
Hendrix, G., 1, *284*
Herrmann, D.J., 104, 223, 229, 235, 238, 255, *281, 286, 293*
Herndon, W.C., 82, *286*
Hersh, H., 41, 42, 101, 102, 104, *280*
Hill, W.C., 179, *289*
Hinton, G.E., 135, 137, 139, 142, *292*
Hirtle, S.C., 38, 39, 243, *289*
Hollan, J.D., 167, *286*
Holyoak, K.J., 101, *285*
Hopcroft, J.E., 98, *279*
Humphreys, M., 120, *286*
Humphreys, M.S., 101, *293*
Hunter, L.E., 223, 224, *292*
Hutchins, E.L., 167, *286*
Hutchinson, J.W., 1, 3, 32, 34, 37, 51, 102, *286, 295*

Hyman, I.E., Jr., 121, *286*
Hymel, S., 47, *279*

I
Isaac, P.D., 105, *286*

J
Jain, A.K., 9, 22, 24, 94, *283*, *286*
James, W., 245, *286*
Jauslin, J.F., 167, *284*
Jenkins, J.J., 101, *286*
Johnson, D.M., 29, 34, 95, 203, *291*
Johnson, M.K., 256, 277, *280*
Johnson, S.C., 10, 20, 21, 22, 25, 28, 29, 45, 139, 232, *286*
Johnson-Laird, P.N., 223, *286*
Jones, W.P., 167, 175, *287*

K
Kahneman, D., 223, *287*
Kaplan, P., 256, *287*
Kaplan, R.M., 1, *296*
Karat, J., 201, *289*
Karweit, N., 48, *285*
Katsuya, A., 232, *290*
Kay, D.S., 184, *287*
Kellogg, W.A., 179, 187, 189, *281*, *287*
Kennard, R.W., 94, *287*
Keppel, G., 117, *287*
Kieras, D.E., 179, 180, *287*
Kintsch, E.H., 201, *287*
Kintsch, W., 1, 31, 199, 200, 256, 259, *288*, *291*, *295*, *296*
Kirkpatrick, S., 81, *287*
Kjeldsen, R., 152, *281*
Klopfer, L.E., 243, *281*
Klovdahl, A.S., 172, *287*
Klun, J.R., 104, *286*
Knapp, M., 132, *279*

Knoebel, A., 89, *287*
Knoke, D., 32, 44, 45, 46, 175, *287*
Koeske, R.D., 32, 38, 41, 102, *281*
Krumhansl, C.L., 3, *287*
Kruskal, J.B., 102, 105, 111, 139, 247, *287*, *288*, 297
Kuklinski, J.H., 44, 45, 175, *287*
Kwasnik, B.H., 169, *280*

L
Lachman, R., 256, *283*
Lamon, M., 277, *288*
Landauer, T.K., 101, 152, 165, *283*, *284*, *288*
Larkin, J., 255, *288*
Lawson, E.D., 49, *288*
Leavitt, H.J., 49, *288*
Lenat, D.B., 227, *286*
Lesk, M., 151, 152, *288*
Lewyckyj, R., 115, *296*
Liebelt, L.S., 201, *289*
Lieblish, I., 31, *288*
Lin, Y., 243, *290*
Linde, L., 179, *288*
Ling, R.F., 20, *288*
Lingoes, J.C., 122, *288*
Lipkis, T.A., 1, *292*
Lockhart, R.S., 277, *288*
Lockhead, G.R., 102, 121, 132, *285*, *286*
Loftus, E.F., 1, 29, 33, 101, 135, *281*
Lorch, R.F., 109, 110, *288*
Lundy, M.E., 3, *286*

M
Mack, R.L., 189, *281*
MacLean, A., 179, *285*
MacQueen, J., 94, *288*
Mannes, S.M., 200, 201, *288*

Author Index

Marriot, F.H., 94, *288*
Marshall, G.R., 34, *288*
Martin, J., 1, *289*
Masson, M.E.J., 179, *289*
Maxwell, K.J., 36, *289*
Mayer, R.E., 179, *289*
Mazur, S., 189, *281*
McClave, J., 55, *289*
McClelland, J.L., 1, 31, 135, 137, 139, 142, *289*, *292*
McCloskey, M., 101, *289*
McDermott, J., 255, *288*
McDonald, D.R., 61, 201, *289*
McDonald, J.E., 40, 42, 43, 61, 94, 135, 173, 179, 180, 191, 200, 201, 227, 228, 238, *282*, *289*
McEvoy, C.L., 121, *290*
McGill, M.J., 176, *292*
McKeachie, W.J., 243, *290*
McKeithen, K.B., 38, 39, 243, *289*
McRae, D.J., 94, *289*
Medin, D.L., 223, 224, *290*
Mendenhall, W., 55, *289*
Mervis, C.B., 29, 34, 95, *291*
Metcalfe, J., 267, 268, 269, 277, *289*
Mettrey, J., 167, *289*
Meyer, D.E., 1, 101, 121, 135, *290*, *293*
Miller, D.T., 223, *287*
Miller, G.A., 124, *290*
Mink, W.D., 101, *286*
Minor, M.J., 32, 44, *280*
Miyamoto, S., 232, *290*
Molander, M.E., 201, *289*
Morris, N.M., 179, *292*
Moran, T.P., 179, 180, *281*, *285*
Morton, J., 179, *285*
Mukherji, B.R., 104, *293*
Murphy, G.L., 223, 224, 225, 228, *290*

N

Nakayama, K., 232, *290*
Nash-Webber, B., 1, *296*
Navathe, S.B., 1, *290*
Naveh-Benjamin, M., 243, *290*
Nelson, D.L., 121, 127, 130, *290*
Nemiroff, P.M., 49, *284*
Newell, A., 180, 227, 255, *281*, *290*
Nicely, P.A., 124, *290*
Nieper, H., 167, *280*
Nilsson, N., 39, *290*
Nisbett, R.E., 243, *290*
Noel, R.W., 201, *289*
Noma, E., 67, *279*
Norman, D.A., 167, 227, *286*, *290*
Norman, R.Z., 31, *286*
Novak, G.S., 1, *290*

O

Oddy, R.N., 167, *280*
Oi, K., 232, *290*
Olson, M.J., 121, 122, 124, 132, *292*
Olson, J.R., 179, *281*
Ortony, A., 5, *290*
Osgood, C., 216, *290*
Owens, J., 256, *290*

P

Paap, K.R., 42, 61, 102, 201, 202, *289*, *291*
Palmer, S.E., 243, *291*
Parad, H.W., 47, *296*
Pasmore, W.A., 49, *284*
Peery, J.C., 47, 48, *291*
Perfetto, G.A., 277, *291*
Perrig, W., 199, *291*
Perry, T.B., 47, 48, *291*

Pichert, J.W., 262, *279*
Polson, P., 179, 180, *287*
Poltrock, S.E., 167, *283*
Poor, D.D.S., 105, *286*

Q

Quillian, M.R., 1, 101, 135, 241, 244, *281*, *291*

R

Ramey, M., 55, *289*
Ramsey, H.R., 173, *291*
Rapoport, A., 1, 32, 102, 121, *284*
Reitman, J.S., 38, 39, 216, 243, 251, *289*, *291*
Reuter, H.H., 38, 39, 216, 243, 251, *291*
Rips, L.J., 33, 101, 102, 104, 108, *291*, *294*
Robertson, S.E., 166, *291*
Rodman, R., 124, *284*
Roistacher, R.C., 47, *291*
Rosch, E., 29, 34, 95, 102, 121, *291*
Roske-Hofstrand, R.J., 42, 61, 102, 202, *291*
Rouse, W.B., 179, *292*
Rubin, D.C., 121, 122, 124, 132, *286*, *292*, *295*
Rubin, J., 94, *284*
Rumelhart, D.E., 1, 31, 102, 135, 137, 139, 142, *289*, *292*
Russell, W.A., 101, *286*

S

Salton, G., 151, 154, 176, *292*
Sampson, G., 164, *292*
Schaeffer, B., 104, *292*
Schank, R.C., 1, 223, 224, *292*
Scheerer, M., 268, *292*
Schkolnick, M., 1, *292*

Schmalhofer, F., 199, *292*
Schmolze, J.G., 1, *292*
Schneider, W., 228, *294*
Schoenfeld, A., 255, *293*
Schorr, D., 277, *294*
Schvaneveldt, R.W., ix, 1, 2, 8, 16, 31, 33, 34, 39, 40, 41, 42, 43, 53, 56, 75, 78, 89, 94, 95, 101, 102, 103, 104, 108, 109, 111, 113, 114, 115, 120, 121, 135, 148, 154, 173, 179, 180, 184, 185, 187, 191, 201, 221, 227, 244, 245, 250, 251, 270, 271, *282*, *287*, *289*, *290*, *293*
Schwartz, R.M., 101, *293*
Sedgewick, C.H.W., 122, *280*
Sexton, G.A., 202, *281*
Shavelson, R.J., 242, *285*, *293*
Shaw, M.E., 31, 49, 50, *293*, *294*
Shepard, R.N., 22, 26, 36, 98, 102, 245, *294*
Shiffrin, R.M., 228, *294*
Shneiderman, B., 167, *294*
Shoben, E.J., 33, 101, 102, 104, 108, *286*, *291*, *294*
Simon, D., 255, *288*
Simon, H.A., 38, 227, 255, *281*, *283*, *286*, *288*, *290*, *294*
Singleton, L.L., 47, *279*
Slavin, R.E., 47, *294*
Sleeman, D., 227, *294*
Smith, E.E., 33, 101, 102, 104, 108, 277, *286*, *291*, *294*
Smolensky, P., 135, 137, 139, 142, *292*
Sneath, P.H.A., 20, *294*
Sokal, R.R., 20, *294*
Somberg, D.R., 47, 48, *283*
Sowa, J.F., 1, 101, *294*
Sparck Jones, K., 152, 153, 166, *284*, *291*
Squires, D.A., 243, *281*
Stanton, G.C., 242, *293*
Stephens, D.L., 251, *294*

Author Index

Stevens, S.S., 39, *294*, 298
Stone, J.D., 201, *289*
Stone, L.A., 94, *287*
Suci, G.J., 216, *290*

T
Takane, Y., 115, *296*
Tannenbaum, P.H., 216, *290*
Thiel, U., 167, *295*
Thorndike, E.L., 132, *295*
Thorndike, R.L., 94, *295*
Thorndyke, P.W., 256, *295*
Thro, M.P., 243, 251, *295*
Tinsley, B.R., 47, *279*
Tolman, E.C., 31, *295*
Torgerson, W., 41, 42, 101, 102, 104, *280*
Townsend, J.T., 124, *295*
Tucker, D.G., 243, *290*
Tucker, R.G., 39, 40, 41, 78, 103, 135, 245, 251, *293*
Tullis, T.S., 201, *295*
Tulving, E., 101, *295*
Tversky, A., 2, 3, 5, 34, 37, 214, 222, *295*

U
Ullman, J.D., 1, 98, *279*, *295*

V
van Dam, A., 200, *295*
van Dijk, T.A., 31, 199, 256, *295*
Vecchi, M.P., 81, *287*

W
Waern, Y., 179, *295*
Wallace, R., 104, *292*
Wallace, W.T., 121, *295*
Wallas, G., 268, *295*

Waltz, D., 1, *295*
Waterman, D.A., 227, *286*, *295*
Weaver, C.A., III., 199, *296*
Webber, B.L., 1, *280*
Weber, E.H., 62, *296*
Weisberg, R.W., 268, 269, 277, *296*
White, H.C., 44, *296*
Whitmarsh, G.A., 34, *281*
Whorf, B.L., 34, *296*
Wiebe, D., 267, 268, 269, 277, *289*
Williams, M.S., 49, *285*
Wilson, T.D., 227, 243, *290*
Wind, Y., 3, *286*
Winograd, T., 225, *296*
Winston, P.H., 1, *296*
Wish, M., 102, 105, *288*
Wood, J.R., 32, 44, 46, *287*
Woods, W.A., 1, 244, *296*
Wright, J.C., 47, 225, 228, *290*, *296*

Y
Yaniv, I., 277, *296*
Yau, S.S., 1, 2, 5, *285*
Young, F.W., 15, *296*
Young, R.M., 201, 228, *285*, *296*

Subject Index

A

abstraction, 1, 43, 173, 210, 221, 238
access, 1, 120, 165, 166, 167, 168, 169, 172, 173, 177, 197, 200, 201, 202
activation, 135, 136, 138, 139, 141, 142, 146, 148
adjacency matrix, 21, 77, 273, 297, 301
algorithm, 6, 11, 25, 28, 29, 53, 55, 59, 60, 79, 94, 98, 102, 175, 176, 222
ambiguity, 152, 153, 160
artificial intelligence, 135, 167, 227
association, associative, 2, 14, 20, 23, 30, 31, 33, 34, 82, 101, 111, 112, 113, 117, 120, 122, 132, 156, 159, 166, 172, 197, 223, 232, 244, 252
asymmetry, asymmetrical, 3, 5, 8, 34, 47, 132, 154, 301

B

block, 45, 95, 96, 98, 99
blockmodel procedure, 44
browse, 166, 198, 202

C

categories, 29, 33, 34, 36, 43, 95, 101, 104, 109, 121, 122, 123, 125, 127, 156, 205, 206, 210, 228, 235
center (of a graph), 47, 265, 274, 301
chunk, 43, 198, 265
city block, 215, 216, 217
classroom, 42, 241, 242, 245, 246, 251
clique, 26, 31, 37, 43, 45, 46, 47, 50, 76, 95, 96, 98, 99, 135, 220, 301
cluster, 3, 9, 15, 20, 24, 28, 29, 43, 89, 92, 93, 94, 95, 96, 97, 98, 99, 122, 146, 168, 178, 186
co-occurrence, 43, 136, 139, 144, 145, 146, 149, 150, 151, 153, 154, 156, 157, 161, 164, 169, 203, 209, 232

complete network, 90, 138, 142, 146, 172, 173, 298, 301, 303
conceptual structure, 167, 173, 175, 177, 191, 213, 224, 225, 226
configural, 81, 241, 244, 246, 250, 251, 253
connectionism, connectionist, 31, 43, 135, 138, 139, 141, 142, 146
context, 20, 43, 101, 108, 150, 158, 159, 161, 164, 173, 213, 215, 222, 223, 224, 225, 226
cutpoint, 45, 95, 301
cycle, 4, 8, 14, 15, 22, 24, 26, 36, 37, 45, 46, 50, 76, 129, 245, 271, 297, 301, 302, 303

D

database, 89, 166, 167, 168, 169, 176, 198
degree (of nodes), 29, 31, 35, 38, 47, 48, 95, 128, 130, 175, 262, 274, 301, 302
density (of graphs), 39, 78, 79, 81, 105, 138, 210, 298, 299, 301
diameter (of graphs), 301
disorientation, 198, 200
document retrieval, 165, 168, 175

E

edge weights, 2, 56, 82, 89, 91, 92, 98, 99
episodic memory, 42, 242
expert system, 40, 166, 225, 227, 228
expertise, 33, 38, 39, 40, 50, 70, 166, 177, 227, 228, 230, 238, 242, 244, 245, 251

313

F

fit (between data and models), 53, 54, 57, 58, 59, 60, 89, 90, 92, 96
fixedness, 268, 269
frequency of occurrence, 150, 151
FUZZYPF, 56, 57, 59, 60

G

geodetic distance, 2
Gestalt, 37, 45
goal, 191, 199, 201, 256, 261, 265, 266
graph structure, 31, 37, 77, 89, 101, 167, 172
graph-theoretic distance, 92, 98, 219, 246, 247, 301, 302

H

hierarchical cluster analysis, 9, 20, 24, 30, 41, 61, 94, 95, 97, 99, 140, 145, 148, 172, 201, 203
human-computer interaction, 179, 201, 227
HyBrow, 202, 203, 206, 209, 210
hypertext, 197, 198, 199, 200, 201, 202, 210

I

indexing, 151, 152, 165, 166, 167, 172, 173, 175, 176
individual differences, 53, 230, 245
inference, 200, 201, 210, 229, 239, 255, 256
information retrieval, 2, 151, 165, 177, 198
interface, 42, 43, 61, 102, 165, 167, 177, 180, 189, 191, 198, 201, 202, 210, 227, 228, 229
introspection, 228, 229
intuition, 32, 33, 43, 102, 267
isomorphic, 28

K

knowledge elicitation, 197, 213, 227, 228, 229, 238, 239, 241
knowledge engineering, 167, 172, 229
knowledge organization, 228, 269, 277
knowledge representation, 1, 63, 165, 167, 243, 244, 250, 253, 266
knowledge structure, 38, 41, 81, 138, 166, 167, 190, 224, 228, 246, 255, 265, 266, 268, 270, 271, 273, 277

L

layout, 204
link membership rule, 6, 7, 63
link structure, 18, 19, 28, 66, 167, 169, 172, 175, 176
link weight, 3, 105, 110, 112, 115, 120, 135, 136, 140, 219, 220, 297, 298, 299, 302, 303

M

meaning, 150, 153, 154, 161, 164, 199, 224, 225, 229, 231, 247, 264, 268
measurement scale, 61
median, 274, 275, 302
memex, 197, 198, 210
mental model, 42, 179, 180, 192, 200, 227
minimal spanning tree, 302
multidimensional scaling, 1, 3, 37, 41, 60, 61, 89, 104, 109, 122, 144, 214, 228, 244

N

natural language, 149, 153, 165, 166, 167, 168, 169, 171, 172, 173, 176, 177, 230, 239
neighborhood, 77, 78, 81, 82, 83, 246
network display, 169, 172
node sublist, 8

Subject Index

P

paired comparisons, 61, 70
path length, 2, 14, 50, 54, 56, 58, 60, 77, 78, 82, 105, 175
performance, 5, 33, 50, 73, 94, 101, 103, 113, 118, 119, 120, 179, 180, 181, 189, 190, 192, 241, 242, 243, 244, 245, 246, 249, 250, 251, 252, 258, 266
perspective, 42, 199, 200, 201, 206, 209
predict, 33, 41, 42, 81, 104, 108, 109, 113, 118, 120, 181, 244, 245, 246, 250, 251
problem solving, 227, 255, 256, 261, 266, 267, 269, 277
protocol, 38, 42, 180, 227, 228, 239
proximity, 1, 2, 5, 18, 21, 24, 42, 61, 90, 94, 111, 148, 167

R

relatedness function, 149, 150, 154, 161, 164, 203
repertory grid, 213, 215, 216, 219, 220, 223, 224

S

schemata, 135, 137, 140, 201, 256, 257, 265, 266
search, 127, 130, 133, 166, 169, 173, 178, 202, 255
semantic memory, 1, 20, 33
sense selection, 149, 150, 154, 158, 164
simulated annealing, 81, 201
situation model, 199
sorting, 184, 203, 229, 231, 232, 234, 235, 238, 239, 242, 245
spreading activation, 135, 152, 167, 277
star, 37, 95, 96, 98
structural equivalence, 44, 45
synonymy, 150, 152, 153
system design, 179, 180, 191

T

text analysis, 172
text comprehension, 199, 255, 256
thesaurus, 152, 166, 167, 168, 169, 171, 172, 173, 176, 177
tree, 14, 29, 39, 75, 105, 172, 230, 243, 297
triangle inequality, 2, 4, 13, 24, 301

U

users, 43, 94, 166, 172, 177, 179, 181, 182, 183, 184, 185, 186, 187, 188, 189, 191, 197, 198, 201, 202, 203, 204, 205, 206, 209, 210

V

vocabulary, 150, 151, 153, 156, 158, 161, 164, 166, 175, 177